Advanced Silicon and Semiconducting Silicon-Alloy Based Materials and Devices

Advanced Silicon and Semiconducting Silicon-Alloy Based Materials and Devices

Edited by

Johan F A Nijs

IMEC, Belgium

CRC Press

Taylor & Francis Group

Boca Raton London New York

CRC Press is an imprint of the
Taylor & Francis Group, an **informa** business

CRC Press
Taylor & Francis Group
6000 Broken Sound Parkway NW, Suite 300
Boca Raton, FL 33487-2742

First issued in paperback 2019

ISBN-13: 978-0-367-40217-4

British Library Cataloguing-in-Publication Data

A catalogue record for this book is available from the British Library.

Library of Congress Cataloging-in-Publication Data are available

**Visit the Taylor & Francis Web site at
http://www.taylorandfrancis.com**

**and the CRC Press Web site at
http://www.crcpress.com**

To Christien, Ronald and Anneleen

Contents

Preface

Crystalline silicon is an "old" material in the semiconductor field. Newer semiconductors have come across such as GaAs and its alloys, InP, II-VI, ... offering a lot of exciting challenges towards scientists and engineers.

Nevertheless, despite the rising R&D stars, crystalline silicon still is used for the major part of the market for semiconductor based applications. With respect to technology, crystalline silicon really offers a lot of advantages such as the specific feature of the easy growth of a silicon dioxide for planar processing and the relatively low defect density at its interface with silicon.

At present some exciting new fields have been opened up for crystalline silicon. Kroemer predicted many years ago the advantages of heterojunctions. However it took a long time to fabricate them in silicon related materials (III-V 's offered this opportunity already for a longer time). So, strained SiGe/Si heterojunction and SiC/Si heterojunctions emerged and along with them new applications.

Alloying as well as heavy doping can considerably alter the basic electronic properties of the material. However the limitation is that defect densities should be kept as small as possible in order not to cause mobility barriers and/or heavy recombination.

Also polycrystalline silicon has some interesting applications. Polycrystalline silicon is used in VLSI applications as a gate contact (CMOS), in order to fabricate a polysilicon emitter (bipolar) or to save place in static RAMS (3-D building of devices), SOI, ... Large grain polysilicon substrates are used as low cost substrates for solar cells. Polysilicon thin film transistors can offer some nice features for active matrix liquid crystal displays.

The chapters of the book are organized in 2 main parts: crystalline silicon and -alloys and polycrystalline silicon. The chapters have been written during 1992 by world's experts in the different fields. Most chapters are giving a literature overview of the field, including author's own work and experience as a special flavour to it. This approach is particularly interesting for a wide audience: from students, over scientists and engineers up till R&D or electronic business managers that want to assess the full potential of the different disciplines that recently have broadened the scope of the silicon semiconductor material.

The editor is and has been personally involved in R&D and R&D management on the majority of the presented fields and has enjoyed during these years the several contacts and collaborations he had and the interesting things he learned from the authors of the chapters in this book but also from so many others. My sincere thanks for this and many thanks to the authors that have put their best knowledge in the careful writing of the chapters.

The editor wants to thank the many scientists and industrials that were involved in critically reviewing the chapters during 1993. Their comments are very valuable and highly appreciated. They also considerably contribute to the value of this book. Their names cannot be disclosed here but they are honoured by their improvements on the quality of the contributions.

Also a special thanking is addressed towards Mrs. C. Deboes for her secretarial help with the editing process.

Finally, I want to thank my wife Christien and my children Ronald and Anneleen for their support and understanding during all the time I could not spend with them while carrying out my R&D activities at IMEC and editing this book. Therefore I want to dedicate this work to them.

Johan Nijs
IMEC, Belgium
November 1993

Introduction
J. Nijs, editor

Crystalline silicon is a material where the silicon atoms are bonded with 4 neighbours and this situation is periodically repeated in exactly the same way. The distance between the Si-atoms is a characteristic of the material, equalling about 5.4 Å. Of course there can be defects of different nature and origin. Dopants can replace the Si-atoms at free lattice points and induce a majority of electrons (n-type) or holes (p-type). High doping levels can to a certain extent alter some physical properties of the material.

Also by alloying silicon with other group IV- elements, different semiconductors can be formed. Very interesting is the situation where heterojunctions are formed between silicon and its alloys. When a silicon-alloy with not too much deviating atomic distance is deposited on silicon and when it is thin enough, this layer can adjust itself in an elastic way to the lattice constant or inter-atomic distance of the silicon substrate. Due to the elastic strain or stress very particular electronic properties can emerge, not existing in the bulk silicon alloy as such. When the deposited layer becomes thicker however, plastic deformation will occur and the deposited layer adjusts itself to its bulk properties via a highly defective transition region.

Polycrystalline silicon consists of different grains of monocrystalline silicon, co-existing next to each other, but with different orientations and joined by so-called grain boundaries. Grain boundaries themselves consist of unsatisfied silicon bonds, defects and are sometimes easily decorated with impurities. The grains can be very small (nm's-0.5 μm) as in the case of low temperature deposited polysilicon on top of silicon oxide or polysilicon deposited on glass e.g. for liquid crystal displays. The grains can be very large as is the case for wafers cut from casted polysilicon blocks, to be used as substrates for solar cell fabrication. In the latter case grain sizes can be from mm's up till cm's.

Organization of the book

This book is organized in 2 parts. The first part deals with (mono-) crystalline silicon and its (strained) alloys. This part consists of 8 chapters, which will briefly be summarized in what follows. Part 2 deals with small and large grain polysilicon and addresses some of its applications. This part contains 4 chapters, totalling to 12 chapters.

In chapter 1 Van Mieghem and Mertens deal with the effects of heavy doping on the properties of crystalline silicon. Heavy dopings are encountered e.g. in emitters of bipolar transistors and solar cells, in source and drain regions of MOSFET's etc. Heavy doping can alter considerably the physical characteristics of the semiconductor material. First the physical mechanisms causing the heavy doping effect are explained. Then experimental results, theories and models will be outlined. Finally the influence of heavy doping effects on the operation of electronic devices is given.

In chapter 2 Claeys and Vanhellemont give an extensive overview of defects in crystalline silicon. Because of the extensive work that has been done worldwide on this subject, the authors had to make choices, because on this subject alone a whole book can be generated. Firstly the structural aspects of the different kinds of lattice defects are discussed, followed by a section on electronic properties. The last section illustrates defect generation during technological fabrication of IC's and the impact of defects on the device performance.

In chapter 3 Kasper and Falco describe the technique of molecular beam epitaxy for metals and for the deposition of silicon and silicon-alloys / heterojunctions. Applications will be outlined for the metals MBE as well as for Si-MBE. Some specific points of attention, such as the cleaning process and the doping problem are addressed. Finally the question is formulated whether MBE is the key technology for future electronic systems.

Low thermal budget CVD-techniques for the deposition of Si and SiGe are described by Caymax and Leong in chapter 4. Prerequisites are described for low temperature growth, followed by a thorough overview of the different low thermal budget CVD-techniques. Consequently the kinetics of growth and doping phenomena are discussed. Also the very interesting possibility of selective growth is discussed. Finally an overview is given of in-situ and ex-situ characterization techniques, that are useful for the study of Si and SiGe-material deposited by the above mentioned techniques.

Consequently in chapter 5 the material properties of (strained) SiGe layers are discussed by Poortmans, Jain, Nijs and Van Overstraeten. Firstly structural and stability issues are addressed, then the bandgap and band-alignments are discussed and finally transport properties are described.

In chapter 6 and 7 Poortmans, Jain and Nijs, respectively Willander discuss the applications of SiGe heterostructures. Chapter 6 deals with the heterojunction bipolar transistor. First the DC-behaviour of the double heterojunction bipolar transistor are outlined . Consecutively the frequency and circuit performance of HBT's with SiGe-base and finally worldwide advanced bipolar technological realizations are shown. In chapter 7 Willander describes different applications such as Si/SiGe field effect transistors, δ-doped field-effect transistors, infrared detectors such as p-i-n detectors, heterojunction internal photoemission detectors, intersubband absorption detectors and finally presents the possibilities of resonant tunneling devices.

In chapter 8 Sugii describes the material properties, growth methods and mechanisms and device applications of a promising new material namely crystalline silicon-carbide. As is the case for SiGe, SiC also can create a heterojunction with silicon. The device applications which are outlined are: the heterojunction bipolar transistor, diodes and field effect transistors and finally light emitting diodes and detectors.

Chapter 9 deals with large grain polycrystalline silicon substrates for solar cells. The advantage in comparison with monocrystalline silicon are that the material is potentially cheaper and perfectly square, leading to a high module fill factor. Martinuzzi and Pizzini firstly describe the different growth techniques for large grain polysilicon (multicrystalline silicon) material. Then the role of defects and impurities on the material properties is outlined. Especially the electrical properties with respect to the solar cell requirements are looked at in detail.

In chapter 10, Migliorato and Quinn describe properties and modeling schemes of polysilicon thin film transistors. First the structural properties and density of states are discussed, followed by a model to describe the influence of the density of states on the device behaviour (current-voltage, capacitance-voltage). Finally the kink-effect and the electrical field at the drain are given specific attention.

In chapter 11, Baert discusses the application and technology of polysilicon thin film transistors for LCD's and projection displays, whereby a lot of attention is paid to direct deposition of polysilicon versus the use of recrystallization. Also the techniques for decreasing the defect density are outlined. Finally a standard coplanar TFT process is described, followed by possibilities to decrease the leakage current on device level.

Finally in chapter 12, Ghannam gives an extensive overview of the use of polycrystalline silicon and its alloys in VLSI applications. First he describes the deposition and structural properties of polysilicon, then technological and electrical properties are given. Moreover an extensive overview of the use of polysilicon in VLSI application and related technologies is given. Finally also the possible use of polycrystalline SiGe in MOSFET's is described. In order to avoid overlap, of course this chapter does not discuss the use of polysilicon for solar cells, neither for active matrix LCD's.

At the end of the book the reader can find the curriculum vitae of editor and authors as well as a keyword index list, guiding him through specific terms used in the chapters of this book.

Figure Acknowledgments

The authors are grateful to the following for granting permission to reproduce figures included in this book.

Academic Press Inc
Akademie Verlag GmbH
American Institute of Physics
American Physical Society
AT&T Bell Laboratories
Editions de Physique
Electrochemical Society Inc
Elsevier Science Publishers BV
Elsevier Sequoia SA
Fujitsu Laboratories Ltd
Institute of Electronic Engineers
Institute of Electrical and Electronics Engineers
Materials Research Society
McGraw-Hill Publishing Co Ltd
Minerals, Metals & Materials Society
Pergamon Press Plc
Philips Research Laboratories
Philips Laboratories
Plenum Publishing Corporation
Sci-Tech Publications
Society for Information Display

The authors have attempted to trace the copyright holder of all the figures reproduced in this publication and apologise to copyright holders if permission to publish in this form has not been obtained.

PART ONE:

Single crystalline silicon and its alloys

1

Heavy Doping Effects in Silicon
P. Van Mieghem and R.P. Mertens

1.1   INTRODUCTION

The doping of a pure semiconductor modifies the material fundamentally. A large number of impurities is inserted in the lattice of the pure semiconductor. These structural changes have important consequences interesting as well from a theoretical as practical point of view. Semiconductors can be doped n-type (p-type) by adding typical donors (acceptors). The type of doping determines the mechanism of conduction. Indeed, donors will add electrons in the conduction band of the pure material and the conduction (at sufficiently high temperature) will be dominated by these donor-electrons. On the other hand, acceptors add holes to the valence band and cause a conduction of holes. Thus, the role of doping is twofold: the high resistivity of the pure semiconductor can be reduced and the type of doping distinguishes between electron (n-type) or hole (p-type) conduction.

The doping concentration can be varied over a large region. When the semiconductor is heavily doped, differences from a lowly doped semiconductor in more than only resistivity are observed. About twenty years ago, anomalies in the current gain of bipolar transistors resulting from a heavily doped emitter were measured. The mechanisms responsible for these supplementary differences, called *heavy doping effects (HDE)*, found an increasing interest. Apart from the bipolar transistor, also the semiconductor lasers seemed to be influenced by HDE. Currently, HDE on semiconductor devices are generally accepted to modify electronic properties. Techniques to dope selected areas very precisely are impressively improved during the last years as demonstrated by the high performance of epitaxial bipolar transistors and primarily by the hetero-junction devices. In research laboratories, a new generation of components, called quantum devices (such as the high electron mobility transistor (HEMT), the resonant tunnel diode (RTS) and superlattices) are investigated. In these devices, quantummechanical effects dominate the operation mechanisms. As the sensitivity of quantum processes exceeds those of the current semi-classical processes, HDE are believed to play an increasing role. HDE can introduce new, remarkable phenomena such as the recently observed Fermi-edge singularities in 2D structure at low temperatures [Schmitt-Rink, 1986; Livescu, 1988; Zhang, 1990].

We will show that HDE consist in both a bandgap narrowing (BGN) and a distortion of the density of states (DOS), called bandtailing. This work confines itself to these two effects. Other topics that involve HDE are omitted. The influence of heavy doping on other semi- conductor parameters such as mobility, lifetime, diffusion coefficients and recombination rate[1] are not considered because the effect of heavy doping on these latter parameters can be related to the important many-body quantity, the self-energy (explained in 1.4.2) [Mahan, 1986]. Numerical techniques for device simulators that include heavy doping models are not discussed but can be found in the work of Bennett and Lowney [1990]. Finally, the presented theoretical discussion is mainly limited to a zero temperature description.

The effects of BGN and the distortion in the DOS have been introduced as a consequence of doping. However, BGN also occurs in non-doped material where a large amount of carriers is piled up. Actually, we will see below that BGN is a many-body effect.

The first chapter of this book is oulined as follows. Section 1.2 explains the physical mechanisms of HDE. Before discussing the theories of HDE in section 1.4, we briefly overview the milestones and most refined theories and experiments in the field of HDE in section 1.3. The last section 1.5 investigates the influence of HDE on some important silicon devices.

1.2   THE PHYSICS OF HEAVY DOPING EFFECTS

Heavy doping effects change the bandstructure of a semiconductor in two particular aspects. First, the bandgap which is a characteristic quantity for each crystalline material, narrows. This effect is called *bandgap narrowing* (BGN). Second, HDE introduce energy levels in the forbidden energy zone resulting in a modification of the density of states (DOS). The latter effect is known as *bandtailing*.

Let us concentrate on a n-type material. Each donor will create an energy level in the forbidden zone that forms the ground state for its weakly bound outer electron. This donor electron circles around the donor atom at a distance of the order of the effective Bohr radius a_B. As long as the donor doping concentration is low, the donor atoms will on the average be distributed far enough from each other so that the donor electron wave functions will neglibly overlap. Thus the impurity levels are well isolated and at sufficiently low temperature, no electronic conduction occurs: the lowly doped semiconductor is an isolator.
Increasing the donor concentration augments the number of impurity states. At a critical doping concentration - the Mott critical doping concentration- the donor

---

[1]We merely refer to del Alamo , Swirhun and Swanson [1985a], del Alamo and Swanson

[1987], Klaassen [1992],   Popp and Weng [1992] (for Si) and to Lundstrom et al.

[1990] (for GaAs) and references herein.

electron wave functions start to overlap significantly such that electronic conduction does not vanish at low temperatures. At and above the Mott critical concentration, the conduction becomes metallic in nature. The overlap of wave functions gives rise to the formation of an impurity band that merges with the conduction band. For still higher doping concentrations, we arrive in the heavily doped regime, where now more than one donor electron moves in the effective Bohr sphere. This large number of donor electrons closely packed to each other will introduce different kinds of interactions that distort both the conduction and valence band.

1.2.1 Electron-impurity interactions

In the heavy doping regime, the large number of electrons will screen the electrostatic Coulomb interaction exerted by the impurity ion on its surrounding donor electron so that the latter becomes more or less unbounded. However, the electron motion is not really unobstructed since the donor ions remain locally electrically active. They polarize the environment. Due to this interaction the individual electron energy is lowered by an amount $æ\Sigma^{ei}$, the electron-impurity self-energy.

The electrons traveling through the crystal can be regarded as balls rolling over a bumpy surface. Some electrons can enter a region where accidently two or more impurity atoms are closer than average to each other. Hence, the potential energy is larger and can even be large enough to capture the electron into a deep energy level. The random distribution of the impurities in the host lattice thus modifies the density of energy states. This effect is called *bandtailing* because a significant number of electrons can be bounded to form energy states lying under the unperturbed band edge.

1.2.2 Electron-electron interactions

Apart from the interactions with donor ions, the electrons interfere among themselves energetically stronger that with the donor ions. In the high density regime, the electrons form a kind of plasma in which the ions slowly oscillate around their equilibrium positions. The very fast motions of the electrons are energetically the most imptant. Figure 1.1. in which the heavy doping effects are quantitatively compared, indicates that electron-impurities effects (curve 8) are indeed less important than the electron-electron interaction and that the statistical bandtail effect (curve 9) contributes least.

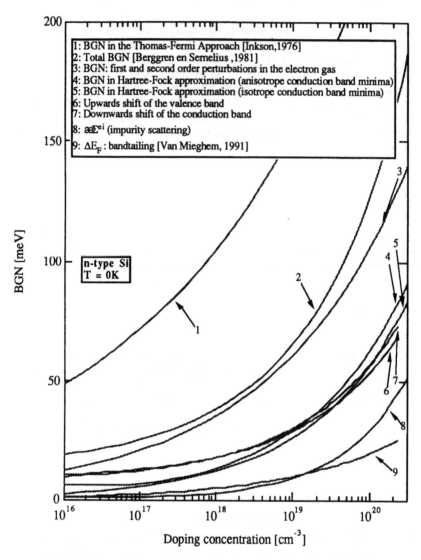

Fig. 1.1 A summary of heavy doping theories (data from Berggren and Sernelius 1981]). The Thomas-Fermi Approximation (1) overestimates BGN. Possible anisotropies of the bandstructure (4-5) contribute insignificantly. The popular random phase method (3) ressembles the most detailed calculations so far, i.e. the theory of Berggren and Sernelius [1981], that gives the total BGN (2) as well as the contributions of the valence- (6) and conduction (7) shift [curve (2) is the sum of (6) and (7)]. The impurity scattering (8) (calculated in the first Born approximation) is comparable to the Fermi-level shift due to bandtailing (9).

The electronplasma interactions can be divided into two classes: an electrical and a purely quantummechanical effect.

1.2.2.1   Correlation effects

The electrical effect - correlation effect - mainly consists in a screening of the charges in the plasma. The long range nature of the bare Coulomb potential is strongly reduced to distances of the order of the mean interparticle separation. This electrical screening initiates the concept of 'correlation hole', an imaginary positive cloud accompanying the electron which causes the same supressed Coulomb potential as the many body screening. As a result of this screening the electron merely interacts with its immediate surrounding neighbours. In addition this screening process involves a rearrangement of a very large amount of electrons. Hence, also collective interactions may occur perturbing the whole electron-plasma. Generally, the correlation effect is described by the dielectric function. It can be shown that the correlation effect in the high density regime is less important than the exchange effect. Therefore it is often viewed as a second (or higher) order perturbation in the self-energy and typically treated in the 'random phase approximation'.

1.2.2.2   The exchange effect

The quantummechanical effect is a consequence of the Pauli exclusion principle. When two electrons with the same energy and spin approach each other, both will change energy through a spin flip. Hence, an apparently enormous force makes identical electrons avoiding each other. Equivalently, an interaction of identical electrons will cause an exchange of energy. This energetic lowering is calculated in terms of the exchange self-energy, $\approx \Sigma^{ex}$. The exchange effect distinguishes the one-electron approximation from the many-body description: non-interacting identical particles can be interchanged without altering the system while interacting identical fermions modify the system's energy by the exchange energy.

The exchange effect can be thought of as an electrical effect. Since the Pauli exclusion principle prevents identical electrons to enter in each other's action sphere, the electron separation is larger than it would be without the Pauli principle. Hence, also the Coulomb interaction is smaller. This reduced Coulomb interaction can be regarded as if the electrons would have less charge at their normal separation distance. Again, we can think of an 'exchange hole' around the electron accounting for the quasi decrease in electron charge.

In summary, heavy doping effects are typically many body effects which can be divided into electron-impurity interactions and electron-electron interactions. The electron-impurity interactions mainly distort the density of states as a result of the stochastic bandtailing effect. The electron-electron interactions and impurity scattering shifts the individual electron states by essentially different amounts. These effects lower the conduction band energy and enhance the valence band energy which causes a narrowing of the bandgap.

1.3   BRIEF REVIEW OF HEAVY DOPING EFFECTS

1.3.1   Theory

The late fifties were very successful for theoretical physics. New formalisms for the solution of quantum mechanical many-body problems have been invented by Feynman, Gell-Mann, Dyson, Brueckner, Pines and others. The first application of this many-body concept to BGN appeared in the work of Wolff [1962]. The early theory of Wolff has been improved successively by Inkson [1976], Mahan [1980] and finally by Berggren and Sernelius [1981]. In the mean time, the statistical problem of the random distribution of impurities in a semiconductor - the bandtail effect - received much attention (see our review [Van Mieghem, 1992]) and actually reached almost perfection for the deep energy region in the work of Halperin and Lax [1965, 1966]. The influence of impurity scattering and the impurity band has been studied in detail by Serre and Ghazali [1983] based on Klauder's multiple scattering theory [Klauder, 1961]. In summary, the current most detailed theories about the understanding of heavy doping effects in 3D are that of Berggren and Sernelius [1981] for BGN, that of Halperin and Lax [1965] and Sayakanit [1980] for the deep bandtails and the Serre and Ghazali multiple-scattering approach [1983] for impurity effects.

During the last years, a number of works investigated the BGN-effect in less than three dimensions (see for references the book by Haug [1990]). The increasing importance of quantum devices (see above), where quantum mechanical effects dominate the operation, are responsible for studies in less than three dimensions.

This section will be ended by connoting some missing elements or weak points in the above mentioned latest theories. As noticed above, the heavy doping effects have been studied as separate entities. The first serious lack is the non-existence of a unified theory both combining BGN and bandtailing. Secondly, all BGN theories are limited to $T=0K$[2]. As phonons come into the picture for finite

---

[2]A temperature dependent formalism to second order in perturbation is presented in the book of Fetter and Walecka [1971]. Calculations about the temperature dependence of the band gap (that only include electron-electron interactions to second order in perturbation) have been presented by Thuselt and Rössler [1985]. A further discussion (also excluding phonons) can be found in Van Mieghem [1991b].

temperatures and the many-body formalism complicates considerably, no accurate calculations of BGN as a function of temperature have been performed. The quantum mechanical bandtail theories are only valid for the deep energy tail. Precise theories for the intermediate states do not exist. In addition, the deep tail theories do not consider many-body interactions and merely assume a Thomas-Fermi approximation (explained in chapter 3). The theory of Serre and Ghazali also describes impurity scattering in the Thomas-Fermi approximation but suffers from occupancy of state corrections and linearizes all integro-differential equations for computational reasons. Finally, all theories assume the validity of the effect mass theorem [Slater, 1949].

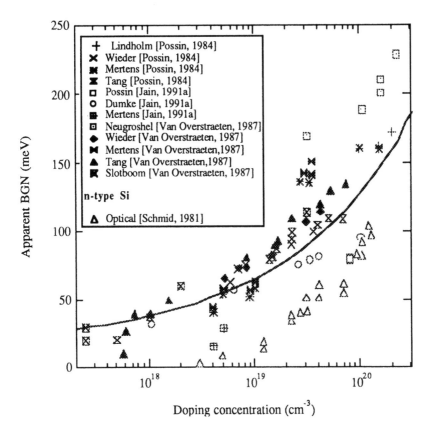

Fig. 1.2a A summary of BGN data on n-type Si found in literature before 1985. The methods determine the apparent BGN from electrical current measurements on a bipolar transistor. Optical absorption data [Schmid,1981] are added for comparison. The line denotes the most accurate theory today of Berggren and Sernelius [1981] at T = 0K. A similar overview can be found in fig. 1 in del Alamo et al. [1985a].

1.3.2 Experiments

Experiments on heavy doping effects can be classified into optical and electrical measurements. In contrast to theory (see Berggren and Sernelius [1981]), no particular measurement results have been taken as a reference.

The optical experiments mainly consist in absorption and photoluminescence measurements (see for an extensive list of references the work of Jain [1991b]). Optical measurements are considered to be more precise and reliable than electrical techniques. Generally, they yield direct information about the distribution of the particles over the energy states (DOS), the shifts in these states (BGN) and the possible impurity effects and lattice defects (indications of the purity of the material). Therefore, most of the theoretical models have been compared with optical data.

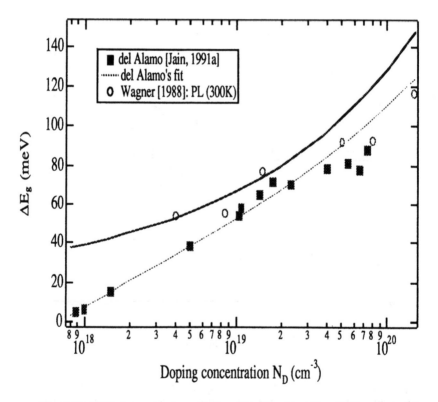

Fig. 1.2b BGN data on n-type Si found in literature after 1985 extracted from optical (PL) and transistor data. Also del Alamo's fit, $\Delta E_g = 18.7 \ln (N_D / 10^{17}) + 25.8 \ln [1 + 0.27 N_D/N_C]$, is shown. The full line denotes the Berggren and Sernelius' theory [1981] at T = 0K. The discrepancy between T = 300K and T = 0K has been discussed by Jain and Roulston [1991a].

Electrical measurements extract information about BGN from current measurements on mainly bipolar transistors (see the review by Mertens et al. [1981]) or capacitance measurements on diodes [Lowney, 1985; Van Mieghem, 1990a, 1990b; Jain, 1988]. Electrical measurements are less accurate than optical measurements and generally measure device parameters which are related to BGN. For example (see also sec. 1.5), in a bipolar transistor expressions for the current contain the quantity $n_{ie} = n_{io}exp[\Delta E_g/2k_BT]$, while capacitance measurements give information about the effective bandgap through the built-in potential. Thus, electrical measurements determine BGN indirectly and strongly depend on the accuracy of the model which connects the measured quantity to heavy doping effects (BGN and DOS distortions). Notwithstanding these disadvantages electrical measurements are important for the design of devices. Although the measured electrical BGN, the apparent BGN, deviates from the true, physical BGN, it is a direct indicator for the heavy doping effects in a particular device.

A comparison of electrical with optical data has created a chaotic situation during many years. A historical overview of the situation before 1985 is sketched in fig. 1.2. The reasons are clear as indicated above: in general, electrical techniques do not precisely yield values of the real (or optical) BGN and second, the electrical measurements do show a significant scattering of BGN data (fig. 1.2a) due to the limited accuracy and differences in both device and measurement technique.

After 1985, much effort has been devoted to bring order in the published data (see e.g. the proceedings of the conference on HDE edited by Landsberg, [1985]). Several old data have been re-interpreted [del Alamo, 1985b]. In particular, Wagner and del Alamo [1988] have shown that optical (absorption and photoluminescence) data agrees well with the Berggren and Sernelius theory (fig. 1.2b and fig. 5 in Lanholt-Börstein [1987] pp. 264). Further, they illustrate that the electrical data on bipolar transistors corresponds well to both theory as optical measurements if the apparent BGN is interpreted properly. A good fit to the existing data for n-type Si, $\Delta E_g = 18.7 \ln (N_D /7 \ 10^{17} \text{ cm}^{-3}) + kT_B \ln \left[1 + 0.27 \exp\{(E_F - E_c)/kT_B\}\right]$ agrees well (at 300K) with optical measurement (fig. 1.2b). From this agreement, they argue that BGN measured by optical techniques is the same as the BGN relevant to device properties. A similar conclusion (with an even better agreement between optical and electrical measurements and theory) was found from capacitance measurements and photoluminescence data by Van Mieghem et al. [1990b] in GaAs. Today, experimental data (absorption, photoluminescence, transistor data and capacitance measurements) agree with each other and with theory with in 25 meV (see fig. 1.2b) when the Fermi-level lies in the conduction (valence) band.

1.4   THEORIES AND MODELS

1.4.1   The Fermi level shift due to heavy doping effects: formal description

In an intrinsic material the Fermi level can be written in terms of the unperturbed density of state (DOS) $\rho_0(E)$ as $E_F[\rho_0(E)]$. As explained before, heavy doping effects (HDE) change the distribution of the individual energy levels in a semiconductor resulting in a DOS $\rho(E) = \rho_0(E) + \Delta\rho(E)$ where $\Delta\rho(E)$[3] reflect the heavy doping influence. Since the Fermi level is a typical macroscopic quantity, we would like to know how it changes due to HDE. More precisely, we are looking for $\Delta E_F = E_F[\rho_0(E) + \Delta\rho(E)] - E_F[\rho_0(E)]$. In general the change in DOS, $\Delta\rho(E)$, is small compared to $\rho_0(E)$ which justifies a low order expansion. Applying functional theory [Feynman and Hibbs, 1965] to first order, we have

$$E_F[\rho_0(E) + \Delta\rho(E)] = E_F[\rho_0(E)] +$$

$$\int_{-\infty}^{\infty} \left.\frac{\delta E_F}{\delta\rho(u)}\right|_{\rho(u) = \rho_0(u)} \Delta\rho(u)\, du\ +\ o(\Delta\rho(E)) \tag{1.1}$$

Assuming electrical neutrality[4] in a heavily doped n-type material and writing the Fermi Dirac distribution function as $f_{FD}(x) = \dfrac{1}{1+\exp[x/k_BT]}$, the general relation connecting microscopic and macroscopic properties is

$$N = n = \int_{-\infty}^{\infty} \rho(E)\, f_{FD}(E-E_F)\,dE \tag{1.2}$$

Functional derivation of (1.2) with respect to $\rho(u)$ yields

---

[3]More generally, $\Delta\rho(E)$ may represent any effect (such as HDE, lattice stress, small magnetic fields, ...) that alters the unperturbed DOS slightly.

[4] The study of HDE has been concentrated on a space charge free semiconductor material. Lowney [1985] has shown that HDE are hard to define in and around space charge regions.

$$f_{FD}(u-E_F) - \left(\int_{-\infty}^{\infty} \rho(E) \frac{df_{FD}(E-E_F)}{dE_F} dE \right) \frac{\delta E_F}{\delta \rho(u)} = 0 \qquad (1.3)$$

Evaluating the change in Fermi level due to HDE around the non-perturbed DOS, $\rho(u) = \rho_0(u)$, results in

$$\left. \frac{\delta E_F}{\delta \rho(u)} \right|_{\rho(u) = \rho_0(u)} = \frac{q^2}{\varepsilon \, \kappa_{TF}^2} f_{FD}(u-E_F) \qquad (1.4)$$

where κ_{TF} denotes the inverse screening length in the Thomas Fermi approximation[5],

$$\kappa_{TF}^2 = \frac{q^2}{\varepsilon} \int_{-\infty}^{\infty} \rho_0(E) \frac{df_{FD}(E-E_F)}{dE_F} dE \qquad (1.5)$$

Combining (1.4) and (1.1) gives

$$\Delta E_F = \frac{q^2}{\varepsilon \, \kappa_{TF}^2} \int_{-\infty}^{\infty} f_{FD}(u-E_F) \, \Delta\rho(u) \, du \qquad (1.6)$$

Expression (1.6) is a general relation for the Fermi level shift that only assumes electrical neutrality[6] and $\Delta\rho(u)$ sufficiently small to justify a first order functional expansion.

Let us consider two easy examples. First, we focus on an energy independent $\Delta\rho(E) = a\theta(E)$, where $\theta(E)$ denotes Heavyside's stepfunction. The resulting Fermi level shift equals

---

[5]Notice that $\kappa_{TF}^2 = \frac{q^2}{\varepsilon} \rho_0(E_F)$ for T = 0K. This implies that κ_{TF}^2 vanishes in non-degenerate semiconductors $(E_F < 0)$ as $\rho_0(E) = 0$ for E < 0. Consequently,

$\left. \dfrac{\delta E_F}{\delta \rho(u)} \right|_{\rho(u)=\rho_0(u)}$ is not defined at T = 0K for non-degenerate semiconductors in a

first order functional expansion.

[6]The presented derivation can be extended to the general neutrality condition, $p + N_D^+ = n + N_A^-$ where the minority charges are no longer negligible as in heavily doped semiconductors and where not all the impurities are ionized.

$$\Delta E_F = \frac{q^2 \, a \, k_B T}{\varepsilon \, \kappa_{TF}^2} \ln\left|1 + e^{E_F/k_B T}\right| \qquad (1.7)$$

For heavily doped semiconductors $E_F > k_B T$, so that (1.7) simplifies to

$$\Delta E_F \approx \frac{q^2 \, a \, E_F}{\varepsilon \, \kappa_{TF}^2} \qquad (1.8)$$

If the constant a does not depend on doping concentration N nor temperature T, we find that $\Delta E_F \sim N^{1/3}$ and that the temperature dependence is almost negligible in the high doping regime. A more realistic example describes a situation where bandtailing is absent and all individual energy levels are shifted by the same amount ΔE_g such that $\Delta\rho(E) = \rho_0(E - \Delta E_g) - \rho_0(E)$. Assuming that ΔE_g is small enough to allow a first order expansion, we may write $\Delta\rho(E) = \frac{d\rho_0(E)}{dE} \Delta E_g$ to find from (1.6) with (1.5) that $\Delta E_F = \Delta E_g$ as it should.

It is interesting that the correct temperature dependence of the Fermi level shift due to HDE can be derived from (1.6). Indeed, two limit values can be obtained for $E_F > 0$:

$$\lim_{T \to 0} \Delta E_F = \Delta E_{F_{max}} = \frac{1}{\rho_0(E_F)} \int_{-\infty}^{E_F} \Delta\rho(u) \, du > 0 \qquad (1.9)$$

and

$$\lim_{T \to \infty} \Delta E_F = 0 \qquad (1.10)$$

As ΔE_F is a continuous function of T, from a physical point of view, ΔE_F is expected to decrease monotonically, but slowly in the temperature interval $(0, \infty)$. A further specification of ΔE_F necessitates the knowledge of $\Delta\rho(E)$.

1.4.2   The many body description of a heavily doped semiconductor

In this section, we introduce the formalism of many-body theory which offers methods to calculate ΔE_g, $\rho(E)$ and $\Delta\rho(E)$. General definitions and equations will be given.

1.4.2.1 Interacting systems and Green's functions

All quantities we will need are conveniently described in momentum space and although a temperature dependent Green's function formalism exists [Fetter and Walecka, 1971; Mahan, 1986], we will start with the simpler zero temperature case.

The Dyson equation (1.12) diagrammatically presented in fig. 1.3 describes the interacting system starting from the non-interacting system and thus gives the formal solution for a HDS. The Green's function for a particle with wave vector $\mathbf{k}$ and energy $æ\omega$ in a non-interacting system with eigen energies $æ\omega_{\mathbf{k}}^{0}$ is

$$G^{0}(\mathbf{k},\omega) = \frac{1}{\omega - \omega_{\mathbf{k}}^{0} + i\eta} = \frac{1}{\omega - æ^{-1}\varepsilon_{\mathbf{k}}^{0} + i\eta} \qquad (1.11)$$

where $\eta < 0$ for occupied and $\eta > 0$ for empty states. Assuming that the interaction is invariant under translations and that the system is spatially uniform, the Dyson equation reads

$$G(\mathbf{k},\omega) = G^{0}(\mathbf{k},\omega) + G^{0}(\mathbf{k},\omega)\Sigma(\mathbf{k},\omega)G(\mathbf{k},\omega) \qquad (1.12)$$

where $G(\mathbf{k},\omega)$ is the Green's function of the interacting system and $\Sigma(\mathbf{k},\omega)$, which is called the self-energy, accounts for the additional energy due to internal interactions. This Dyson's equation can be solved explicitly as

$$G(\mathbf{k},\omega) = \frac{1}{[G^{0}(\mathbf{k},\omega)]^{-1} - \Sigma(\mathbf{k},\omega)} \qquad (1.13)$$

or after substitution of (1.11)

$$G(\mathbf{k},\omega) \equiv \frac{1}{\omega - \omega_{\mathbf{k}} + i\eta} = \frac{1}{\omega - æ^{-1}\varepsilon_{\mathbf{k}}^{0} - \mathrm{Re}(\Sigma(\mathbf{k},\omega)) - i\mathrm{Im}(\Sigma(\mathbf{k},\omega))}$$

$$(1.14)$$

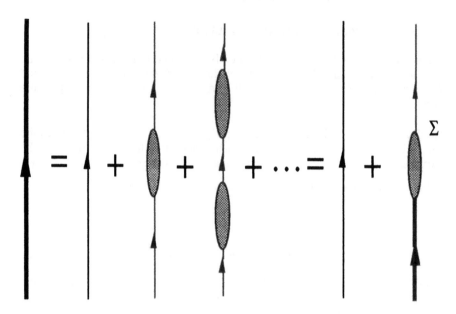

Fig. 1.3 General structure of $G(\mathbf{k},\omega)$ (Dyson's equation). The light line denotes $G^0(\mathbf{k},\omega)$ whereas the heavy line denotes $G(\mathbf{k},\omega)$. A substitution of the heavy line yields an infinite-order expansion of $G(\mathbf{k},\omega)$ in terms of $\Sigma(\mathbf{k},\omega)$ and $G^0(\mathbf{k},\omega)$.

Notice, that in the latter expression $i\eta$ is now irrelevant and that the solution for ω of $\mathrm{Im}(\Sigma(\mathbf{k},\omega)) = 0$ defines the chemical potential μ (or Fermi level E_F): energy states above μ are empty ($\mathrm{Im}(\Sigma(\mathbf{k},\omega)) < 0$) while energy states below μ are occupied ($\mathrm{Im}(\Sigma(\mathbf{k},\omega)) > 0$). The density of states is defined as

$$\rho(E) = \frac{1}{V_0} \sum_\alpha \delta(E - E_\alpha) \tag{1.15}$$

where E_α is an energy eigenvalue belonging to the eigenfunction $\phi_\alpha(\mathbf{r})$ of some particular Hamiltonian and V_0 denotes the macroscopic volume. This function $\rho(E)$ is a (real positive) non-analytic function of the real variable E and, for a unit volume, is written in terms of Green's functions using the identity for $\eta \to 0^+$,

$$\frac{1}{E - E_\alpha \mp i\eta} = P\left(\frac{1}{E - E_\alpha}\right) \pm i\pi\delta(E - E_\alpha) \tag{1.16}$$

as

$$\rho(E) = \mp\frac{1}{\pi} \mathrm{Im}\left[\sum_\mathbf{k} G(\mathbf{k}, E \pm i\eta)\right] = \mp\frac{1}{\pi} \mathrm{Im}[\mathrm{Tr}\, G(\mathbf{k}, E \pm i\eta)] \tag{1.17}$$

where Tr denotes the trace of the matrix. The latter is the general definition because usually, the Green's function is not a diagonal matrix. With (1.14), we find

$$\rho(E) = \sum_{k=0}^{k_r} \left\{ \frac{1}{\pi} \frac{\text{Im}(æ\Sigma(\mathbf{k},E))}{[E - \varepsilon_k^0 - \text{Re}(æ\Sigma(\mathbf{k},E))]^2 + [\text{Im}(æ\Sigma(\mathbf{k},E))]^2} \right\}$$

(1.18)

The quantity between {} is called the spectral density $A(\mathbf{k},E) = \mp \frac{1}{\pi} \text{Im}(G(\mathbf{k},E \pm i\eta))$, which is the probability that an electron having energy E is in state $|\mathbf{k}\rangle$; the DOS is then expressed as $\rho(E) = \text{Tr}(A(\mathbf{k},E))$. The interaction clearly has two effects. It broadens all states so that, for a given ω, a range of momenta contribute to ρ, and produces a level shift of amount $\text{Re}(æ\Sigma)$. From (1.14) and (1.18), we see that the HDS description will crucially depend on the self-energy of the system.

1.4.2.2   Band Gap Narrowing (BGN)

The previous section has sketched the formalism to describe electron gas properties or the electron behaviour in a particular energy band of a HDS. This knowledge will be applied to one of the most important consequences of many-body interactions: the experimentally observed shrinkage of the energy gap[7],

$$\Delta E_g = E_g^0 - E_g$$

(1.19)

The expression for the Green's function (1.14) suggests that the interacting electron system may be interpreted as described by non-interacting quasi-particles. The corresponding quasiparticle dispersion for the conduction band is.

$$E_c(\mathbf{k},\omega) = \varepsilon_c^0(\mathbf{k}) + æ\Sigma_c(\mathbf{k},\omega)$$

(1.20)

In the case of the valence band, however, the unperturbed energy $\varepsilon_v^0(\mathbf{k})$ already contains a Hartree-Fock exchange contribution because the band is completely

---

[7]Contributions to this shift due to the direct influences of impurity centres such as local strain, etc. (which are independent of carrier concentration n) are not considered.

occupied. When adding free carriers, either electrons or holes, one is only interested in the shifted self-energy. The relevant quasiparticle energy is

$$E_V(k,\omega) = \varepsilon_V^o(k) + \text{æ}\Sigma_V(k,\omega) - \text{æ}\Sigma_V^{HF}(k) \qquad (1.21)$$

This was first noted by Inkson [1976]. The shift of the bandgap relative to the band edges is

$$\Delta E_g = \text{æ}\Sigma_V(0, \text{æ}^{-1}\varepsilon_V^o(0)) - \text{æ}\Sigma_V^{HF}(0) - \text{æ}\Sigma_C(0, \text{æ}^{-1}\varepsilon_C^o(0)) \qquad (1.22)$$

In general, the energies involved are complex (see (1.14)). Since the imaginary parts of the energies (corresponding to lifetime broadening effects) are normally much smaller than the real parts, the real part of the energies appearing in (1.22) has been used. The minimal energy to remove (or add) a particle is reduced by an amount given by (1.22). In optical experiments, excited particles must end up above the Fermi-level (at zero temperature) and often the effective reduction in energy gap is measured at the Fermi level yielding,

$$\Delta E_F = \text{æ}\Sigma_V(0, \text{æ}^{-1}\varepsilon_V^o(0)) - \text{æ}\Sigma_V^{HF}(0) - \text{æ}\Sigma_C(k_F, \text{æ}^{-1}\varepsilon_C^o(k_F)) \qquad (1.23)$$

It is interesting to note that the shift in band gap is expressed in terms of quasi-particle self-energies, rather than arising from a total energy shift due to many-body effects. A total energy shift ensues from all energy bands while it is only the difference between valence and conduction band shifts which will be measured [Berggren 1981, Abram 1984]. This point of view [Inkson 1976] differs from the early theories based on Wolff's work [Wolff 1962]. The widely used interpretation of BGN as a rigid shift of valence and conduction band towards each other is seen to be incorrect and would require wave vector independent self-energies. However, the rigid band shift picture is a good approximation in 3D[8]. The screening of the Coulomb potential localizes the interaction in coordinate space or equivalently, extends over a broader range of wave-vectors in k-space and flattens the k-vector dependence of the self-energy. The mathematical complexity prevents analytical expressions for (1.22) and (1.23). Accurate calculations (at T = 0K) [Berggren 1981] exhibit a $n^{1/3}$ dependence (see fig. 1), which is often used in experiments to extract BGN.

---

[8]In 2D and especially at low temperature, enhanced interactions round the Fermi-level lead to Fermi-edge singularities and exhibit a strong k-dependence of the self-energy [Schmitt-Rink 1986, Livescu 1988, Zhang 1990].

1.4.3   BGN theories

1.4.3.1   The semi-classical model of Lanyon and Tuft [1979]

Lanyon and Tuft have developed a model for BGN which takes into account the electrostatic energy of interactions between a minority carrier and the majority carriers surrounding it. This energy reduces the thermal energy required to create an electron-hole pair. In contrast with preceding theories that calculated HDE only in the conduction or valence band, their model calculates a pair energy similar to the excitonic binding energy of bound electron-hole pairs in semiconductors at low temperature. Their idea is, in fact, a reduction of the activation energy of creation of minority carriers rather than a true reduction of the energy gap, E_g.

They calculated, semi-classically, the pair energy and equate it to the bandgap reduction. They state that the background majority carrier concentration affects the degree to which the electron-pair is localized. At low majority carrier concentrations the pair is widely separated, resulting in very little binding energy whereas high background doping concentrations the pair is in close proximity with large binding energy. This binding energy has been calculated as the difference in electrostatic energy between a screened Coulombic field E(r) and the bare Coulomb field, $E_0(r) = \dfrac{q}{4\pi\varepsilon r^2}$. The screened Coulomb field was computed in a Thomas-Fermi approximation [Ashcroft and Mermin,1981; Mahan, 1986], resulting in $E(r) = -\nabla V(r) = \dfrac{q}{4\pi\varepsilon r}\left(e^{-\kappa r}\left(1/r + 1/\kappa\right)\right)$

The carrier pair self-energy is taken as the difference between the energy stored in these fields,

$$\Delta E_g = \frac{\varepsilon}{2} \int_{\text{all space}} \left(E_0^2(r) - E^2(r) \right) dr$$

which yields

$$\Delta E_g = A \frac{q^2}{4\pi\varepsilon} \kappa \qquad (1.24)$$

They obtained the value 3/4 for the quantity A. Since the significance of A can not be taken to seriously as the model is only approximative, A (of the order of 1) can be used as fit parameter [Van Mieghem et al., 1990].

The good agreement of this simple model with experiment (see e.g. our measurement of BGN on abrupt diodes [Van Mieghem et al., 1990]) is surprising because of several reasons. Apart from the semiclassical approach instead of applying many-body theory, the model neglects exchange and correlation interactions, electron-impurity interactions, bandtailing and impurity band broadening. Moreover, Mahan [1980] objected that Lanyon and Tuft assumed an infinite efffective mass. He further showed that due to this localized hole the hole self-energy is considerably overestimated. In addition, Jain and Roulston [1990] mentioned that (1.24) does not give correct values for other semiconductors.

1.4.3.2   The variational model of Mahan [1980]

Mahan [1980] has calculated all contributions to BGN in n-type Si and n-type Ge with the omission of bandtailing. The principle idea in calculating BGN ressembles that of Lanyon and Tuft in the sense that the energy gap is defined as the minimum energy to create an electron-hole pair. In contrast to Lanyon and Tuft, Mahan first considers the minimum energy required to add an electron (which is the chemical potntial), he then determines the minimum energy to produce a hole (Σ^h) and defines the difference between these two quantities as the optical energy gap[9] $E_{go} + E_F + \Sigma^h$. First, he considers the downwards shift of the conduction band, ΔE_c, and states that

$$\Delta E_c = \Sigma^{ee}_{HF} + \Sigma^{ee}_{corr} + \Sigma^{ei} \qquad (1.25)$$

The first contribution in (1.25) reflects the exchange self-energy (in the Hartree-Fock approximation [Van Mieghem, 1991b]),

$$\Sigma^{ee}_{HF} = \frac{q^2}{2\pi^2\varepsilon} \Lambda \frac{k_F}{2} \qquad (1.26)$$

where Λ is a factor to account for anisotropic effective mass. The correlation term, Σ^{ee}_{corr}, was neglected, because Lundqvist [1967] showed that correlation term was much smaller than the exchange term. Moreover, Mahan assumes that the wavevector dependence of $\Sigma^{ee}_{HF} + \Sigma^{ee}_{corr}$ is small and that (1.26) is a reasonable expression. The electron-impurity interaction is somewhat more complied due to the distribution of screening charge around the ion. The first Born approximation (assuming non-uniform screening charge distribution) in the Thomas-Fermi approach (which assumes a uniform screening charge distribution) is thus logically inconsistent while, as Mahan argued, also inaccurate. The uniformity of the screening charge around the donor can be modeled in two extreme ways: a localized screening cloud (leading to the density derivative of Lanyon and Tuft

---

[9]The reference of E_F is the bottom of the unperturbed conduction band, while the reference for Σ^h is the top of the unperturbed valence band.

model (1.24) or, in Mahan's notation, $\Sigma^{ei} = \delta\mu_{i'} = -\frac{1}{8}\frac{q^2}{4\pi\varepsilon}\kappa)$ and a truly uniform electron density with no density variation due to donor screening (leading to $\Sigma^{ei} = \delta\mu_i = -0.481\frac{q^2}{4\pi\varepsilon}n^{1/3}$). Both extremes models, $\delta\mu_{i'}$ and $\delta\mu_i$, give very similar results for Si doped between 10^{18} cm^{-3} and 10^{20} cm^{-3}. Mahan then performed a more rigorous variational calculation which accounts for the inhomogeneity in the electron screening charge around the impurity and found

$$\Delta E_{cSi} = -12.58\left(n/10^{18}\right)^{1/3} - 3.1 \qquad [meV] \qquad (1.27)$$

The upwards shift of the valence band was described by,

$$\Delta E_v = \Sigma^{hi} + \Sigma^{he} + \Sigma^h_{corr} \qquad (1.28)$$

The first self-energy, Σ^{hi}, reflects the hole (a minority carrier in n-type material) scattering with the screened donor. The second self-energy, Σ^{he}, denotes the interactrion of the hole with conduction band electrons (without back reaction on the hole), while the last term corresponds to the hole correlation energy origination from the electron-hole interaction who act back upon the hole. Mahan found that the first two terms exactly cancel in the first Born approximation, leaving the correlation self-energy. The latter was calculated in the plasmon-pole approximation. Similar to the electron case, Mahan then calculated variationally the electron wave function in the electron-hole system combining the hole energy terms and arrived at

$$\Delta E_{vSi} = -13.1\left(n/10^{18}\right)^{1/4} + 6.1\left(n/10^{18}\right)^{1/3} [meV] \quad (1.29)$$

The BGN was then found by adding (1.27) and (1.29). However, the impurity interactions Σ^{hi} and Σ^{ei} cancels each other in Mahan model because he assumed a regular impurity sublattice. As explained below, Berggren and Sernelius [1981] showed that this is incorrect. Thus, the Mahan model does not treat the impurity contribution properly.

In the low density limit, Mahan correctly finds the exciton binding energy. However, the variational approach diminishes the transparancy of Mahan's model. The overall agreement with experiments is very satisfying.

1.4.3.3	The model of Berggren en Sernelius

In 1976, Inkson [1976] pointed out that HDE affects all energy bands and that a calculation of the many-body influences in only one band, neglecting the other bands, is definitely incorrect. This point of view was already an improvement over the early work of Wolff [1962] who was among the first to apply the Feynman diagrams to HDE in semiconductors. Inkson used a simplified many-body approach (invoking the Thomas-Fermi approximation) resulting in the simple expression for BGN,

$$\Delta E_g = \frac{q^2}{2\pi^2 \varepsilon} k_F \left(1 + A(k_F, \kappa) \right) \tag{1.30}$$

with	$A(k_F, \kappa) = \frac{\kappa \pi}{2k_F} - \frac{\kappa}{k_F} Atan\left(\frac{k_F}{\kappa} \right)$

a correction on the Hartree-Fock approximation. However, his calculations overestimated BGN, as shown by Abram et al. [1978]. Abram et al. improved Inkson's model by taking the Lindhard dielectric function instead of that of Thomas-Fermi and have extensively compared their results with experiment.

The most detailed calculation up to now has been performed by Berggren and Sernelius [1981]. They have rigorously applied many-body theory up to second order and at zero temperature and zero wavevector k (although they later gave explicit results at $k = k_F$ [Berggren and Sernelius, 1985]). They also showed that the influence of conduction band anisotropy is almost negligible (in n-Si). A general Green's function description for the complex valence band was deduced (without considering the spin-orbit splitt-off band), but they limited the numerical calculation to zero wavevector for simplicity. Their total result for BGN is somewhat larger than that of Abram et al. (who omitted the supplementary energy shift due to impurity scattering) but smaller than Inkson's result (1.30).

The final part of their paper discusses in detail comparison with Mahan's calculation. The overall agreement was reasonably good except that the contribution of the impurity scattering was larger. In Mahan's work the impurities were assumed to form a regular sublattice in the host semiconductor. As a result of this unrealistic assumption, the contribution of hole scattering (valence band) and electron scattering (conduction band) turn out to cancel each other. On the contrary, Berggren and Sernelius who assumed a random distribution of impurities found that they added. However, the inpurity scattering contributes less to BGN than the other effects as follows from fig. 1.1. In fact, both most refined theories did not include properly stochastic effects giving rise to exponential bandtails.

The Berggren and Sernelius results (see fig. 1.1) show that BGN varies as $N^{1/3}$-law for sufficiently high doping concentrations. This agrees well with experiments.

1.4.3.4 The Jain-Roulston model

The model of Jain and Roulston [1991a] combines theoretical simple models (based on characteristic lengths) and empirical results. The proposed model assumes that BGN consists of energetic contribution of the majority carrier exchange ΔE^{ex}_{maj}, minority carrier correlation ΔE^{cor}_{min} and both majority ΔE^{i}_{maj} and minority ΔE^{i}_{min} carrier impurity scattering,

$$\Delta E_g = \Delta E^{ex}_{maj} + \Delta E^{cor}_{min} + \Delta E^{i}_{maj} + \Delta E^{i}_{min} \qquad (1.31)$$

which closely relies on Mahan's results for exchange and correlation and on Berggren and Sernelius for the impurity part. They have worked out (1.31) in dimensionless parameters as the effective Rydberg energy $R = \dfrac{13.6\, m}{\varepsilon^2}$ (eV), the effective Bohr radius $a_B = \dfrac{4\pi\varepsilon\hbar^2}{me^2}$ and $r_s \equiv \dfrac{r_0}{a_B}$ with the interparticle distance $r_0 = \left(\dfrac{3}{4\pi n}\right)^{1/3}$ to obtain

$$\frac{\Delta E_g}{R} = 1.83\,\frac{\Lambda}{N_b^{1/3}\, r_s} + \frac{0.95}{r_s^{3/4}} + \left[1 + \frac{R_{min}}{R}\right]\frac{1.57}{N_b\, r_s^{3/2}} \qquad (1.32)$$

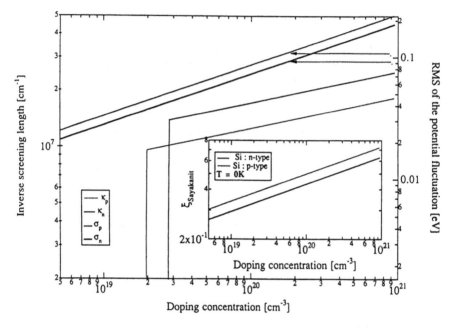

Fig. 1.4 The inverse screening length κ, the second moment of the potential fluctuation and in the inserted graph, the Sayakanit dimensionless parameter, $\xi_{Sayakanit} = \dfrac{\sigma^2}{E_\kappa^2}$, where $E_\kappa = \dfrac{\hbar^2\kappa^2}{2m}$ versus doping concentration for n- and p-type Si at zero temperature and calculated in the Thomas-Fermi approximation. Notice the abruptness in inverse screening length occurring when the Fermi level enters the band (at T = 0K) (see footnote 3). Only if $E_F > 0$, the parameters shown make sense. The inserted graph eases interpretation of the folowing two figures below.

For p-type semiconductors, the two parameters in their model, N_b and Λ, were determined from fittings of experimental values of BGN with (1.32) and we refer to the parameter table they gave in their paper. The merit of this model is that it is both simple and accurate and well suited for a first order study of HDE in still unstudied semiconductors, such as SiGe and SiC.

1.4.4 Bandtail theories

Bandtails in the DOS originate from a stochastical effect. The random distribution of impurity ions cause potential fluctuations over the semiconductor lattice. If locally a potential fluctuation is enough pronounced to allow for a supplementary energy state lying in the forbidden energy zone, an electron can be captured into this energy state modifying the DOS.

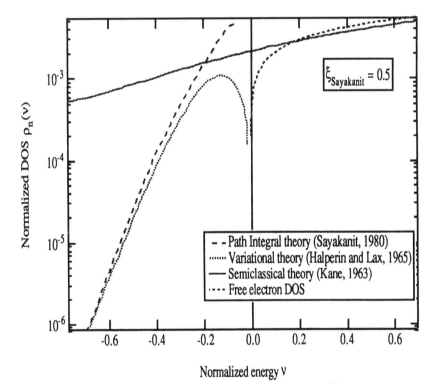

Normalized energy v

Fig. 1.5 Important bandtail theories for the parameter $\xi_{Sayakanit} = 0.5$ (data extracted from Sayakanit et al. (1980)) in normalized quantities:

$$v = \frac{E}{E_K} \text{ and } \rho_n(v) = \frac{E_K \xi_{Sayakanit}^2}{\kappa^3} \rho(E) = a(v) \exp[\frac{b(v)}{2\xi_{Sayakanit}}]$$

where $a(v)$ and $b(v)$ are dimensionless functions of the dimensionless energy v.

In the past, much effort has been devoted to calculate this tailed DOS. Although a detailed review of the most important bandtail theories is beyond the scope of this article, but can be found in Van Mieghem [1992a], we can summarize as follows. The deep energy tail in the DOS has been studied in great detail by a variety of quantummechanical techniques. All converge to the same asymptotic form first established by Halperin and Lax [1965] (see fig. 1.5). The high energy region is well described by semiclassical theories [Van Mieghem, 1991a; Kane, 1963], because a semiclassical approach can be demonstrated to converge to the exact high density limit.

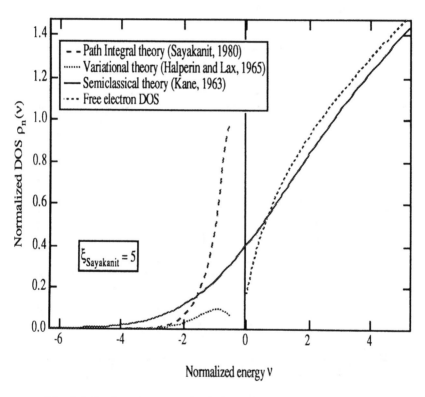

Fig. 1.6 Graph analogous to fig. 1.5 but on a linear scale and for
ξSayakanit = 5.

The intermediate energy region poses some problems because the form of the potential fluctuation - unfortunately unknown - is here rather important. The quantummechanical deep tail theories assume a most probable shape of the potential fluctuation and a Gaussian distribution function (facilitating further computations). The variational deep tail theories (Halperin and Lax [1965] and Efros [1971,1974]) only consider the lowest energy state in a potential fluctuation and neglect possible higher energy states. More recent deep tail theories such as the path integral approach [Sa-yakanit, 1980] and the replica method [Cardy,1978; John et al.,1988, 1990; Van Mieghem, 1992a] are for computational reasons limited to quadratic actions leading to Gaussian integrals. Physically, a confinement to quadratic actions means the assumption of a Gaussian distribution function although a broad class of correlation functions can be chosen. All the available theories neglect many-body effects and assume the effective mass theorem to be valid. These shortcomings result in the lack of theories descibing the intermediate energy region properly. Various interpolation schemes between the deep tail and high energy region are proposed but none of them garantees a good physical description. An accurate solution of the bandtail problem is certainly a formidable task. Because of the lesser importance of the bandtail effect in HDE, imperfections in the bandtail theories do not induce in general serious errors.

For majority carriers in heavily doped semiconductors, semiclassical theories yield sufficiently accurate results. Minority carrier properties (such as their amount, mobility, lifetime diffusion coefficient,...), on the other hand, may strongly depend on the shape of the tailed DOS.

1.5   INFLUENCE OF HDE ON SILICON DEVICES

1.5.1   Modification of the minority   carrier transport equation

In most semiconductor devices the doping concentration depends on the position. Therefore, in the general case, the bandgap is also position dependent. This necessitates a modification of the well known drift-diffusion equation. As demonstrated by Kroemer [1957], a position dependent bandgap gives rise to quasi-fields that have to be added to the conventional electrostatic field. The quasi-field, acting on the minority carriers, strongly affects the device performance and depends on the effective intrinsic carrier concentration n_{ie} defined as

$$n_{ie} = n_{io}exp(\Delta E_g/2k_B T) \qquad (1.33)$$

In (1.33) ΔE_g is the apparent bandgap narrowing (see sec. 1.3.2). From a practical viewpoint bandgap narrowing is most important in a heavily doped neutral region of a minority carrier device. In such a region the expression of the minority carrier quasi-field, E_{quasi}, becomes

$$E_{quasi} = \frac{k_B T}{q} \left[\frac{1}{n_{ie}^2} \frac{dn_{ie}^2}{dx} \right] = \frac{1}{q} \frac{d\Delta E_g}{dx} \qquad (1.34)$$

In n^+ neutral material the hole current j_p can be written as [Van Overstraeten et al., 1973]

$$j_p = q p \mu_p \frac{k_B T}{q} \left[\frac{1}{n_{ie}^2} \frac{dn_{ie}^2}{dx} - \frac{1}{N_D} \frac{dN_D}{dx} \right] - q D_p \frac{dp}{dx} \tag{1.35}$$

1.5.2   Influence of HDE on a bipolar transistor

In a VLSI npn bipolar transistor the emitter is normally transparant for the holes injected from the base. Equation (1.35) can then be integrated over the emitter region. If the surface recombination velocity at the emitter surface equals S, the emitter injected hole density can be expressed as [Van Overstraeten et al., 1973]

$$j_p = \frac{q n_{io}^2 \exp(q V_{BE}/k_B T)}{\int\limits_{emitter} (N_D/D_p) (n_{io}/n_{ie})^2 \, dx + (N_{D,s}/S) (n_{io}/n_{ie,s})^2} \tag{1.36}$$

In (1.36) V_{BE} is the applied base-emitter voltage, N_D the position dependent donor concentration, D_p the hole diffusion constant in the emitter; $N_{D,s}$ and $n_{ie,s}$ refer to the surface donor concentration and the intrinsic carrier concentration at the surface respectively.

Formula (1.36) clearly shows the influence of HDE on I_p. As I_p is the largest component of the base current, the common emitter current gain will be proportional to the denominator of (1.36). Using the expressions for BGN derived in the previous sections and typical emitter doping profiles, we obtain average values for $(n_{ie}/n_{io})^2$ larger than 10. If S is higher than 10^4 cm/s, the first term in the denominator of (1.36) normally dominates and BGN results in common emitter current gains that are at least one order of magnitude smaller. BGN also explains the temperature dependence of the current gain.

The gain-bandwidth product f_T of a bipolar transistor is also influenced by BGN [Van Mieghem et al., 1992b]. In the absence of collector resistance effects the maximum value of f_T is given by

$$\frac{1}{2\pi \, f_{T;max}} = \tau_e + \tau_b + \tau_c \tag{1.37}$$

In (1.37) τ_e, τ_b and τ_c refer to the neutral emitter, neutral base and collector depletion layer delay times respectively. Without BGN effects τ_e is negligible as it is proportional to the minority carrier emitter charge and therefore to the weighted average of n_{io}^2/N_D over the emitter. This value of n_{io}^2/N_D in the emitter is indeed very small due to the high emitter doping level. With BGN the emitter charge becomes important and cannot be neglected anymore with respect to τ_b and τ_c.

Also the fall-off of f_T at low current levels is influenced by BGN because the emitter-base space charge layer capacitance increases due to a lower value of the built-in voltage [Van Overstraeten et al., 1973].

In general, BGN causes a poorer performance of a bipolar transistor. On the other hand recently [Suzuki et al., 1989] the concept of a pseudo-HBT (hetero junction bipolar transistor) has been introduced. In this transistor BGN has been used to enhance the device performance. By using a structure with a very thin base that is more heavily doped than the emitter, BGN effects will occur in the base more strongly than in the emitter. In this way the beneficial effects of a heterojunction emitter-base structure are obtained.

1.5.3   Influence of HDE on a solar cell

In a modern silicon solar cell two heavily doped regions are present: the n^+ diffused emitter and the p^+ backsurface field region. BGN enhances the injection of holes in the emitter and from electrons in the p-p^+base (see (1.35)).This results in lower values of the open circuit voltage and therefore also of the efficiency of the solar cell.

1.5.4   Influence of HDE on a Si MOSFET

A MOSFET is a majority carrier device and therefore its performance is affected to a smaller extent by BGN than minority carrier devices. However doping levels in source and drain are normally sufficiently high to cause BGN. The main effect of BGN in the source is to lower the current gain of the parasitic bipolar transistor. Consequently, the importance of bipolar breakdown in short-channel MOS transistors is overestimated if BGN in the source is not included.

BGN effects not only occur in the source region but also in the inversion layer under conditions of strong inversion. In that case BGN has to be treated as a two-dimensional problem. Calculations [Girisch et al.,1989] indicate that BGN in such a 2D system is compensated by an effective energy-gap widening resulting from the upward shift of the electro-chemical potential caused by band filling. As a result the "effective" energy gap is found to be almost concentration independent over a large electron-density range.

At very low temperature, potential fluctuations that cause bandtailing can be responsible for obstructing the transport in the channel [Hafez et al., 1990].

1.6   SUMMARY

This first chapter has briefly overviewed the field of heavy doping effects in Si. The physical mechanisms have been emphasized and the most important theoretical achievements as known today are discussed. Heavy doping effects arise from electron-electron and electron-impurity interactions and results mainly from the many-body nature of these interactions.

The influence of HDE on Si devices may be summerized roughly as

$\Delta E_g \sim N^{1/3}$

$\Delta E_g \sim 10 \% \ E_{go}$ of Si

ΔE_g almost temperature independent if the semiconductor is degenerate.

The most important examples of HDE on Si devices are illustrated.

Acknowledgements. We would like to express our gratitude to Professor S.C. Jain for useful discussions.

1.7   REFERENCES

Abram, R. A., G. J. Rees and B. L. H. Wilson, 1978, *Heavily doped semiconductors and devices*, Advances in Physics, 27, 6, 799-892.

Abram, R. A., G. N. Childs and P. A. Saunderson, 1984, *Band gap narrowing due to many-body effects in silicon and gallium arsenide*, J. Phys. C: Solid State Phys., 17, 6105-6125.

Ashcroft, N. W. and N. D. Mermin, 1981, *Solid State Physics*, Saunders College, Philadelphia.

Bennett, H. S and J. R. Lowney, 1990, *Physics for numerical simulation of silicon and gallium arsenide transistors*, Solid State Electron., Vol. 33, No. 6, 675-691.

Berggren, K.-F. and B. E. Sernelius, 1981, *Band-gap narrowing in heavily doped many-valley semiconductors*, Phys. Rev. B, 24, 4, 1971-1986.

Berggren, K.-F. and B. E. Sernelius, 1985, *Very Heavily Doped Semiconductors as a "Nearly-Free-Electron-Gas" System*, Solid-State Electronics, 25, 1/2, 11-15.

Cardy, J. L., 1978, *Electron localisation in disordered systems and classical solutions in Ginzburg-Landau field theory*, J. Phys. C, 11, L321.

del Alamo, J. A., and R. M. Swanson, 1987, *Modeling of minority carrier transport in heavily doped silicon emitters*, Solid-State Electronics, Vol. 30, Nos. 11, pp 1127-1136.

del Alamo, J. A., S. Swirhun and R. M. Swanson, 1985a, *Measuring and modeling minority carrier transport in heavily doped silicon*, Solid-State Electronics, Vol. 28, Nos. 1/2, pp 47-54.

del Alamo, J. A., S. Swirhun and R. M. Swanson, 1985b, IEDM, Washington, DC (IEEE, New York), p. 290

Efros, A. L., 1971, *Theory of electron states in heavily doped semiconductors*, Sov. Phys. - JETP, 32, 3, 479-483.

Efros, A. L., 1974, *Density of states and interband absorption of light in strongly doped semiconductors*, Sov. Phys.-Usp., 16, 6, 789-805.

Fetter, A. F. and J. D. Walecka, 1971, *Quantum Theory of Many-Particle Systems*, McGraw-Hill, New York.

Feynman, R. P. and A. R. Hibbs, 1965, *Quantum Mechanics and Path Integrals*, (McGraw-Hill, N. Y.)

Girisch, R., R.Mertens and O.Verbeke, 1990, *Energy-gap change in silicon n-type inversion layers at low temperature*, Solid State Electronics, Vol.33, No.1, 85-91.

Hafez, I. M., G. Ghibaudo, and F. Balestra, 1990, *Assessment of interface state density in silicon metal-oxide-semiconductors at room, liquid-nitrogen, and liquid-helium temperatures*, J. Appl. Phys. 67,4, pp. 1950-1952.

Halperin, B. I. and Melvin Lax, 1966, *Impurity-Band Tails in the High-Density Limit. I. Minimum Counting Methods*. Phys. Rev., 148, 2, 722-740.

Halperin, B. I. and Melvin Lax, 1967,*Impurity-Band Tails in the High-Density Limit. II. Higher Order Corrections*, Phys. Rev., 153, 3, 802-814.

Haug, H. and S. W. Koch, 1990, *Quantum Theory of the Optical and Electronic Properties of Semiconductors*, World Scientific, Singapore, pp. 97-111.

Inkson J. C., 1976, *The effect of electron interaction on the band gap of extrinsic semiconductors*, J. Phys. C, 9, 1177-1183.

Jain, S. C. and D. J. Roulston, 1991a, *A simple expression for the band gap narrowing (BGN) in heavily doped Si, Ge, GaAs and Ge_xSi_{1-x} strained layers*, Solid-State Electronics, Vol 34, No. 5, pp. 453-465.

Jain, S. C., R. P. Mertens and R. J. Van Overstraeten, 1991b, *Bandgap Narrowing and its effects on the Properties of Moderately and Heavily Doped Germanium and Silicon*, Academic Press, 82

Jain, S. C., R. P. Mertens, P. Van Mieghem, M. G. Mauk, M. Ghanam, G. Borghs and R. J. Van Overstraeten, 1988, *Effects of Bandgap Narrowing on the Capacitance of Silicon and GaAs pn junctions*, Proc. IEEE Bipolar Circuits and Technology Meeting, pp. 195.

John, S. and C. H. Grein, 1990, *Instantons, Polarons, Localization and The Urbach Optical Absorption Edge in Disordered Semiconductors*, Reviews of Solid State Science, Vol. 4, No. 1, 1 - 59 (World Scientific).

John, S., M. Y. Chou, M. H. Cohen and C. M. Soukoulis, 1988, *Density of states for an electron in a correlated Gaussian random potential: Theory of the Urbach tail*, Phys. Rev. B, 37, 12, 6963-6976.

Klaassen, D. B. M., 1992, *A unified mobility model for device simulation - II Temperature dependence of carrier mobility and lifetime*, Solid State Electron., Vol. 35, No. 7, 961-967.

Klauder, J. R., 1961,*The modification of electron energy levels by impurity atoms*, Annals of Physics, 14, 43 - 76.

Kroemer, H., 1957, *Quasi-electric and quasi-magnetic fields in non uniform semiconductors*, RCA Rev., vol. 18, 332-342.

Landsberg, P. T., 1985, editor of the proceedings of the conference on *Heavy doping and the metal-insulator transitions in semiconductors*, Solid-State Electron., vol. 28, No 1/2.

Lanholt-Börnstein, 1987, New Series, Group III, Volume 22, subvolume a, edited by O. Madelung, p. 264.

Lanyon, H. P. D. and Richard A. Tuft, 1979, *Bandgap Narrowing in Moderately to Heavily Doped Silicon*, IEEE Trans. Electron Devices, ED-26, 1014-1018.

Livescu, G., David A. B. Miller, D. S. Chemla, M. Ramaswamy, T. Y. Chang, Nicholas Sauer, A. C. Gossard and J. H. English, 1988, *Free Carrier and Many-Body Effects in Absorption Spectra of Modulation-Doped Quantum Wells*, IEEE Journal of Quantum Electronics, 24, 8, 1677-1689.

Lowney, J. R., 1985, *Band-gap near in the space-charge region of heavily doped silicon diodes*, Solid-State Electron., Vol. 28, Nos. 1/2, 187 - 191.

Lundqvist, B. I., 1967, Phys. Kond. Mater. 6, 193, 206.

Lundstrom, M. S., M. E. Klausmeier-Brown, M. R. Melloch, R. K. Ahrenkiel and B. M. Keyes, *Device-related material properties of heavily doped gallium arsenide*, Solid State Electron., Vol. 33, No. 6, 693-704.

Mahan, G. D., 1980, *Energy gap in Si and Ge : Impurity dependence*, J. Appl. Phys., 51, 5, 2634-2646.

Mahan, G. D., 1986, *Many-Particle Physics*, Plenum Press, N. Y.

Mertens, R. P., R. J. Van Overstraeten and H. J. De Man, 1981, Advances in Electronics and Electron Physics, vol. 55, Academic Press, pp 77

Popp, J. and J. Weng, 1992, *Measurement of transport parameters in heavily-doped emitters of bipolar transistors*, Solid State Electron., Vol. 35, No. 7, 999-1003

Sa-yakanit, V. and H. R. Glyde, 1980, *Impurity-band density of states in heavily doped semiconductors: A variational calculation.* Phys. Rev. B, 22, 12, 6222-6232.

Schmid, P. E., 1981, Phys. Rev. B23, 5531.

Schmitt-Rink, S., C. Ell and H. Haug, 1986, *Many-body effects in the absorption, gain and, luminescence spectra of semiconductor quantum-well structures*, Phys. Rev. B, 33, 2, 1183-1189.

Serre, J. and A. Ghazali, 1983, *From band tailing to impurity-band formation and discussion of localization in doped semiconductors: A multiple-scattering approach*, Phys. Rev. B, 28, 8, 4704.

Slater, J. C, 1949, *Electrons in Perturbed Periodic Lattices*, Phys. Rev., 76, 11, 1592-1601.

Suzuki, K., T.Fukano, T.Yamazaki, S.Hijiya, T.Ito and H.Ishikawa, 1989, *Pseudo-HBT with Polysilicon Emitter Contact and an Ultrashallow Highly doped Base by Photoepitaxy*, IEDM Technical Digest, 811-814.

Thuselt, F. and M. Rösler, 1985, *Gap Shift in Doped Semiconductors at Finite Temperatures*, phys. stat. sol. (b) 130, 661- 673.

Van Mieghem, P., 1991b, *Heavy Doping Effects in Semiconductors*, Ph.D. Thesis, (K. U. Leuven, Belgium.)

Van Mieghem, P., 1992a, *Theory of Bandtails in Heavily Doped Semiconductors*, Rev. Mod. Phys., Vol. 64, No. 3, pp. 755-794.

Van Mieghem, P., G. Borghs and R. Mertens, 1991a, *Generalized semiclassical model for the density of states in heavily doped semiconductors*, Phys. Rev. B, 44 (23),12822-12829.

Van Mieghem, P., R. P. Mertens and R.J. Van Overstraeten, 1990a, *Theory of the junction capacitance of an abrupt diode*, J. Appl. Phys. 67(9), 4203-4211.

Van Mieghem, P., R. P. Mertens, G. Borghs and R.J. Van Overstraeten, 1990b, *Band-gap narrowing in GaAs using a capacitance method*, Phys. Rev. B., 41 (9), 5952-5959.

Van Mieghem, P., S. Decoutere, G. Borghs and R. Mertens, 1992b, *Influence of Majority Bandtails on the Performance of Semiconductor Devices*, Solide-State Electronics Vol. 35, No. 5, pp. 699-704.

Van Oversraeten, R., H. De Man and R. Mertens, 1973, *Transport equations in heavily doped silicon*, IEEE Trans. Electron Devices, ed-20, 290-298.

Wagner, J. and J. A. del Alamo, 1988, *Band-gap narrowing in heavily doped silicon: A comparison of optical and electrical data*, J. Appl. Phys. 63 (2), 425-429.

Wolff, P. A., 1962, *Theory of the band structure of very degenerate semiconductors*, Phys. Rev., 126, 2, 405.

Zhang, Y., De-Sheng Jiang, R. Cingolani and K. Ploog, 1990, *Fermi edge singularity in the luminescence of modulation-doped GaInAs/AlIn As single heterojunctions*, Appl. Phys. Lett., 56, 22, 2195-2197.

2

Defects in Crystalline Silicon

C. Claeys and J. Vanhellemont

2.1 INTRODUCTION

This chapter gives a brief overview of different aspects of defects in crystalline silicon. As restrictions in total length do not allow to give a complete in-depth review, preference has been given to a discussion of some basic aspects of the structural and electrical properties of the defects in order to obtain a better fundamental understanding. The first section discusses the structural aspects of lattice defects, while the second section focusses on their electronic properties. The third section is aimed to illustrate the defect generation and behaviour during integrated circuit processes, and also includes a short discussion on the impact of crystallographic defects on the electrical device performance.

2.2 STRUCTURE OF LATTICE DEFECTS

2.2.1 Semiconductor silicon: crystal structure and physical properties

The equilibrium phase of crystalline semiconductor silicon under normal pressure and temperature conditions has the diamond cubic structure as schematically shown in figure 2.1. This structure is face-centered cubic (fcc) with two atoms per unit cell. The lattice constant a is 0.5431063 nm at room temperature and the density of atoms C_{Si} is 5.028 10^{28} m^{-3}. Other phases of silicon which are commonly used during integrated circuit processing are amorphous and crystalline and polycrystalline silicon mostly deposited by chemical vapour deposition (CVD) techniques. Under certain conditions small clusters of hexagonal silicon, which is a high pressure phase of silicon, have also been observed in silicon substrates after low temperature processing. A number of other high pressure silicon phases obtained under extreme pressure conditions are also reported in the literature, some of them having metallic or even superconducting properties (Gupta and Ruoff 1980).

An exhaustive overview of the electrical, mechanical and optical properties of doped and undoped silicon, together with solubility and diffusivity data of most common dopants and metals in silicon can be found in the reference work "Properties of Silicon" (Ning 1987). The following paragraphs give a short overview of the structure and properties of lattice defects in silicon. Some basic aspects of the electronic properties of lattice defects are given in section 2.3.

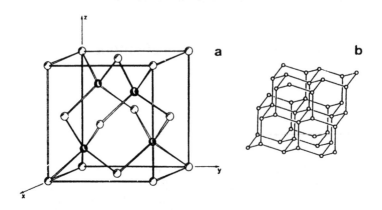

Figure 2.1 a) Schematic representation of the diamond cubic structure with the <001> crystallographic axes. b) View along the [001] direction with the (111) planes horizontal.

2.2.2 Intrinsic and extrinsic point defects

Intrinsic point defects

A crystal lattice is formed by a periodic three dimensional arrangement of atoms. In this periodic structure two types of intrinsic point defects can be formed, i.e. self-interstitials and vacancies. In the simplest model self-interstitials are atoms of the same element which forms the crystal lattice, but in positions in-between the lattice sites. The diamond structure has a low packing density with an occupancy of only 34% in the hard sphere model, thus favouring the formation of self-interstitial point defects rather than vacancies. The silicon lattice thus contains much less vacancies in thermal equilibrium than e.g. metals. The observation that most of the extended defects in processed silicon are of the interstitial type is a clear illustration that self-interstitials are the dominant intrinsic point defect in silicon. In fcc lattices the possible positions for interstitials are $(\frac{1}{4},\frac{1}{4},\frac{1}{4})$ in tetrahedral and $(\frac{1}{2},\frac{1}{2},\frac{1}{2})$ in octahedral cavities in the lattice, as illustrated in figure 2.2. Vacancies are lattice sites where an atom is missing. Both type of defects are schematically illustrated in figure 2.3. There still exists some controverse on the physical structure of the self-interstitial "point" defect in silicon. The discussion is mainly about the position of the self-interstitial atom: tetrahedral, hexagonal bond centre or even about a more sophisticated description of the self-interstitial as dumb-bell configuration of two silicon atoms with a double bond in-between them, the whole located on a lattice site (Mainwood *et al* 1978, Frank 1975). As one of the double bonds of the dumbbell can easily be opened the self-interstitial can be charged.

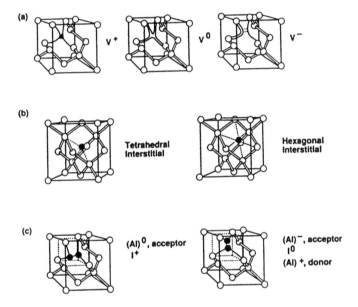

Figure 2.2 Illustration of the position of the cavities in the silicon lattice. Schematic representations are given of (a) a vacancy in three different charge states, (b) tetrahedral and octahedral cavities, (c) dopant ($A \neq I$) or silicon ($A = I$) interstitial defects involving two atoms A and I.

A large number of physical phenomena in silicon such as diffusion of extrinsic point defects, temperature dependence of the yield stress, oxidation kinetics... are strongly correlated with the presence and behaviour of intrinsic point defects, so that for more than three decades a continuous effort has been going on in order to obtain a better understanding of their properties. Due to the difficulties to address intrinsic point defects - their properties can only be studied indirectly through their influence on other macroscopical measurable physical parameters - a large uncertainty and dispute still exists with respect to their properties. An extensive review on point defect properties and their influence on dopant diffusion in silicon has recently been published by Fahey *et al* (1989).

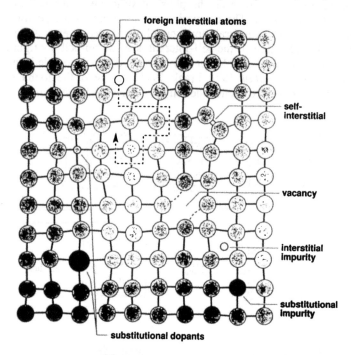

Figure 2.3 Schematic representation of intrinsic (vacancy V, self-interstitial I) and extrinsic point defects in a simple cubic crystal lattice (Gösele and Tan 1991).

The number of intrinsic point defects in thermal equilibrium depends on the temperature and on their formation energy and can be estimated using basic statistical physics. For a constant pressure, the equilibrium density of point defects C_x^* (x = V for a vacancy and x = I for a self-interstitial) is given by

$$C_x^* = \frac{N_{Si}}{1 + e^{\frac{\Delta G_x}{kT}}} \tag{2.1}$$

with N_{Si} the number of silicon lattice sites and ΔG_x the Gibbs free energy for the formation of the point defect. Due to their long lifetime, self-interstitials and vacancies are already present in supersaturation in silicon crystals after solidification from the melt. Their grown-in concentrations strongly depend on the cooling rate of the ingot and on the subsequent heat treatments. These grown-in point defects and their interactions with extrinsic defects lie on the basis of the formation of so-called swirl defects which are observed in Float Zone silicon. More details on the different types of grown-in defects in starting silicon is given in section 2.4.

Since the early experiments by Watkins in 1965 a substantial amount of information on intrinsic point defects found in the literature is based on low temperature electron irradiation experiments, which are analysed with Electron Spin Resonance spectroscopy. A review on this topic has been published by Watkins *et al* (1978). One of the main conclusions is that the intrinsic defects can have several charge states due to sequential trapping and detrapping of charge carriers generated by ionisation processes. A strong interaction with extrinsic point defects, i.e. dopant atoms, has also been observed. Recently, Romano-Rodríguez and Vanhellemont (1992) obtained additional insight into the different interactions between irradiation induced point defects and dopant atoms by using a high-voltage transmission electron microscope for in-situ 1 MeV irradiations. Theoretical calculations are often hampered by the controversial values for the migration energies and formation enthalphies that have been reported. Therefore further careful experimental and theoretical work is required to improve the understanding of point defect behaviour during irradiation.

Extrinsic point defects

Extrinsic point defects can be present either in interstitial or in substitutional positions, as schematically shown in figure 2.3. As the size of the extrinsic point defects differs from that of silicon lattice atoms, they can cause a macroscipical measurable contraction or expansion of the silicon lattice. Celotti *et al* (1974) have used powder and single-crystal X-ray techniques for a precise measurement of the lattice parameters, and obtained the values 0.117, 0107 and 0.091 for the tetrahedral radii of silicon, phosphorus and boron respectively. The lattice contraction coefficient β for substitutional impurities with covalent radius r_{ext} introduced in the silicon matrix with N_{Si} lattice sites and lattice atomic radius r_{si} is given by

$$\beta = \frac{1}{N_{Si}} \left[1 - \left(\frac{r_{ext}}{r_{si}} \right)^3 \right] \qquad (2.2)$$

To understand the influence of substitutional impurities on defect nucleation, this stress component has to be added to the other process-induced stress components present in the material as shown by Vanhellemont and Claeys (1991a). As will be discussed in section 2.4, in heavily doped silicon the strain field may be sufficiently large to introduce so-called misfit dislocations.

There exists a large variety of extrinsic point defects in silicon. In general, it is possible to classify them into four main groups, depending on their importance for silicon processing :

- oxygen, carbon and nitrogen. These atoms have a strong impact not only on individual processing steps, but also on the overall device performance and yield aspects. The can be either inherently present in the starting material or introduced during the device processing.

- dopants such as phosphorus, arsenic, boron, antimony, gallium ... These atoms are essential for the device fabrication and are therefore introduced in a well controlled manner during either the crystal growth or the epitaxial and CVD deposition processes, and by ion implantation or by indiffusion.

- transition metals (TM's). These metals can be introduced by contamination during the device processing (e.g. from metal tweezers, chucks, gas tubing ...) and even by some wafer cleaning procedures. After silicon processing the detectable metals mostly belong to the 3d group, although a limited amount of work has also been published on 4d and 5d TM's.

- chalcogen atoms. In recent years intensive investigations started on the behaviour of chalcogen related donors in S, Se and Te doped silicon. More details on the properties of extrinsic point defects can be found in the review by Claeys and Vanhellemont (1987).

Solubility of point defects

If one assumes that the formation energy of a point defect X is given by E_X^f one can write for the thermal equilibrium concentration C_X of point defects X (X = vacancies V, self-interstitials I or an extrinsic point defect P in a substitutional or interstitial position)

$$C_X = c_X \, e^{-\dfrac{E_X^f}{kT}} \quad \text{with} \quad c_X = C_{Si} - C_X \qquad (2.3)$$

In most cases $C_X \ll C_{Si}$ so that $c_X \approx C_{Si}$. The solubility of intrinsic point defects is less well known than that of extrinsic point defects as it is only accessible through indirect experiments. The available data for the equilibrium concentration of self-interstitials and vacancies show thus a large scatter as illustrated in figure 2.4a. Recent solubility data for metals and dopants are shown in figures 2.5a and 2.6.

Diffusion of point defects

A concentration (C_X, X= I, V, P) gradient will cause a flux J_X of point defects described by the phenomenological relation known as Fick's law :

$$J_X = -D_X \text{ grad } C_X \qquad (2.4)$$

J is the number of point defects crossing a unit area per time unit, while D is the diffusion constant expressed as a surface per unit time. The diffusion constant depends on the temperature and can generally be written as

$$D_X = D_{0X} \, e^{-\frac{E_a}{kT}} \qquad (2.5)$$

with E_a the activation energy of the diffusion process.

With respect to the diffusivity of self-interstitials an even larger uncertainty exists as reflected by the larger spread of the data shown in figure 2.4b. For the moment there is no agreement on the exact mechanism of self-diffusion (and also of diffusion of point defects in silicon in general). Three different schools exist proposing :

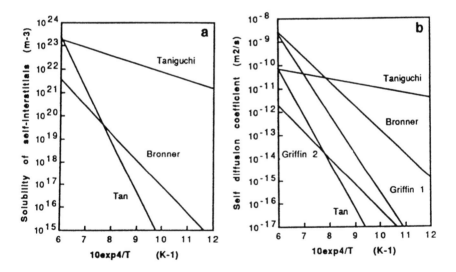

Figure 2.4 (a) Bronner and Plummer (1987a) solubility data of intrinsic point defects in silicon. The arrows indicate the upper bounds calculated from gold diffusion experiments. (b) The diffusivity of the self-interstitials in silicon as a function of the temperature. The arrows indicate the upper (a) and lower (b) bounds calculated from gold diffusion experiments.

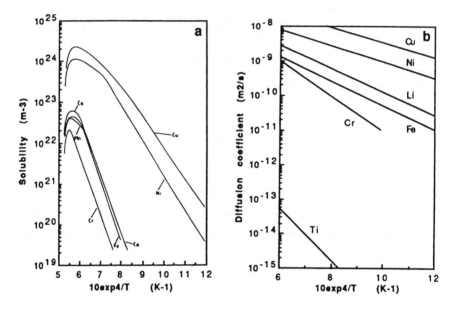

Figure 2.5 Solubility (a) and diffusivity (b) data of metals in silicon as a function of the temperature, after Graff (1982).

- a pure vacancy,
- a pure interstitial and
- a combined self-interstitial-vacancy mechanism.

Although the nature of the predominant intrinsic point defects in diffusion processes, i.e. self-interstitial or vacancy driven, is still controversial, it is becoming more and more accepted that both types of intrinsic point defects not only coexist in silicon but also participate simultaneously in many physical processes. Seeger and Chik (1968) claim that the self-diffusion data below 1173 K are consistent with a vacancy mechanism, while at higher temperature a self-interstitial mechanism is dominant. This leads to the following equation for the self-diffusion coefficient D^{SD}

$$D^{SD} = D_I C_I^* + D_V C_V^* \qquad (2.6)$$

where $D_I(D_V)$ is the diffusivity and C_I^* (C_V^*) the thermal equilibrium

concentration of self-interstitials (vacancies). A large number of estimates can be

found in the literature for the values of D_I and D_V, and the estimates differ by

many orders of magnitude. In the case of experimentally determined self-diffusion coefficients, the self-diffusion coefficient can be written in the form

$$D^{SD} = d^{SD} e^{-\frac{E^{SD}}{kT}} \qquad (2.7)$$

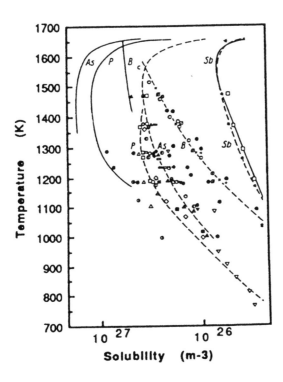

Figure 2.6 Steady state solubility of dopants in silicon, after Borisenko and Yudin (1987). The full lines represent the old Trumbore data from 1960. The dashed lines are calculated using the corrected thermodynamic parameters.

In the temperature range between 1123 K and 1673 K the reported values for the activation energy range between 4.1 and 5.1 eV. The pre-exponentional factor d^{SD} however, may range between 5.8 and 9000. Diffusion data for metals in silicon are given in figure 2.5b.

Lifetime of intrinsic point defects

A self interstitial and a vacancy can recombine by the formation of an undisturbed lattice site O of silicon, according to the reaction

$$I + V \Leftrightarrow O \tag{2.8}$$

Expressing thermal equilibrium leads immediately to the mass action law commonly used in solid state physics

$$C_I C_V = \text{constant} = C_I^* C_V^* \tag{2.9}$$

Hu (1985) and Tan (1986) critically reviewed the problems of recombination of vacancies and interstitial and they pointed out, however, that this law is not generally valid. They showed that a supersaturation of self-interstitials can coexist with a supersaturation of vacancies. A reaction barrier preventing the spontaneous recombination of V and I is introduced in order to explain the observations.

Another recombination mechanism of intrinsic point defects is the disappearance at the silicon surface, which is always less than a few hundred microns away. As integrated circuit processing often creates large stress fields at the wafer surface, this will play an important role both as source and sink of intrinsic (and extrinsic) point defects. Quantitative data on the recombination of self-interstitials with vacancies are scarce. Well accepted, however, is that the lifetime increases strongly with decreasing temperature. Tan and Gösele (1985) calculated the recombination lifetime τ_r and found that the value increases from about 1 day at 1273 K to several weeks at 1173 K. This result strongly indicates that there exists only a weak coupling between the concentration of vacancies and self-interstitials and that for the thermal treatments commonly used for integrated circuit processing they can even be considered as completely independent of each other. The situation becomes even more complex as it is known that intrinsic point defects can be electrically charged and can form complexes of two or more point defects. This topic is addressed in the following paragraph.

It has been reported (Corbett and Bourgoin 1975) that both vacancies and self-interstitials can exist in different charge states. For vacancies four different charge states, i.e. neutral (V), single negatively charged (V$^-$), double negatively charged (V$^=$) and single positively charged (V$^+$), have been postulated although some of these charge states are still controversial. Basically this implies that intrinsic point defect clusters can exist in silicon, and their configurations have been studied by using Electron Paramagnetic Resonance spectroscopy (EPR). There still exists a dispute on the stability of self-interstitial/vacancy pairs in view of their reaction barrier (Gösele *et al* 1982), and the complexity of the problem is mainly caused by i) the amphoteric nature of the point defects, ii) the influence of the Fermi level on the type of ionised pair formed, iii) the screening effect by the dielectric constant, and iv) the impact of the presence of ionised extrinsic acceptor and or donor impurities.

2.2.3 Clustering of point defects: pairing

The formation of defect clusters is strongly dominated by the charge state of the defects. Coulombic pairing occurs if opposite charged defects are attracted towards each other. This pairing process of point defects X_1 and X_2 will be controlled by the law of mass action and can be represented as

$$[X_1^+ X_2^-] = K_{pair} [X_1^+] [X_2^-] \qquad (2.10)$$

with K_{pair} the so-called pairing constant given by (Lambert and Reese 1968)

$$K_{pair} = k\, e^{\frac{E_b}{kT}} \qquad (2.11)$$

k is a constant and E_b is the binding energy between the point defects.

Pairs of intrinsic point defects

The first step of the formation of isolated self-interstitials and vacancies is the creation of so-called Frenkel pairs. A Frenkel pair can be generated by e.g. a knock on collision of an incident electron with a silicon lattice atom which moves the lattice atom from its equilibrium position. The displaced atom can be described as a close neighbouring self-interstitial/vacancy pair. Most of these

Frenkel pairs are unstable and recombine in a very short time resulting in an undisturbed lattice site. In a small fraction of them the separation of the Frenkel pair is large enough so that the self-interstitial and the vacancy can be considered as independent from each other. These intrinsic point defects can then form pairs and even larger clusters. The dominant stable intrinsic point defect pair which is observed in silicon is the di-vacancy which can also occur in different charged states depending on the position of the Fermi level. This defect is stable up to 400°C.

Intrinsic-extrinsic point defect clusters

The most studied intrinsic-extrinsic point defect clusters are the so-called A-centre (oxygen-vacancy complex, [VO]) and E-centre (phosphorus-vacancy complex, [PV]). These defects are the dominant type of clusters formed in irradiated n-type silicon. The A-centre has an acceptor level 0.17 eV below the conduction band caused by an unpaired electron in the antibonding orbital. Based on EPR analyses, Watkins (1975) has determined the structure of both the A-center and the E-center and came up with the models shown in figure 2.7. The oxygen-vacancy pair contains one oxygen impurity and has an infrared vibrational band v_3 at 828 cm^{-1}. The unpaired electron is shared between two of the silicon atoms, giving rise to the g-tensor and hyperfine interaction in the EPR spectrum. For the E-center, the unpaired electron is located in the broken bond of a Si atom neighbouring a vacancy. For a correct interpretation of experimental spectra, one has to take into account that Jahn-Teller distorsion has a pronounced effect on the electronic degeneracy of the defect centre. The E-centre has an acceptor level at about 0.40 eV below the conduction band.

Extrinsic point defect clusters

A large variety of electrically active defect clusters are known which consist of two extrinsic point defects. A well known and technologically important example is the iron-boron (FeB) pair which is often observed in boron doped silicon when iron contamination is unintentionally introduced during the processing [Kimerling and Benton 1983]. Other well known clusters are inherent to the substrate itself and are formed during irradiation with e.g. high energy electrons. Kimerling *et al* (1989) performed a detailed DLTS study of electron irradiation induced defect clusters in silicon and found that the dominant vacancy related defects in phosphorus doped silicon, i.e. the A- and the E-center, have a self-interstitial related counterpart which has a coincident deep level but with a different annealing behaviour. One of their important conclusions is that the silicon self-interstitial has a parallel of every known vacancy reaction with the reaction products often more stable. As the self-interstitial is much more mobile it can not be detected directly but only through its reaction products with extrinsic point defects. A typical interaction is the so-called kick-out mechanism through which a self-interstitial atom kicks-out a substitutional extrinsic point defect thus rendering it mobile. This mechanism commonly occurs with boron and carbon atoms thus leading indirectly to important electrically active point defect clusters such as interstitial boron/substitutional boron (B_iB_S), interstitial boron/interstitial oxygen (B_iO_i) and interstitial carbon/substitutional carbon (C_iC_S).

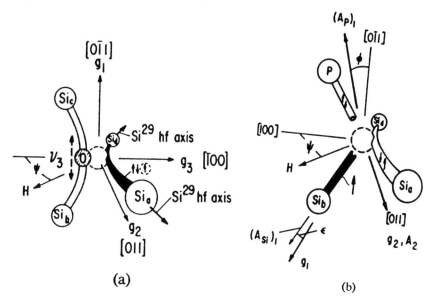

Figure 2.7 Model for the A-centre (a) and E-centre (b) based on EPR spectroscopy, after Watkins (1975).

2.2.4   Line defect : dislocation

As many excellent textbooks, e.g. Weertman and Weertman (1965), Hull (1965, Kovacs and Zsoldos (1973), Nabarro and Ravi (1981), exist on the basic properties of dislocations only the most important notions, which are essential for the understanding of process induced defects and gettering, will be briefly recalled. The structure of dislocations in the diamond structure has been discussed in detail by Amelinckx (1979). Alexander (1986) published an extensive review on dislocation dynamics and the plastic properties of silicon.

Burgers vector of a dislocation

A dislocation is completely defined by its Burgers vector and by its direction. The Burgers vector can be obtained by drawing a Burgers circuit which is any closed loop around a dislocation going from atom to atom as illustrated in figure 2.8. If the same path is followed in a dislocation free crystal and the loop does not close, the vector required to close the loop is called the Burgers vector **b**. For an edge dislocation, respectively a screw dislocation the Burgers vector is perpendicular, respectively parallel to the dislocation line. When the closing vector is a lattice vector the dislocation is called perfect, in the other case one speaks of a partial dislocation. Dislocations must form either closed loops or must end at the crystal surface. When dislocations, e.g. b_1 and b_2, interact and form nodes, conservation of the Burgers vector must occur, e.g. $b_3 = b_1 + b_2$.

Dislocations can easily move by glide in the (glide or slip) plane formed by the Burgers vector and the dislocation line. This motion is conservative which means that no additional intrinsic point defects are absorbed. When a dislocation loop is not lying in its glide plane, movement by glide is impossible. In that case the loop can only grow by a non-conservative (climb) mechanism which requires the absorption of intrinsic point defects.

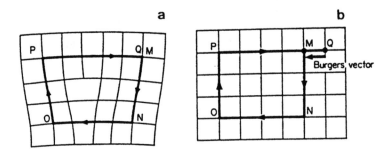

Figure 2.8 Schematic representation of a Burgers vector circuit around a dislocation (a). In (b) the same circuit is drawn in a perfect crystal. The closure vector is the Burgers vector.

Edge and screw dislocations

Self-interstitials present in supersaturation in areas with concentrated stress can easily form line defects or dislocations. These line defects are fully characterised by their direction and their Burgers vector. The dislocation energy is proportional to the square of the Burgers vector. In silicon the shortest vector between two lattice positions is $\frac{a}{2}$ <110>. Dislocations with this Burgers vector will thus have the lowest formation energy and are called perfect dislocations as the corresponding displacement (Burgers) vector is part of the lattice. Dislocations with displacements which cannot connect two lattice positions are called imperfect or partial. The lowest energy slip plane which is expected for the $\frac{a}{2}$ <110> type of Burgers vector is {111} as confirmed by experimental observations. The most current dislocation types are 60°, screw and 90° dislocations in order of occurrence. These dislocation types are named after the angle made by the Burgers vector and the dislocation line. For the 60° dislocation e.g. the $\frac{a}{2}$ <110> Burgers vector makes an angle of 60° with the dislocation line which thus must lie along a <011> direction. These <011> directions correspond with the lowest Peierls energy in the {111} planes. For a screw dislocation the Burgers vector is parallel to the dislocation line. In principle each plane containing the screw dislocation is a glide plane although in practice the two common {111} glide planes for a screw dislocation with Burgers vector $\frac{a}{2}$ <110> are again preferred. The atomic structure of both the 60° and screw dislocation was studied in detail by Hornstra (1958) and is shown in figure 2.9. Each dislocation can be decomposed into two basic dislocation types, i.e. pure edge and pure screw. Figure 2.10 illustrates schematically these two basic dislocation types. An edge dislocation can be considered as resulting from the insertion of a half plane of atoms. The " extra half plane " consists of two (110) planes of silicon atoms. A screw dislocation can be described as a single-surface helicoid somewhat like a spiral staircase.

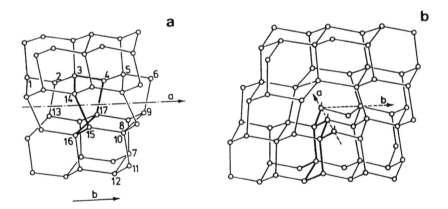

Figure 2.9 The atomic structure of a screw (a) and a 60° dislocation (b) in silicon, after Hornstra (1958).

An experimental observation of the atomic structure of both 60° and 90° dislocations is illustrated by the high resolution transmission electron microscopy (HREM) images in figure 2.11. In this figure the white dots represent the <110> oriented columns of pairs of silicon atoms. The glide planes with the lowest energy are (111) and (100). As the (111) related glide systems are activated at the lowest energy, plastic deformation of silicon will mainly occur by glide and cross-glide of a $\frac{a}{2}[110]$ type dislocations in {111} glide planes. hardening of the substrate then occurs by the interaction of glide dislocations from these different glide systems, resulting in the formation of dense dislocation networks which are sessile.

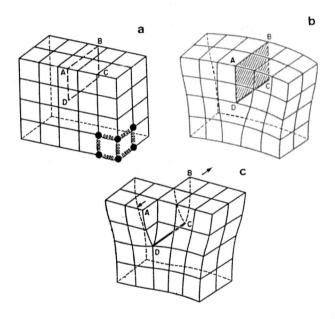

Figure 2.10 Schematic representation of a simple cubic lattice (a). An edge dislocation is formed by the insertion of an extra half plane of atoms ABCD (b). A screw dislocation is obtained by displacing two parts of the crystal relative toe each other in the direction AB (c) (After Hull 1965).

Forces on a dislocation

The different forces which can act on a dislocation are due to the interaction of the stress field of the dislocation core with another stress field such as:

　* external applied forces giving rise to an external climb and glide force. In integrated circuit processing a large number of stress inducing step are used

such as the deposition or growth of thin films with structural parameters different from those of the silicon substrate, ion implantation, or in-diffusion of dopants, heating and cooling processes ... The force F exerted by a stress field (σ_i, τ_{ij} with i, j = x, y, x) on a dislocation is given by the Peach and Koehler formula (Weertman and Weertman 1965)

$$F = t \times G = \begin{vmatrix} e_x & e_y & e_z \\ t_x & t_y & t_z \\ G_x & G_y & G_z \end{vmatrix} \qquad (2.12)$$

with

$$G_x = \sigma_x b_x + \tau_{xy} b_y + \tau_{xz} b_z$$
$$G_y = \tau_{yz} b_x + \sigma_y b_y + \tau_{yz} b_z$$
$$G_x = \tau_{zx} b_x + \tau_{zy} b_y + \sigma_z b_z$$

t is the unit vector parallel to the dislocation line, the Burgers vector b is defined in the conventional FS/RH way by making a circuit around the dislocation clockwise when looking in the sense of t.

* other dislocations will exert attracting or repelling forces on the dislocation depending on their Burgers vector and habit plane. When dislocations of the same type lie in the same glide plane, pile-up phenomena will occur in the neighbourhood of concentrated stress fields or of obstacles. The force between two dislocations can also be calculated using the Peach and Koehler formula with as stress field this of the second dislocation (Kovacs and Zsoldos 1973). For an edge dislocation along the z-axis one obtains:

$$\sigma_x = - \frac{\mu b}{2\pi(1 - v)} \frac{\sin\theta (2 + \cos 2\theta)}{r}$$

$$\sigma_y = - \frac{v\mu b}{\pi(1 - v)} \frac{\sin\theta \cos 2\theta}{r} \qquad (2.13a)$$

$$\tau_{zy} = - \frac{\mu b}{2\pi(1 - v)} \frac{\cos\theta \cos 2\theta}{r}$$

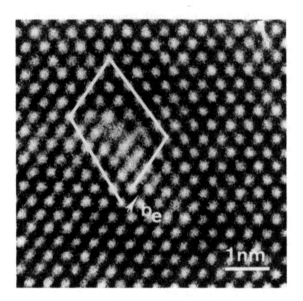

Figure 2.11 HREM images of the structure of undissociated 60° (a) and 90° (b) dislocations (courtesy H. Bender).

For screw dislocations one obtains

$$\tau_{xz} = -\frac{\mu b}{2\pi} \frac{y}{x^2 + y^2}$$

$$\tau_{yz} = -\frac{\mu b}{2\pi} \frac{x}{x^2 + y^2} \qquad (2.13b)$$

For a mixed dislocation the stress field will a linear combination of both expressions.

* a curvature of the dislocation will lead to a force caused by the line tension which is similar to the surface tension of a liquid or a soap bubble. A good approximation of the line tension force is given by

$$F_1 = T \frac{d^2 r}{dr^2} = F_1 l \qquad (2.14)$$

with T the line tension and

$$F_1 = \frac{T}{R} = \frac{\mu_s b^2}{8\pi R(1-v)} \ln \frac{R}{r_0} [(2-v) + 3v \cos 2\theta] \quad \text{with} \quad (2.15)$$

$$R = \frac{[1 + (\frac{dz}{dx})^2]^{1.5}}{\frac{d^2 z}{dx^2}} \qquad \text{and} \qquad r_0 = 5b$$

μ_s is the shear modulus and v is the Poisson ratio. θ is the angle between the Burgers vector and the direction of the dislocation.

* the stress field associated with the dislocation core will attract impurities resulting in the formation of a " Cottrell atmosphere " or even in the

formation of clusters on the dislocation line causing a frictional force hampering dislocation glide.

The previous forces were all external of nature. Some internal forces connected with the crystal structure and the presence of intrinsic point defects also occur :

* a friction force of Peierls-Nabarro force opposing dislocation movement by glide is due to the interaction of the dislocations stress field with the periodic internal stress field of the lattice. An estimate of this force F_{np} is given by Kosevich (1979) as

$$F_{np} = \frac{2\mu_s d_{112}}{1-\nu} \, e^{-\frac{\pi(3-2\nu)d_{111}}{2(1-\nu)d_{112}}} \, \sin\frac{2\pi x}{d_{112}} \tag{2.16}$$

with d_{ijk} the distance between the lattice planes (ijk).

* a supersaturation of intrinsic point defects will cause a chemical or climb force which is directed perpendicular to the glide plane and which is the driving force for dislocation growth by climb. The climb force can be written as

$$F_{ni} = \frac{b}{\phi l^3} \, kT \ln \frac{N_v}{N_v^0} \tag{2.17}$$

with N_v^0 the thermal equilibrium and N_v the actual number of vacancies.

Interaction between dislocations and point defects

The interaction between point defects and dislocations is an important topic as it allows to understand key phenomena in integrated circuit technology such as the decoration of dislocations by metallic impurities making them electrically active, enhanced dopant diffusion along dislocation lines and last but not least its application for gettering purposes. A number of interactions exist between dislocations and point defects of which the most important are :

* linear elastic interaction
* electrical interaction
* chemical or Suzuki interaction related to the change of chemical potential close to the dislocation and thereby changing locally the solubility of the point defects.

Describing the point defect as an elastic inclusion allows to calculate the elastic interaction energy which is composed of two components, the principle one being the first-order interaction while the second one is the inhomogeneity interaction energy. The size interaction is due to the difference in size between the inclusion and the space in which it is forced into the silicon lattice. The inhomogeneity interaction energy is related to the difference between the elastic constants of the inclusion and of the silicon. The first-order elastic energy Φ_{int} between a point defect and an edge dislocation with Burgers vector $b(b,0,0)$ is according to Teodosiu (1982) by

$$\Phi_{int} = -\frac{\mu_s b(1-v)\,dV}{3\pi(1-v)}\frac{\sin\theta}{r} \qquad (2.18)$$

r, θ are the cylindrical coordinates of the point defect with the positive direction of the dislocation chosen as the z-axis. dV is the change of volume of the spherical inclusion after insertion in the silicon lattice. The inhomogeneity interaction is smaller and decreases with the square of r. This expression is interesting because it shows that the sign of the interaction energy depends on the angle θ, showing that a dislocation can getter as well impurities with $dV > 0$ as with $dV < 0$.

It is clear that a dislocation can both repel and attract a point defect depending on their relative position. Dislocations can thus getter all types of extrinsic point defects which give rise to tensile or compressive strains in the silicon lattice. The location (inside or outside the dislocation loop) where they are gettered (and thus precipitate during the cooling) will be different however, and depends on the species. Sumino and Imae (1983) calculated the locking stress τ_l caused by particles (decoration) on the dislocation core

$$\tau_l = \frac{N}{b^2}\left\{\,E_0 - kT\ln\frac{LNv_v}{\Gamma}\,\right\} \qquad (2.19)$$

with N the density of locking particles on the dislocation, E_0 the maximum energy of the interaction between the dislocation and the point defect, L the length of the dislocation, v_v the frequency of the dislocation, Γ the release rate of the dislocations and b the Burgers vector length.

Dislocation movement and growth

Growth of dislocation loops can occur by climb and/or glide. Growth by climb occurs by the absorption of self-interstitials. This growth (movement) of dislocations is called non conservative.

Growth by glide does not require the presence of a supersaturation of self-interstitials and is therefore called conservative growth (movement). All experimental results on the motion of straight 60° or screw dislocations lying along the <110> directions show that it is a thermally activated process. For stresses well above the critical shear stress the dislocation velocity v as been calculated by Alexander (1986) as

$$v = v_0 \left(\frac{\tau}{\tau_0}\right)^m e^{-\frac{Q}{kT}} \tag{2.20}$$

v_0 is a material constant which depends strongly on the dopant type and concentration, m is weakly temperature dependent. The theoretical models of dislocation motion are based on the assumption of the formation and propagation of double kinks.

Once a dislocation is formed, its movement and growth can easily be understood with the concepts discussed in the previous paragraphs. The exact nature of the nucleation mechanism of dislocations (and stacking faults) is however, not so well established at the moment. Different models have been proposed such as: vibrating string, nucleation by climb ... A very important parameter for the nucleation of dislocation loops is the self-interstitial concentration as this strongly influences the climb force and thus climb by growth. This aspect has been discussed in detail by Vanhellemont and Claeys (1988a).

Plastic deformation

Plastic deformation of silicon requires the presence of large numbers of dislocations. In order to reach the stage of macroscopic plastic deformation, dislocation sources have to be activated as well as multiplication mechanisms.
A typical examples of a dislocation multiplications mechanism is the Frank-Read source which is schematically shown in figure 2.12. Figure 2.13 shows an experimental observation of a Frank-Read like dislocation source activated by film edge stresses in a (111) silicon substrate. Multiplication of dislocations is a much easier process than nucleation as it is based on glide and cross-glide mechanisms. The modern crystal growth techniques allow the growth of silicon crystals with grown-in dislocation densities well below 10^8 m^{-2}. A clear correlation exists between the upper yield stress and the starting dislocation density. Above a critical number of grown-in dislocations, pinning effects by impurity atoms (such as e.g. oxygen) will have a strong influence on the yield stress of the material.

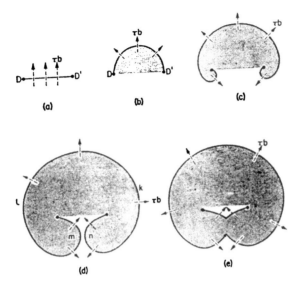

Figure 2.12 Schematic representation of the Frank-Read mechanism for the multiplication of dislocations. Slip has occurred in the shaded area [after Read 1953].

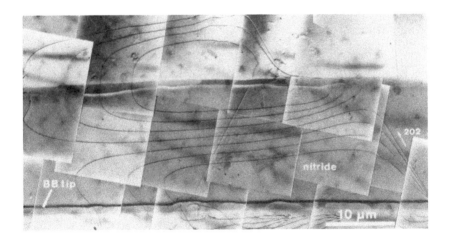

Figure 2.13 Plan view HVEM micrograph showing a single ended Frank-Read source activated by thin film stress [Vanhellemont and Claeys 1988a].

2.2.5 Planar defects: twins and stacking faults

A stacking fault is a planar defect which interrupts the normal lattice sequence. In a ABCABC stacking of closed-packed planes there are two possible equivalent positions to put one layer on top of the other. An atom of a closed packed A layer van be posed either in a B or in a C position and both positions are geometrically equivalent. In a fcc lattice like that of silicon two types of stacking faults are theoretically possible and referred to as extrinsic or intrinsic. They are schematically represented in figure 2.14. An extrinsic stacking faults can be considered as an extra (111) lattice plane inserted in the silicon lattice. It is clear that a supersaturation of self-interstitials will favour the formation of this type of stacking fault. Similarly an intrinsic stacking fault is formed by a missing (111) lattice plane. All the stacking faults observed in silicon are extrinsic in nature except for small stacking fault tetrahedra which are intrinsic of nature. These defects are however, only observed after ion implantation with very high fluences thus creating a local supersaturation of vacancies (Coene *et al* 1985). As the diamond structure is formed by covalent bonds, only the nearest neighbour interaction is important. Only fourth-nearest neighbours are affected by the removal or insertion of a double layer of silicon atoms in (111) planes. Due to this the stacking fault energy is relative low compared to the surface energy, explaining why in silicon e.g. twinning during crystal growth occurs quite easily. Perfect dislocations can easily dissociate and form extended dislocations. In some cases the hexagonal form of silicon is also observed after low temperature treatments.

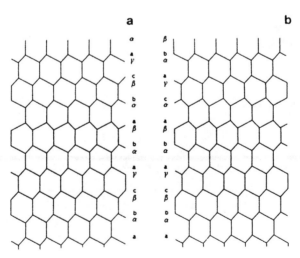

Figure 2.14 Schematic representation of (a) an extrinsic and (b) an intrinsic stacking fault in fcc structures viewed along a<110> direction in the stacking fault plane.

Stacking faults are bordered by partial dislocations, the most common one being the Frank type extrinsic stacking fault with $\mathbf{b} = \frac{1}{3} <111>$. Stacking faults can only grow by climb (they are sessile), i.e. by the absorption of intrinsic point defects. As the stacking fault energy is linearly dependent on the stacking fault surface, it increases faster than the energy of a perfect dislocation with the same size as this depends only on the dislocation length. The total energy of a stacking fault is the sum of the surface energy σ and the line energy of the surrounding partial dislocation. Small stacking faults thus have a lower energy than a perfect dislocation loop with the same size. With increasing size a critical radius is reached above which the energy of the perfect loop becomes smaller than that of the stacking fault and un unfaulting mechanism yielding a perfect dislocation loop becomes energetically favourable. A typical unfaulting reaction involves the simultaneous formation of and reaction with two partial dislocations and can be written as

$$\frac{1}{3} <\bar{1}11> + \frac{1}{6} <1\bar{1}2> + \frac{1}{6} <12\bar{1}> \quad \Rightarrow \quad \frac{1}{2} <011> \qquad (2.21)$$

The imperfect dislocations with Burgers vector $\frac{1}{6} <112>$ are called Shockley partials and nucleate in the stacking fault plane which is also their glide plane. After nucleation they glide very quickly over the stacking fault and react with its bounding partial dislocation. Similarly, multiple stacking faults nucleated at the same spot will react and unfault when this becomes energetically favourable. Experimental observations of both an extrinsic and an intrinsic stacking fault are shown in figure 2.15.

Another type of planar defects which are commonly observed in silicon are twins. Twinning of the silicon lattice is a common process during low temperature epitaxial growth as the energy difference between the two possible orientations is small. Plastic deformation at low temperatures (< 873 K) is mostly associated with the formation of large numbers of twins as dislocation glide is extremely difficult. Under certain conditions hexagonal silicon is formed which can be considered as a pile up of microtwins.

Extended dislocations

As mentioned in the previous paragraph, 60° dislocations can dissociate and form extended dislocations. During this process the perfect dislocation splits up in two Shockley partial dislocations with in-between a stacking fault. Using weak beam transmission electron microscopy the splitting up can be measured quite accurately and from the spacing between the two partials the stacking fault energy is calculated to be of the order of 60-70 mJm^{-2}. From the accurate measurement of the spacing of separated screws, Föll and Carter (1979) deduced

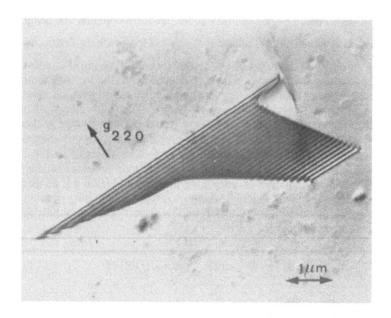

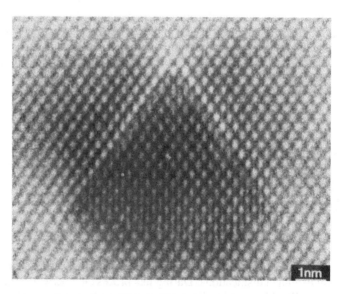

Figure 2.15 a) Interstitial type (extrinsic) "sailboat" stacking fault observed at the silicon surface. b) Vacancy type stacking fault tetrahedron observed in high dose phosphorous implanted silicon (Coene *et al* 1985).

that the extrinsic stacking fault energy is lower than that of intrinsic faults. In most cases however, dislocations seem not to be dissociated in silicon as also suggested by the very frequent observation of cross-section glide mechanisms leading to defect multiplication.

2.2.6 Volume defects: precipitates

During heat treatments which make them mobile, extrinsic point defects present in supersaturation will tend to agglomerate with the formation of precipitates. These precipitates can be either amorphous or crystalline. Their size will be determined by parameters such as anneal time and temperature, the concentration of the precipitating species and the number of stable precipitate nuclei which can grow. The shape of the inclusions is mainly determined by the minimum of the interface and strain energy. The anisotropy of the elastic constants both of the precipitate phase and of the silicon will thus play an important role. In case of coherent precipitation of crystalline phases in silicon the precipitate will take a shape to minimize the misfit between the two lattices.

Intrinsic point defects can also agglomerate with the formation of precipitates. A supersaturation of vacancies can annihilate by the formation of voids which can be considered as vacancy precipitates. Recently, Bender and Vanhellemont (1988) presented evidence that silicon self-interstitials precipitate in rod-like hexagonal silicon precipitates during low temperature treatments of oxygen-rich material which can be considered as precursors of dislocations formed at higher temperatures (Vanhellemont *et al* 1989).

Critical size of precipitates

Precipitates can only grow if their size is larger than a critical value. Precipitation of dissolved species will result in a local increase of the strain due to the volume differences between the precipitated phase and the amount of consumed silicon. The strain can be released by the emission or absorption of intrinsic point defects as schematically illustrated in figure 2.16. Theoretical calculations are facilitated by the assumption of a spherical precipitate consisting of a phase M_yP_z, M being a matrix atom and P being an atom of the precipitating species, the *precipitants*. The critical size of the precipitate is then determined by the condition that the free energy change for an increase of volume dV becomes zero or dG = 0. Smaller precipitates will dissolve while larger ones will grow. An analytical expression for the critical radius has been derived by Vanhellemont and Claeys (1987 and 1992) for isotropically elastic material

$$r_c = \frac{2\sigma}{\frac{E_x kT}{\Omega_p} \ln \frac{C_p}{C_p^*} [\frac{C_V}{C_V^*}]^\beta [\frac{C_I}{C_I^*}]^\gamma - 6\mu\delta\varepsilon} \qquad (2.22)$$

with σ the interface energy between the precipitate and the surrounding lattice. Ω_p and Ω_M are the precipitate volume, respectively the matrix volume containing y matrix atoms. Strain relief by the absorption of β vacancies and γ self-interstitials per precipitated atom is also taken into account.

$$E = (1 - \varepsilon)^{-3}$$

$$\delta = \{ \frac{\Omega_p}{[1 + x(\beta + \gamma)]\Omega_M} \}^{-\frac{1}{3}} - 1$$

$$\varepsilon = \frac{\delta}{1 + \frac{4\mu}{3K}}$$

with δ the linear misfit, ε the "constrained strain", K the bulk compressibility of the precipitate, μ the shear modulus of the matrix M, $x = \frac{z}{y}$ the number of precipitant P atoms per precipitate unit. Hereby it is taken into account that in order to obtain a spherical precipitate with radius r one has to insert a precipitate with radius $(1 - \varepsilon)^{-1}r$ and that the change in strain energy for a volume dV is given by $(6\mu\delta\varepsilon)dV$. The hydrostatic pressure p on the spherical precipitate after insertion in the matrix is $p = 4\mu\varepsilon$. Substitution of the equilibrium pressure at room temperature for the possible polymorphs of the precipitate P allows to calculate the total intrinsic point defect interaction ratio $\alpha = \beta + \gamma$

$$\alpha = \frac{1}{x} \{ [p (\frac{1}{4\mu} + \frac{1}{3K} + 1)]^{-3} \frac{\Omega_p}{\Omega_M} - 1 \} \qquad (2.23)$$

Neglecting the influence of the strain energy and intrinsic point defect concentration leads to the well known expression

$$r_c = \rho \frac{T_s^p}{T_s^p - T} \qquad (2.24)$$

with $\quad \rho = \dfrac{2\sigma\Omega_P}{xkT_P}$ and T_s^p the theoretical temperature at which the amount P

would be soluble.

Equation 2.22 clearly demonstrates that the critical radius not only depends on the supersaturation of precipitating atoms and vacancies, but also on the supersaturation of self-interstitials. It also allows to understand the influence of the temperature on the shape of the precipitate through the strain factor in the denominator: there will be tendency to have a plate-like precipitate for lower temperatures due to the anisotropy of silicon and the difficult generation of intrinsic point defects. At higher temperatures there will be a tendency to form more spheroidal precipitates as the strain component decrease so that the interface energy becomes more important. More detailed expressions for the critical dimensions of precipitates should take into account the shape of the precipitate and also the strain field which results from it (Vanhellemont and Claeys 1991b). The more simple expression 2.22 has the merit that it gives in a transparent analytical form the influence of the different parameters on the critical size of the precipitates.

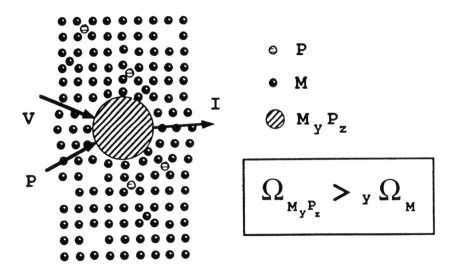

Figure 2.16 Schematic representation of the growth of a spherical precipitate by the absorption of vacancies and precipitant atoms and by the emission of self-interstitials.

Real processes can be much more complex and can involve more than one precipitating species, such as e.g. the co-precipitation of carbon and oxygen in silicon or the co-precipitation of two different metal contaminants in silicon. Vanhellemont and Claeys (1991b) derived a generally valid expression to determine the critical shape of such precipitates with volume V and surface S

$$\frac{dV}{dS} = \frac{\sigma}{\frac{kT}{\Omega_P} \sum_X \gamma_X \ln \frac{C_X}{C_X^*} - E_s + E_i + E_r} \tag{2.25}$$

with C_X and C_X^* the actual concentration and the thermal equilibrium

concentration and γ_X the number of point defects X absorbed per precipitate

molecule (X = precipitant P, vacancy V or self-interstitial I). E_s and E_i are respectively the strain energy in the precipitate and the surrounding matrix, and the elastic interaction energy of one precipitate with the strain field of the other precipitates. The latter is only important when the average distance between the precipitates is becoming of the same order of magnitude as the critical size of the precipitates. E_r is the relaxation energy associated with the formation of extended defects such as misfit dislocations, punched out dislocation loops and stacking faults. Whether or not dislocation loop nucleation will take over the role of the emission of self-interstitials in the stress relieving process is depending on the total strain build-up around the precipitate. The practical use of this general expression will be illustrated in section 2.4.

Precipitate nucleation and growth kinetics

When a supersaturated solution is formed the system will try to lower its total free energy by precipitation of the excess of dissolved species. Two main cases can be considered:

- initially there are no precipitates present. The nucleation of precipitates can then satisfactory be described by the classical nucleation theory of Zeldovich (1942), as shown by Inoue *et al* (1981) for oxygen precipitation.
- there exists already an initial distribution of precipitates. In that case the precipitation will be governed by the dissolution of the subcritical nuclei and by a further growth of the supercritical ones.

In both cases the further diffusion limited growth of the stable nuclei follows to a large extent the theory of Ham (1958).

2.3 ELECTRONIC PROPERTIES OF DEFECTS

2.3.1 Donor and acceptor properties

The breakthrough of the silicon integrated circuit technology is based on the ability to produce high purity substrates and to control their electrical properties by doping with impurities such as boron, arsenic, phosphorus, gallium .... In this section we will only treat impurities as defects when they are not intentionally introduced in the lattice. It was already mentioned in section 2.2.3 that intrinsic point defects such as vacancies and interstitials can have different charge states and therefore they will also have acceptor or donor-like properties. A neutral vacancy is characterised by a donor (0/+) trapping behaviour, while a single negatively charged vacancy has an acceptor-like behaviour (-/0). In general one can state that an electrically active defect centre results in the generation of of a set of localised energy levels in silicon band gap. Mostly only the ground state with ionisation energy E_T is considered. Depending on the position in the band gap, one speaks about shallow or deep levels. Shallow donors have an energy level a few meV below the minimum of the conduction band, while shallow acceptors have an energy level a few meV above the maximum of the valence band. Theoretically these shallow levels can be analysed by using the effective mass theory. An important parameter of a defect level is its relative position to the Fermi level E_F, as this will determine the probability of the level to be occupied by an electron. The classical treatment is based on the Fermi-Dirac statistics, whereby the probability $f(E_T)$ of a level to be occupied by an electron is given by

$$f(E_T) = \frac{1}{1 + \frac{1}{g} e^{\frac{E_T - E_F}{kT}}} \qquad (2.26)$$

with g the ground-state degeneracy factor (g = 2 for shallow donors, g = 1/4 for shallow acceptors). At room temperature, for typical doping levels, $E_A \ll E_F$ and $E_D \gg E_F$ so that the acceptor levels are occupied by an electron and thus negatively charged, while the donor levels are empty. In the case of defect centres, an amphoteric character is often observed which means that the centre can act both as an acceptor and as a donor depending on the position of the Fermi level. This schematically illustrated in figure 2.17. More detailed information on the calculation of the exact charge state of the different defect levels can be found in the review by Milnes (1973). However, if the energy difference between two subsequent states is much larger than kT, then the behaviour of the defect centre in neutral material will be determined by one of the ground-states depending on the position of the Fermi level.

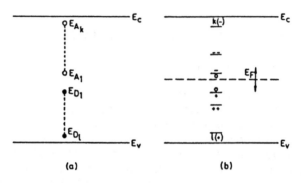

(a)   (b)

Figure 2.17 (a) Schematic representation of an amphoteric defect centre with k acceptor levels E_{A1}....E_{Ak} and l donor levels E_{D1}...E_{Dl}. (b) Corresponding charge state representation. The defect centre will change its charge state from neutral to positive when the Fermi level rises below E_{D1}, it will becomes negative when the Fermi level rises above E_{A1}.

The presence of acceptor or donor levels will have a direct impact on the resistivity of the material. The number of free carrier can be calculated for the general case of k acceptor levels E_{Ai} (i=1...k) with homogeneous density N_{Ai} and l different donor levels E_{Dj} (j=1...l) with homogeneous density N_{Dj} by using the charge balance requirement

$$n + \sum_{i=1}^{k} N_A\, f(E_A) = p + \sum_{j=1}^{l} N_D\, (1 - f(E_D)) \qquad (2.27)$$

and the equilibrium condition

$$np = n_i^2 = N_C N_V\, e^{-\frac{E_G}{kT}} \qquad (2.28)$$

with N_C and N_V the density of states in the conduction and valence band, E_G the silicon band gap, and n_i the intrinsic carrier density corresponding with undoped silicon.

The resistivity of n-type material is defined as

$$\rho = (q\mu_e n)^{-1} \qquad (2.29)$$

with μ_e the electron mobility. It should be remarked that a defect centre has not only a direct influence on the resistivity by changing the density of free carriers, but in some cases also indirectly by affecting the free carrier mobility. If different scattering mechanisms are active, the mobility is determined by adding reciprocally the mobilities that would exist if each process were active separately. Depending on the number of ionised defect centres N_I, the free carrier mobility can be reduced due to the scattering mechanism associated with these ionised impurities. According to Conwell and Weisskopf (1950) this term is given by

$$\mu_I = \frac{2^{7/2}\varepsilon^2(kT)^{3/2}}{\pi^{3/2}m^{*1/2}q^3Z^2N_I} \frac{1}{\ln[1 + (3\varepsilon kT\, r\, q^2 N_I^{1/3})^2]} \approx \frac{\sqrt{m^* T^{2/3}}}{N_I} \qquad (2.30)$$

with m^* the effective mass, ε the dielectric constant, qZ the charge associated with the defect and r the the Hall factor which is impurity concentration and temperature dependent.

Thermal and new donors

Since the early days it is well known that oxygen containing silicon is prone to the generation of thermal donors unless a special soak treatment is implemented in the crystal fabrication process. The thermal donor generation is strongly temperature dependent, i.e. reaches a maximum for 450°C anneals and dissapears for temperatures around 650°C. A large variety of models have been proposed to explain the core structure of these donors, based on the agglomeration of either oxygen atoms, or silicon self-interstitials. Recently, Claeys and Vanhellemont (1989) have reviewed these different models. The YLID-model, based on the saddle point configuration of interstitial oxygen diffusion, the Ourmazd-Schröter-Bourret (OSB) model with an oxygen cluster containing in its centre a silicon atom pushed out into a quasi interstitial position along the <001> direction, and the Mathiot's model where the thermal donor has two self-interstitials arbitrary in the bond-centered position on a O_2 complex, are schematically shown in figure 2.18. For a general treatment of thermal donors in silicon, one has to take into account the role of doping atoms and the influence of the presence of respectively carbon, nitrogen and hydrogen atoms. Recently, Heijmink *et al* (1992) have used photoluminescence and EPR/ENDOR studies and found some evidence that the type of acceptor doping has no influence on the thermal donor generation kinetics. In addition to the "standard" thermal donors, Kamiura *et al* (1989) obtained experimental evidence that prolonged annealing of phosphorus doped CZ silicon also results in a family of oxygen-related shallow donors (0.04 and 0.09 eV from the conduction band). Although these shallow donors can be correlated with donor levels revealed by ENDOR analysis, the NL-10 band, the proposed core structure of thermal donors remains somewhat hypothetical.

Kanamori and Kanamori (1979) were the first ones to report that after annealing of the thermal donors, further annealing at temperatures above 650°C may result in the formation of so-called "new donors". These new donors have a maximum generation rate around 850° C and remain stable up to about 1000°C. Although there exists an even larger uncertainty about the exact nature of these new donors, a so-called " SiO_x Interface Model " has been postulated (Hölzlein *et al* 1984 and Pensl *et al* 1989). This model correlates the donor activity to trap states originating from either interface states at the surface of oxide precipitates and bound states in Coulombic wells of a fixed positive charge located in the precipitates.

The thermal and new donor formation is not only important in oxygen containing bulk silicon, but can also influence the electrical properties of Silicon-On-Insulator (SOI) material. It has been reported that an increased donor activity can also occur in SOI material fabricated by a high dose oxygen implantation followed by a high temperature anneal step, i.e. so-called Separation by IMplanted OXygen substrates (SIMOX). These donors can be generated both in the superficial silicon film and in the silicon substrate due to the excess oxygen concentration present after the SIMOX process and subsequent activated during the device processing. Even in SIMOX structures fabricated on high resistivity substrates, thermal donor concentrations of about 10^{19} m^{-3} have been observed.

Beside oxygen, also nitrogen can have a donor activity. This is especially important when silicon nitride crucibles are used for the growth of oxygen lean crystals. However, also nitrogen-oxygen donors have been observed in CZ crystals deliberately doped with nitrogen by added Si_3N_4 to the melt (Suezawa *et al* 1986), and by giving undoped CZ silicon a high temperature anneal in a nitrogen ambient (Hara *et al* 1989). These donors have been studied by EPR, photothermal ionisation spectroscopy (PTIS), infrared spectroscopy, and photoluminescence (PL). Although up to nine independent shallow double donors have been observed, there still exists some dispute about the exact role of nitrogen in the donor activity. According to Griffin *et al* (1989) nitrogen is only playing a catalytic role. For crystals grown in a nitrogen ambient, Yang *et al* 1991 observed a thermal acceptor activity which they correlate to a kind of Si-O-N complex. However, no other reports on such thermal acceptors are available. Finally, it should also be mentioned that hydrogen in silicon is also showing a donor activity. Recently, Zhong *et al* (1992) presented experimental evidence for a hydrogen-related donor state at 0.118 eV from the conduction band. For their experiments they used neutron transmutation doped Float Zone silicon crystals grown in a hydrogen atmosphere.

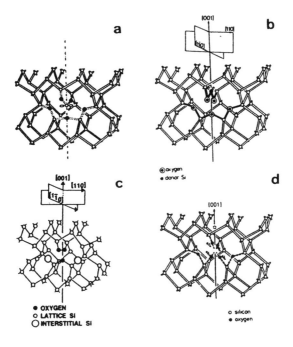

Figure 2.18 Schematic representation of the YLID (a), OSB (b), interstitial Mathiot model (c), and an updated OSB-model (d) proposed to explain the structure of the thermal donors in silicon.

2.3.2 Minority carrier lifetime

In silicon the minority carrier lifetime is strongly influenced by the presence of trapping levels in the forbidden band gap. The interaction between a carrier and a defect centre, is caused by either a thermal or an optical mechanism which involves respectively phonons or photons to deliver or dissipate the energy associated with the trapping or emission of a carrier. Only for high doping levels or deep cryogenic temperatures the Auger mechanism, which involves a third carrier, has to be taken into account. Milnes (1975) has reviewed in detail the different mechanisms dominating the minority carrier lifetime. For most of the defect centres in standard silicon, the minority carrier lifetime can be calculated by using the Shockley-Read-Hall model for the thermal recombination/generation mechanisms. Basically this approach results in the following equation for the carrier lifetime τ, assuming that the electron and hole lifetimes are equal and that the defect density is relatively low

$$\tau = \frac{1}{n + p + \Delta n} \left[\frac{p + p_1 + \Delta n}{c_n N_I} + \frac{n + n_1 + \Delta n}{c_p N_I} \right] \qquad (2.31)$$

with p_1 and n_1 the hole and electron concentration when the Fermi level coincides with the defect level, Δn and Δp the excess carrier concentrations, and c_n and c_p the electron and hole capture probabilities. As recombination rates add together, and the recombination rate is inversely proportional to the lifetime, the lifetimes due to different defect centres will add reciprocally. The position of the defect level is an important parameter, and energy levels close to the middle of the band gap will dominate.

The defect levels are often characterised by determining their thermal electron trapping (c_n) and re-emission rate (e_n) given by

$$c_n = \sigma_n \, n \, <v_n>$$

$$e_n = \frac{\sigma_n \, <v_n> \, N_C}{g} \, e^{-\frac{E_T}{kT}} \qquad (2.32)$$

with $<v_n>$ the average thermal velocity and σ_n the electron capture cross section of the defect level. The capture cross section varies between 10^{-8} m^{-2} (attractive centres) to 10^{-12} m^{-2} (neutral traps) and even to 10^{-16} m^{-2} (repulsive traps).

A defect centre is electrically completely characterised by its energy level, its density and its capture cross section. A common technique to determine these trap parameters is Deep Level Transient Spectroscopy (DLTS). To facilitate the defect identification by its DLTS signature, so-called defect libraries have been developed as e.g. done by Benton (1990) for transition metals in silicon.

Low-frequency noise analyses of silicon transistors are also gaining more and more interest as a diagnostic tool for defect spectroscopy (Murray *et al* 1991, Scholz and Roach 1992) of completely processed devices. The basic equations for characterising (N_T, E_T and σ_T) the defects are (Murray *et al* 1991)

$$\ln \left[\frac{T^2}{f} \right] = \frac{E_C - E_T}{k} \left(\frac{1}{T} \right) + \ln \left[\frac{\pi}{A\sigma_T} \right] \qquad (2.33)$$

$$N_T = \frac{2S_{vpeak}(\omega)\, C_{ox}^2\, WL\, (4\pi f)}{q^2 \Delta W_D\, K} \left(\frac{W_D}{W_D - x_D}\right)^2 \qquad (2.34)$$

$$\text{with} \quad K = \ln\left(\frac{\alpha + V_{GT}}{\alpha}\right) - \frac{V_{GT}}{\alpha + V_{GT}}$$

$$\alpha = \frac{kT}{q}\left(1 + \frac{C_D}{C_{ox}}\right)$$

f is the measuring frequency, S_{vpeak} the spectral density at the maximum of the gate referred noise power spectral density versus temperature, W_D the width of the depletion layer, C_D and C_{ox} respectively the depletion layer and the gate oxide capacitance per unit area, WL the width-length product of the transistor gate, and x_D is the location in the depletion layer where the carrier capture and emission time constant are equal. Equation (2.33) points out that by using an Arrhenius plot of ln (T^2/f) versus 1/T, the defect energy level and the capture cross section are given by respectively the slope and the intercept of the line. Equation (2.34) can then be used to calculate the defect density. It should be remarked that only defects in the vicinity of x_D, the location in the depletion layer where the assumptions are correct, will contribute to the noise peak. The location can be varied by changing the applied bias conditions.

This spectroscopical approach has also a strong potential to identify silicon film defects in SOI material (Simoen and Claeys 1993).

2.3.3. Optical properties

As defects have energy states in the forbidden band gap, these levels will also affect light absorption properties because of the probability of optically induced carrier transitions between these levels and the conduction or valence band. For this reason photoconductivity is often used as a very sensitive technique to study the electronic properties of defects (e.g. Bube 1969). The absorption properties strongly depend on the photon energy. For wavelengths corresponding with the band gap energy, electronic transitions between localised defect states and the band edges will dominate. However, even for longer wavelengths non-electronic absorption bands may be observed. A typical example is the 9.1 μm infrared absorption band in silicon, which is directly correlated to the interstitial oxygen content of the material.

Corbett *et al* (1961 and 1964) have extensively used infrared spectroscopy to study the A-centre in silicon, and were able to conclude that this defect is giving a peak in the absorption spectrum around 830 cm^{-1} ($\approx$ 12 µm). By performing different annealing experiments they also identified other oxygen related infrared absorption bands in electron irradiated silicon. Bean *et al* (1970) have performed an in-depth classification of the different vibrational bands observed in electron irradiated silicon and concluded that complexes involving both oxygen and carbon are formed, as well as those involving either carbon or oxygen alone. Intrinsic point defects play an important role during annealing cycles used to study the interactions between the different bands. It is out of the scope of this chapter to discuss in detail the large variety of irradiation induced defects and to review the recent insights into the different defect models. An overview of radiation induced defects has recently been assembled by Claeys and Vanhellemont (1992).

Oxygen in silicon has been studied for more than three decades and depending on thermal heat treatments and the experimental conditions, a large variety of oxygen related lattice defects may be formed as discussed in section 2.1. Recent reviews on oxygen in silicon have been published by Bender and Vanhellemont (1992, 1993). The infrared absorption spectra related with different precipitates have also been used to gain more insight in their composition. As an illustration of the potential of this approach, it can be mentioned that Hu (1980) has compared experimental and theoretical infrared spectra to identify plate-like oxygen precipitates, and observed that they give rise to an absorption peak at 1230 cm^{-1}.

Also the influence of carbon on oxygen precipitation can be studied by infrared spectroscopy. A recent demonstration has been given by Sun *et al* (1990) who pointed out the intermediate role of the substitutional carbon/self-interstitial [$C_S Si_I$] complex to enhance oxygen precipitation processes and to introduce changes in the precipitate morphology. A detailed review of the potentials of infrared spectroscopy for defect studies in silicon has been compiled by Newman (1988). The infrared spectra of figure 2.19 illustrate the potential of the technique to study the interaction between oxygen and carbon. Prolonged annealing reduces the interstitial oxygen peak (at 517 cm^{-1}), but does not result in the growth of C_S-O_i complexes.

Finally it should be mentioned that photoluminescence (PL) has also successfully been applied to study defects in silicon. When electron-hole pairs are created in the material, the carrier recombination processes often produce light emission referred to as luminescence. This phenomenon can beneficially be used to study the energy levels of dominant recombination centres. Originally applied to obtain information on chemical impurities, it also resulted in a better understanding of irradiation induced defects (e.g. Spry and Compton 1968). During the last decade PL studies have been used to characterise respectively

swirl defects (Nakashima and Shiraki 1978), dopant-impurity complexes impacting the free- bound and multi-excitons complex related fine structure (Nakayama *et al* 1980), Si-O complexes (Tajima *et al* 1980), thermally-induced microdefects (Katsura *et al* 1982), thermal donors (Tajima *et al* 1983), and nitrogen-oxygen donors (Steele *et al* 1990). Wijaranakula (1991) performed photoluminescence analysis to investigate the interactions between extended dislocation loops and respectively interstitial (oxygen), substitutional (tin) and interstitial-type (Cu) impurities. Some of his observations are shown in figure 2.20, giving PL spectra of samples containing a low oxygen content, a high oxygen content, and tin doping respectively. The interaction between the edge dislocation component and the impurities produces optical defect centres giving rise to the D1 line at 0.81 eV, while the interaction between the screw component and impurities results in the D2 line at 0.89 eV. D-line luminescence is also often observed in plastically deformed silicon (e.g. Queisser 1982), although the deformation introduced dislocations seems not to be a prerequisite (Wijaranakula *et al* 1990). Lelikov *et al* (1989) made a classification of dislocation energy levels according to the length of their Burgers vector and found a clear correlation between the Burgers vector and the line position in the PL spectrum. Their analysis is based on the concept of a one-dimensional dislocation related exciton bound in the dislocation strain field.

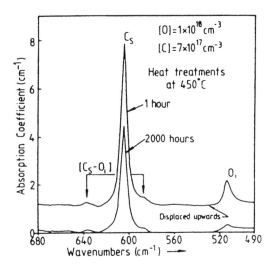

Figure 2.19. Infrared spectra of a sample with a high oxygen and carbon content, after different thermal anneals at 450°C (after Newman 1988).

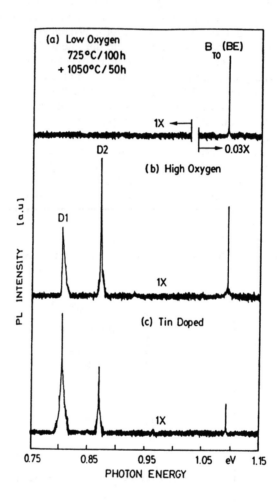

Figure 2.20 Photoluminescence spectra of a low oxygen concentration (a), a high oxygen concentration (b) and tin doped (c) samples respectively (after Wijaranakula 1991).

2.4 DEVICE PROCESSING RELATED DEFECTS

2.4.1 Grown-in defects

The grown-in defects are directly related with the crystal growth technique and appear either in a striated or non-striated distribution over the silicon wafer. A typical distribution pattern, revealed after preferential etching of the wafer surface, is shown in figure 2.21. These microdefects are generally classified into A- and B-defects or so-called swirl defects, and D-defects which appear in an arbitrary pattern. The swirl defect kinetics strongly depends on the presence of extrinsic point defects; boron, phosphorus and carbon impurities enhance the swirl formation, while oxygen has a retarding effect. The nature of these swirl defects has extensively been studied and possible models can be found in the reviews by Chikawa *et al* (1978), Abe *et al* (1973) and de Kock (1973). It is generally accepted that the larger A-defects (between 0.5...3 μm) are interstitial type microdefects, while the nature of the smaller B-defects (< 0.1 μm) is much more difficult to identify. Originally Föll and Kolbesen (1973) assumed that these B-defects are agglomerates of self-interstitials and trapped impurities like e.g. carbon, while de Kock and Wijgert (1980) more favour the vacancy type nature. Up to now, the exact nature has not been determined unambiguously. However, it is generally accepted that the nucleation of the A-defects is based on the collapse of the B-defects. Concerning the D-defects, first observed by Roksnoer and Van den Boom (1981), experimental evidence has recently been given by Abe and Kimura (1990) on their vacancy type nature in agreement with the original hypothesis by Voronkov (1982) as being octahedral voids.

For completeness, it should also be mentioned that in the literature reference is sometimes made to so-called C-type defects. These defect striations have been observed by X-ray topography of copper decorated FZ-silicon, and sometimes in dislocation-free CZ silicon (de Kock *et al* 1979). These defects are smaller than B-defects and are present in regions close to the B-defect ones. De Kock *et al* (1974) noticed that elimination of these defects is obtained by either low (< 0.20 mm/min) or high (> 5.00 mm/min) pulling rates of carbon-lean crystals.

The formation of the grown-in defects is closely related with the thermal gradients, caused by crystal rotation and melt convection, and the stress distribution in the crystals. A-defects are mainly formed when the thermal gradient is large and the silicon under tensile stress, while D-defects are created in regions with a low thermal gradient and a compressive stress (Abe and Kimura 1990). Swirl defects can be found in both Float Zone and Czochralski silicon, although a small addition of hydrogen to the pulling atmosphere seems to be sufficient to suppress them in FZ silicon.
In many cases the grown-in defects are at the basis of the heterogeneous nucleation of process-induced defects.

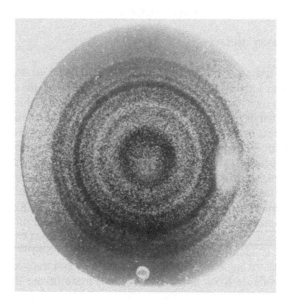

Figure 2.21. Typical grown-in defects pattern, observed after preferential etching of the silicon.

The dominant grown-in point defect in silicon grown by the Czochralski technique is interstitial oxygen which is present with a concentration ranging between 2×10^{17} cm^{-3} and 2×10^{18} cm^{-3}. As a result of this large grown-in oxygen super saturation, oxygen will precipitate with the formation of a silicon oxide phase during all treatments performed at temperatures above 300°C. During treatments at temperatures in the range between 300 and 500°C, the precipitation of the interstitial oxygen is accompanied by the formation of so called "thermal donors". The formation of silicon oxide particles occurs with a strong volume increase, i.e. the precipitated phase has a much larger volume than the amount of silicon which is consumed during the oxide formation. As discussed in paragraph 2.2.6, as a result of this excess strain large numbers of self-interstitials or dislocation loops can be formed thus strongly influencing other defect formation mechanisms such as discussed in the next paragraph. Figure 2.22 shows a schematic view of the behaviour of oxygen and carbon in silicon as a function of the temperature. The indicated temperatures are only indicative as there is a gradual transition between the different configurations. Recent reviews on the behaviour and properties of oxygen in silicon were published by Claeys and Vanhellemont 1989, Bender and Vanhellemont 1992 and Bender and Vanhellemont 1993.

2.4.2 Stress-induced defect generation

In silicon IC technology, different sources of mechanical stress can become active and may result in defect nucleation and growth. In general, these stresses can be divided into well-defined categories, depending on their origin and/or correlation to a processing sequence. An extensive review of stress related problems in silicon technology has recently been published by Hu (1991a). A brief discussion on different stress related phenomena will be given in this section.

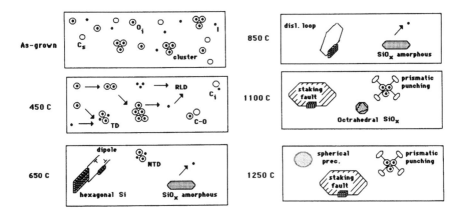

Figure 2.22. Schematic representation of the behaviour of oxygen and carbon in silicon during thermal treatments at different temperatures (Claeys and Vanhellemont 1989).

Thermal stresses

The stress which has been studied since the early days of silicon technology is the one caused by temperature gradients across the wafer, i.e. the thermal stress. This stress component is especially important during heating or cooling the wafers and may result in the generation of so-called slip patterns. During heating the temperature gradients produce a tensile stress in the central wafer region and a tangential compressive stress in the outer regions, while the opposite occurs for the cooling cycle. If the stress is exceeding a critical level, i.e. the silicon yield stress, dislocations nucleate and may eventually introduce plastic deformation. As shown by Vanhellemont and Claeys (1988a), the yield stress strongly depends on the concentration of intrinsic point defects. Due to the oxygen precipitation process, the yield stress is lower in CZ silicon compared to FZ silicon. The external climb force needed for yielding at temperature T is given by

$$F_{ne}(T) = F_n^y(T) - \frac{b_e k}{\phi l^3} \, T \ln \frac{C_I^0(T)}{C^*} - \frac{b_e k}{\phi l^3} \, T \ln \left[1 + \frac{\Delta C_I(T)}{C_I^*} \right] \quad (2.35)$$

with b_e the edge component of the Burgers vector, C_I and C_I^* the actual and

equilibrium concentration of silicon self-interstitials, $\Delta C_I = C_I - C_I^0$ with C_I^0

the concentration at time $t = 0$, ϕ a constant depending on the crystal structure

and l the interatomic spacing between the (111) glide planes. $F_y^n(T)$ is the yield

force in the case that the concentration of self-interstitials is equal to the equilibrium concentration. The generation of self-interstitials, e.g. during the oxygen precipitation process, will result in an excess of self-interstitials and thus to a decrease of the yield stress. The nucleation of dislocations will preferentially occur during the cooling process as the long lifetime of the self-interstitials results in a supersaturation which decreases with increasing temperature. For inert anneals of CZ silicon one also has to take into account the initial dissolution of the subcritial oxide precipitates for the calculation of the concentration of self-interstitials. As illustrated in section 2.2, the critical radius of oxygen precipitates is directly influenced by the concentration of point defects.

For flat wafers the tangential and radial stress components vary with the distance to the centre of the wafer, but not along the wafer thickness. This is no longer true for wafers with a bow, which is for nowadays silicon typical between 1 and 3 μm, where the stresses will be more compressive at the centre of the concave side and less compressive or tensile at the centre of the convex side (Leroy 1980). This observation in combination with the knowledge of the parameters influencing the yield stress, allow to conclude that dislocation nucleation will preferentially occur near the centre of the wafer and at the edges preferentionally in regions with a high oxygen precipitation. Slip dislocations will easily be generated in {111} planes along the <110> directions. Depending on the magnitude of the tangential stress component, different glide systems are preferred so that typical slip patterns are obtained. Models to describe slip generation have been developed by Hu *et al* (1976) and by Bentini *et al* (1984). At high stress levels plastic deformation may occur, resulting in severe wafer warpage. This effect has a strong degrading influence on the lithographic yield of submicron technologies. To restrict or avoid thermal stresses during processing, temperature ramping techniques are standard used nowadays. This ramping can be achieved by either controlling the insertion and withdrawal rate of the wafers into and from the furnace hot zone or ramping up and down the furnace temperature after the hot processing step. It should be remarked that during "rapid thermal processing", typically used for low temperature budget processes such as e.g. shallow junction formation or silicidation, a tight control of the thermal stresses is also essential (Vanhellemont and Claeys 1993).

Film stresses and film-edge induced stresses

The integrated circuit fabrication process involves the growth or deposition of a large number of different thin films, such as e.g. thermal silicon dioxide, silicon nitride, polycrystalline or amorphous silicon, silicides... Depending on the growth/deposition technique and the experimental conditions these surface films introduce compressive or tensile stress in the silicon substrate. The origin of the stress can be e.g. the lattice mismatch between the film and the silicon or the intrinsic stress grown-in in the film. Due to the importance for advanced processing technologies the first stress cause will be treated in another paragraph. Some typical values, taken from the review by Hu (1991a), for intrinsic stress levels in different films are respectively -0.2 to - 0.3 GPa for thermal oxides, 0.1 to 0.4 GPa for CVD oxides, -0.1 to -0.3 for polysilicon (strongly dependent on the grain size), 0.6 to 1.2 GPa for Si_3N_4, and 1- 3 GPa for silicides such as e.g. $TiSi_2$ and $CoSi_2$. The actual temperature also has a strong influence on the stress level. Near film edges the stress strongly increases and may in many cases exceeds the silicon yield stress. Whether or not the subsequent dislocation nucleation occurs heterogeneous or homogeneous strongly depends on the number grown-in lattice defects. The latter nucleation process requires a higher stress level. The motion of existing dislocations is strongly influenced by the presence of so-called pinning centres or locking agents such as e.g. dissolved oxygen or nitrogen. These mechanical strengthening phenomena have been studied in detail by Sumino (see e.g. Sumino 1981 and 1985). For the homogeneous yielding process, based on the direct condensation of self-interstitials, Vanhellemont and Claeys (1988a) have modified equation 2.35 and derived an expression for the critical film thickness h_c required for dislocation nucleation taking into account the influence of a varying self-interstitial concentration

$$h_c(T) = h_c^0(T) - \Phi \; \frac{b_e k}{\phi l^3} \; T \ln \left[\; 1 + \frac{\Delta C_I(T)}{C_I^*} \; \right] \qquad (2.36)$$

with $h_c^0(T)$ the critical film thickness for yielding at $t = 0$ and Φ a constant which can be calculated from the experimental data.

The interactions between dislocations and the stress field at the edge of a thin film was originally studied by Hu (1975). The defect generation near silicon nitride film edges, which is of crucial importance for the most commonly used device isolation techniques, have been systematically studied for a large variety of experimental conditions. This resulted in a classification of the different types of observed defects, e.g. 60° dislocations parallel to the nitride edge, dislocation half-loops or so-called Hu-loops, triangular half-loops, Frank-Read dislocation sources ... in function of both the silicon substrate and the nitride film orientation. Some typical defect structures, observed by transmission electron microscopy, are shown in figures 2.13 and 2.23. Vanhellemont *et al* (1987a and 1987b) and Vanhellemont and Claeys (1988b) have derived a theoretical model for the calculation of the type and the shape of the generated dislocations as a function of the relative orientation of the film edges and the silicon substrate lattice. The criterion for homogeneous dislocation nucleation is based on the fact that for each geometrical configuration, the dislocation with a Burgers vector maximising both the internal and external glide forces will nucleate first after reaching the yield stress. The model also predicts that in the case of periodic patterns, the stress increases for smaller spacings between the film bands as observed experimentally (Vanhellemont and Claeys 1991a). As only the magnitude and the nature, tensile or compressive, of the intrinsic film stress has to be known, the same model can be applied for any type of film. Good results have also been obtained for the defect study in silicidation processes (Van den hove *et al* 1988) or for defect nucleation in trench isolation structures (Fahey *et al* 1992). Quantitative information on the stress distributions at film edges can be obtained with sub-micron lateral resolution using micro-Raman spectroscopy (De Wolf *et al* 1992) or with nm resolution by using electron diffraction techniques (Vanhellemont *et al* 1992).

Stresses associated with " embedded structures"

The stress calculations are much more complex for cases dealing with "embedded structures" in the silicon matrix. A typical example is the trench isolation structure used in submicron process technology. For such processes a trench is etched in the silicon substrate and subsequent thermally oxidised before being filled again with either polysilicon or a CVD SiO_2 layer.

According to Hu (1990) there are three main driving forces for stress development near trenches: i) the volume expansion associated with the thermal oxidation of the trench, ii) the different thermal expansion coefficients for silicon and the trench filling material, and iii) the intrinsic stress in the trench filling materials. Other examples in which similar types of stress problems may be encountered are isolation processes based on recessed oxide structures, V-groove trenches, raised source-drain approaches, filling of contacts and vias in multilevel metallisation schemes, buried silicides, implanted buried layers... The stresses are caused by both the internal stress fields associated with the embedded films and the lattice mismatch between the matrix and the structure. For a theoretical modelling of the stresses it becomes essential to go over a 3-dimensional approach. Hu (1989) developed a three-dimensional analytical model for solving the problem of a parallelepipidic "thermal" inclusion in a semispace and has applied this successfully for the stress calculations of isolation trenches (Hu 1990). A methodology based on the plane strain finite element approach, has been proposed by Chidambarro *et al* (1991) for stress calculations taking into account the complete trench fabrication process. Although, depending on the cases studied, there are differences with the analytical approach both are giving similar trends for the stress variation in view of the technological parameters. The resolved shear stresses depend on the alignment of the trench compared to the orientation of the silicon substrate, on the length, width and depth of the trench, and on the spacings between different trenches. The resulting stress level can be sufficient either to initiate propagation of already present dislocations or to start dislocation nucleation. Here also a combination of micro-Raman spectroscopy and electron diffraction techniques and correlation with simulations based on finite element calculations can be used to experimentally determine the two dimensional distribution of stresses around the embedded structure (Vanhellemont *et al* 1993). Kolbesen *et al* (1991) used TEM to study the dislocations generated by trenches and observed that these dislocations are glissile on one of the four {111} planes and have a Burgers vector $\frac{a}{2}$ <110> inclined to the silicon surface. As trench processes are used for the fabrication of storage capacitors in Dynamic Random Access Memories, the dislocation density and electrical activity has to be controlled in order to reduce the refresh failures. As reported by Stiffler *et al* (1990) the stress levels in DRAM structures can be reduced by using higher oxidation temperatures allowing oxide viscous flow, by aligning the trench structures in the <100> direction instead of the <110> direction, by inserting a thin nitride layer at the trench sidewalls, and by optimising the design concept.

Stresses associated with heteroepitaxial layers

Heteroepitaxial structures are nowadays extensively studied due to the high potential of Si-Ge systems for microelectronic devices, as outlined in some other chapters in this book. Although both silicon and germanium are of the diamond type, their different lattice constant of respectively 0.543 an 0565 nm leads to a cell mismatch of about 4.2%. The relation between the silicon lattice constant a and the concentration Cp of a solute species P follows Vegard's law

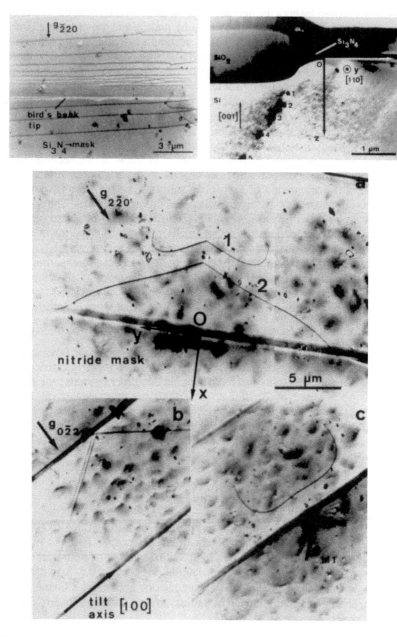

Figure 2.23 Typical defect structures observed along nitride film edges: a) and b) plan view and cross-section image of 60° dislocations at <110> film edges; c) Triangular half-loop observed along <100> edges (Vanhellemont *et al* 1987a, 1987b, Vanhellemont and Claeys 1988).

$$\frac{\Delta a}{a} = \beta \, C_P \qquad (2.37)$$

with β the lattice contraction (when negative) coefficient. According to the review paper by Hu (1991a) the lattice concentration is of the order of -4.5 to -5.6 10^{-24} cm^3/atom, -1.8 10^{-24} cm^3/atom, 6.2 to 7.1 10^{-25} cm^3/atom, and negligible for respectively boron, phosphorus, germanium and arsenic. This means that for increasing Ge concentrations in the Si-Ge system, the silicon lattice constant increases. The accommodation of the mismatch of heterosystems can occur either elastically by increased lattice strain or plastically by the generation of misfit dislocations. For relatively thin layers mismatch accommodation mainly occurs by strain, while the probability for dislocation generation increases for thicker layers. For embedded structures, e.g. oxide precipitates, the emission of point defects or the generation of prismatic dislocation loops is involved. For structures with a free surface, misfit dislocation generation starts at the surface and the formed dislocation nucleus might grow by glide further away from the interface.

The physical basis for stress calculations in heteroepitaxial systems is formed by the so-called thermodynamic equilibrium theory proposed by van der Merwe (1963). The total energy of a systems consists of the strain energy in the film and the interface energy associated with the lattice incoherence between the film and the structure. By deriving formulas for both energy components and minimising the total energy, van der Merwe obtained an expression for the critical film thickness h_c for initiating the generation of misfit dislocations

$$h_c = \frac{(1 - 2\nu)a}{4\pi(1 - \nu)^2 f} \left[\ln\frac{1 - \nu}{2\pi f} + 1 \right] \qquad (2.38)$$

f is the mismatch of the the spacing between the epitaxial layer and the substrate, measured at the interface and expressed as the fraction $\Delta a/a$. The predictions of (2.38), together with some experimental results obtained by Kasper and Daembkes (1987) for Si-Ge structures grown by Molecular Beam Epitaxy (MBE) are schematically shown in figure 2.24. It can be noticed that the critical film thickness is higher than the theoretically predicted one and can even be further increased by lowering the MBE growth temperature.

Another approach for calculating the critical film thickness has been proposed by Matthews and Blakeslee (1974) and is based on considering the misfit dislocations explicitly by taking into account the line tension force of the misfit dislocation by balancing the driving force from the epitaxial strain. More recently, People and Bean (1985 and 1986) combined the two previous calculation approaches by deriving a formula based on the original free energy minimization concept of van der Merwe, but now taking into account explicitly the energy of the misfit dislocations. Their calculated values are in better agreement with the experimental observations. Recently, Willis *et al* (1990) demonstrated however that the theory of van der Merwe and that of Matthews and Blakeslee give identical results if one uses the same expression for the dislocation energy. A discussion of these different approaches is outside the scope of this paragraph. Recently also Hu (1991b) pointed out that the different approaches should essentially result in the same critical film thickness if the right assumptions are made. He derived the following expression for the critical film thickness h_c

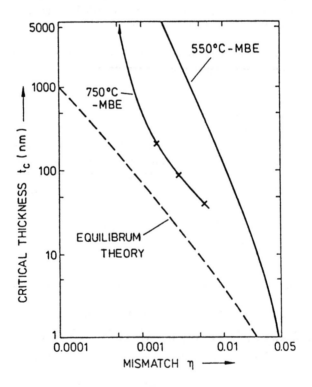

Figure 2.24 Critical film thickness for misfit dislocation generation versus the lattice mismatch for Si-Ge structures grown by MBE (Kasper and Daembkes 1987).

$$h_c = \frac{b}{8\pi f \sin\theta \cos\phi} \left(\frac{1 - \nu \cos 2\theta}{1 + \nu}\right) (\ln\frac{h_c}{r_0} - 1) + K \qquad (2.39)$$

$$K = \frac{1 - \nu}{2G(1 + \nu)f \cos\phi} \left(\frac{E_c}{b \sin\theta} + \gamma\right)$$

with θ the angle between the dislocation line and its Burgers vector, ϕ the tilt angle between the dislocation line and the film plane, r_0 the core radius of the dislocation and E_c the dislocation core energy. The K term, related to the increase of the critical film thickness by taking into account the dislocation core and surface energy adds about 20-30% for practical situations and is very often neglected.

Jain *et al* (1992) took also into account the elastic interaction energy between the misfit dislocations to calculate the energy of an array of dislocations and derived a more correct expression for the critical thickness. As the total energy is reduced by including the interaction energy, the critical film thickness is increased. Mostly the difference is quite small but for strained $Si_{(1-x)}Ge_x$ layers with large concentrations of germanium grown on silicon, it can increase to about 25%.

Stresses associated with extrinsic point defects

Somewhat related to the previous paragraph is stresses associated with extrinsic doping impurities in silicon. In section 2.2 it was mentioned that extrinsic point defects cause a contraction or extension of the silicon lattice. The lattice contraction coefficient for substitutional impurities with a covalent radius was defined in equation (2.2). In heavily doped regions the stress level may be sufficient for the generation of misfit dislocations. The generation of doping-induced dislocation networks has more than three decades ago nearly simultaneously been reported by Queisser (1961) and Prussin (1961). This subject has recently been reviewed by Hu (1991a). Queisser used the so-called Shockley model, whereby the generation and propagation of a dislocations is considered as resulting form the driving force to minimize the free energy. The inwards gliding of a dislocation over a depth x from the surface will reduce the strain energy of the diffused layer by

$$E_S = \frac{\sqrt{2}G(1 + \nu)}{1 - \nu} x \, \varepsilon \, b \sin\theta \cos\phi \qquad (2.40)$$

with ε the strain in the diffused layer. The preferred dislocations are 60° dislocations in {111} glide planes. By comparing the energy reduction with the elastic energy required for putting the dislocation at a distance x form the interface, Hu (1991a) derived the following formula for the critical doping concentration C_{crit} required to start the nucleation of misfit dislocations

$$C_{crit} = \frac{b}{2^{5/2}\pi\beta(1+\nu)\sin\theta\cos\phi} \ln\left[2^{3/2}\frac{x_d}{r_0}\right] \qquad (2.41)$$

with x_d equal to the diffusion length $2\sqrt{Dt}$.

In Prussin's model, the energy of the dislocation is not taken into account. When the stress field exceeds the critical value for dislocation glide, the misfit dislocations will glide away from the surface towards a depth where the doping concentration has dropped below the value corresponding with the critical glide stress. Many experimental data reported in the literature are in disagreement with Prussin's model, whereas the critical dose predicted by the Shockley model is often obtained. It should also be remarked however, that for small diffusion areas as encountered in advanced technologies the above given model is no longer valid, as these areas can then be considered as a type of "embedded " regions where the strain field is enhanced. In that case more complex calculations are required as discussed in a previous paragraph.

The Vegard's equation given in (2.37) can also be used to explain the strain compensation effects reported by Yeh and Joshi (1969). They noticed that by a simultaneous doping of boron and tin, the the strain energy of the system is reduced. This compensation effect is in the last decade extensively studied for the Ge-B system, as by an optimization of the doping ratio a complete strain compensation can be obtained, so that no process-induced dislocations are generated in the heavily doped regions. The addition of Ge has no influence on the electrically active doping concentration. In addition to the suppression of misfit dislocations, it has also been reported that the co-doping of Ge and B with a germanium-to-boron ratio between 3 and 7 has a beneficial influence on the mechanical strength of CZ wafers due to role as locking agents to immobilize dislocations (Fukuda and Ohsawa 1991). Strain compensation effects only occur for the co-doping with elements which do not interact with each other, by other mechanisms (e.g. Coulomb pairing). This implies that donor and acceptor elements must be excluded due to their possibility to form ionic pairs.

2.5. Metallic contamination and gettering

Since the early days of silicon processing, it is well known that metallic impurities have a strong degrading influence on the device performance. Most of the metallic contaminants such as Au, Fe, Ni, and Cu introduce generation-recombination centres resulting in a reduction of the minority carrier lifetime, an enhanced junction leakage current, and a lowering of the breakdown voltage. In addition metal contamination has a strong influence on the harmful properties of grown-in or process-induced defects by rendering them electrically active. For the advanced processing technologies with deep submicron geometries, even low concentrations of metallic impurities have detrimental effects on the device yield. Therefore this section review some important aspects of metallic contamination in silicon. The gettering of metallic contaminants will also briefly be addressed.

Origin of metallic impurities

The starting silicon may contain trace amounts of metals due to the polishing of the wafers, but nowadays most silicon manufacturers are offering nearly contamination free material. For the state of the art processing lines, metallic contamination reported in the early days as originating from the use of metal tweezers and metallic wafers chucks has disappeared due to the well controlled and optimized wafer handling technology. Diffusion furnaces at high temperatures can cause some contamination if no precautions are taken by using ultra pure furnace tubes and quartz material. Special care has also to be taken with susceptors in e.g. epi and CVD systems. A more common source of contamination may be related to the ion implantation step in case of beam misalignment leading to sputtering effects. In most of the above situations, if metallic contamination occurs it will have a dramatic influence on the process yield and will therefore relatively easily be discovered. The standard procedure to avoid the occurrence of metallic contamination is implementing on a regular basis monitoring techniques. Nowadays, there exists a variety of rather fast and accurate characterization tools such as e.g. Total reflection X-Ray Fluorescence, haze analysis, Surface Photovoltage Decay, thermal wave..., which allow a nearly on-line monitoring of the contamination problems.

The situation is becoming however much more difficulty for the control of very thin gate oxides, where metal concentration in the 10^9 cm^{-2} range already seriously degrades the oxide integrity. Therefore a huge amount of investigations are concentrating at present on the contamination control of materials such as gases and chemicals. Especially the standard wafer cleaning procedures can be an important source of metal contamination. Not only the purity grade of the chemicals used in cleaning solutions is important, but also the mixture of the solutions and the sequence in which the different parts of the cleaning procedure are performed. In addition to a low metallic impurity content, attention also has to be given to the particles removal rate of the cleaning solution and to the organic contaminants. An important aspect is also the impact of the cleaning process on the micro-roughness of the silicon surface, as a higher surface roughness increases the sticking probability of the metals to the surface.

Metal precipitation phenomena

Most of the metallic contaminants have a high diffusion coefficient so that in short times and at relative low temperatures large wafer areas can be contaminated by localised sources. Near the silicon melting point the diffusion constant is of the order of 10^{-3} to 10^{-4} cm^2/s (Weber 1983). This property allows the use of rapid thermal processing for the study of metallic contaminations as e.g. done by the so-called haze test (Alpern *et al* 1989). For most of the transition metals, an increasing diffusion coefficient is noticed for increasing atomic number. Most of the metals have a low solubility, which is decreasing very steeply for lower temperatures (see figure 2.5) so that supersaturation is easily obtained. Graff *et al* (1988) pointed out that in general, the solubility is influenced by parameters such as the cooling rate, the interstitial oxygen content of the substrate, and the atomic number of the metal. The solubility properties in combination with the high diffusivity implies that fast precipitation processes will occur. As outlined in section 2.2.6, the critical radius of the precipitates is controlled by both the concentration of point defects and the strain field around the precipitates. Therefore for isolated precipitation processes, i.e. the precipitation of a single metallic species in the silicon substrate, lattice matching between the precipitate and the silicon lattice is of key importance. To understand the influence of lattice strain on precipitation it is useful to consider the ratio ζ of the volume of the unstrained precipitate to the volume of silicon which is consumed (Vanhellemont and Claeys 1989)

$$\zeta = \frac{\Omega_{Si_yP_z}}{y\Omega_{Si}} \tag{2.42}$$

P is the impurity which precipitates into Si_yP_z. If $\zeta < 1$, the precipitation will preferentially occur in tensile strained regions as the lower strain energy results in a decrease of the critical radius. Most metals will precipitate in a silicide phase. The ζ factor for some of the most important metal silicides is given in Table 2.1. This table allows to conclude that Fe, the most common metal impurity in silicon, will precipitate as $FeSi_2$ in intrinsic gettered regions, as near oxide precipitates there is a tensile strain region. The same methodology can be used to understand the co-precipitation process of two or more metals. This is illustrated by the study performed by Vanhellemont *et al* (1990) concerning metal precipitation in high resistivity silicon material. In this material the semicoherent precipitation of rod-like α- or β- $FeSi_2$ phases has been observed. Figure 2.25 illustrates the needle-like iron silicide precipitates observed in boron implanted regions. These defects are lying in the <110> directions and consist of either tetragonal α-lebolite or orthorombic β-lebolite. In some cases the co-precipitation with $CuSi_x$ is also observed, as this is favoured by the strain compensation resulting in an overall strain reduction for the total system.

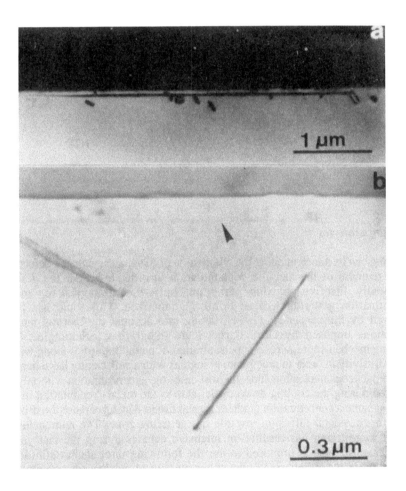

Figure 2.25 Needle-like iron silicide precipitates observed in boron implanted silicon (Vanhellemont *et al* 1990). In (a) a needle parallel to the silicon surface is observed together with end of range defects, while in (b) both an inclined and a surface parallel (in extinction: arrow) precipitate are shown.

Table 2.1 ζ-factor for metal precipitates in silicon (after Bronner and Plummer 1987)

Silicide	Ω_p (10^{-23} cm^{-3})	ζ
CuSi	4.15	2.01
Fe$_3$Si	4.52	2.26
FeSi	2.26	1.13
FeSi$_2$	1.88	0.94
Ni$_3$Si	4.31	2.16
Ni$_2$Si	3.27	1.64
NiSi	2.43	1.22
NiSi$_2$	1.97	0.96

Gettering strategies

Gettering can be aimed at either the elimination of the process-induced defects or at the removal of the metallic contamination in order to render the defects electrically inactive. In this paragraph only the strategies for metal contamination gettering will be briefly summarized. This topic has been reviewed by Simoen *et al* (1991), taking into account the thermal budget restrictions imposed by Ultra Large Scale Integration technologies. The gettering mechanism requires a supersaturation of metal impurities along with a high diffusivity in order to precipitate in regions with a sufficiently large density of heterogeneous nucleation sites. In most cases the supersaturation will only be achieved during the cooling down cycle. Due to the metal precipitation in the gettering zone a concentration gradient is established which becomes the driving force for the metal diffusion towards the gettering zone. The nomenclature related to gettering has resulted in intensive debates during the last years. Schröter *et al* (1991) proposed to use the following three main definitions. *Relaxation-induced gettering* is the above mentioned mechanism whereby the heterogeneous nucleation sites allow supersaturated regions to relax. This process would be retarded or even inhibited if no nucleation sites were present. In some cases no supersaturation is needed to obtain a gettering action. In that case the gettering action is initiated either by enhanced metal solubility or by stabilization of new metallic compounds in the gettering region. This gettering mechanism is referred to as *segregation-induced gettering*. Finally, for processes based on a coupling between substitutionally dissolved metals and silicon self-interstitials the term *injection-induced gettering* is used. This mechanism is typical for the well known phosphorus diffusion gettering. It is also possible to differentiate between so called internal and external gettering in view of the creation of the heterogeneous gettering sites (Gillis *et al* 1990). *Internal gettering* is provided by intrinsic or extrinsic defects created in the interior of the material, such as e.g. the oxygen precipitation during high temperature treatments. In the case of *external gettering*, the gettering sites are generated by an external processing step such as e.g. surface abrasion, ion implantation, phosphorus diffusion.

The three major steps in a gettering process are (i) the dissolution or release of the contaminants in the active device area, (ii) the diffusion of the contaminants, and (iii) the stable capturing of the contaminants at predefined gettering sites. Very often the main problem is the stability of the last step of the gettering process during the complete processing cycle. For the first step, the dissolution or release of the contaminants, can be understood by equation (2.42). For precipitates with a ζ-factor < 1, an injection of silicon self-interstitials can be absorbed by the precipitates so that no initiation of the dissolution process will take place. This means that if these kind of precipitates are formed in a relatively early phase of the fabrication process, they will remain stable during the subsequent processing and no dissolution will occur. For a ζ-factor >1 the injection of self-interstitials will lead to a dissolution of the precipitates and the subsequent getter steps can be performed.

For substitutional metallic impurities, the diffusion process occurs either by a dissociative mechanism, whereby the substitutional metal atom occupies an interstitial lattice position and a vacancy is created, or a kick-out mechanism. The latter has been proposed by Gösele *et al* (1980) for Au diffusion in silicon and is based on the interaction of the substitutional metal atom with a silicon self-interstitial pushing the metal atom in an interstitial lattice position. In both cases there is an interaction with intrinsic point defects which are thus required in order to have an effective gettering.

For the capturing of the metal impurity at the getter site, different mechanisms such as solubility enhancement, Coulombic pairing or capturing by extended defects can play an important role. The solubility enhancement is based on the fact that the solubility of the metal is enhanced by a shift of the Fermi-level so that the concentration of other charged point defects has to be taken into account. This phenomenon has been treated in detail by Kang and Schroder (1989). A typical example of a case where Coulombic pairing is important is the formation of iron-boron [FeB] complexes due to the interactions between the interstitial Fe donors and the substitutional B acceptors. These complexes have been studied in detail by Kimerling and Benton (1983). The pairing reaction occurs as

$$[Fe_i]^+ + [B_s]^- \quad \Leftrightarrow \quad [Fe_i B_s] \qquad (2.43)$$

This reaction indicates that the charge state of the metal impurity is important. In n-type silicon the interstitial Fe would be neutral so that no pairing can occur, even if there would be compensating boron atoms present. Rotsaert *et al* (1991) have demonstrated that at room temperature the [FeB] pair formation is possible for boron concentrations as low as 3×10^{12} cm^{-3}. The thermal stability of the [FeB] is rather restricted and dissociation is observed for short time anneals at temperatures of 200 °C. Storage at room temperature again favours the pair formation, and reported time constants for this process are between 250 and 300 min. Based on DLTS measurements, a strong recombination level at 0.4 eV and a shallow level at 0.1 eV above the valence band are reported for respectively Fe_i and [FeB].

If solubility enhancement and pairing occur simultaneously, then the gettering efficiency can be expressed as

$$\frac{[P_1]}{[P_1]_i} = 1 + \alpha [P_2] \; (1 + K_{pair} [P_2]) \qquad (2.44)$$

with $[P_1]$ the concentration of the metal to be gettered, $[P_1]_i$ the solubility of the metal in intrinsic silicon, $[P_2]$ the concentration of the opposite charged pairing species, α a positive constant, and K_{pair} the pairing constant which is exponentially dependent on the pair binding energy. This formula indicates that for the Fermi-level effect the solubility enhancement is proportional to the doping concentration, while it is proportional to the square of the doping concentration when ion pairing occurs. For each of the different metal atoms, it is possible to determine which silicon doping impurity gives the highest gettering efficiency. The discrepancies sometimes reported in the literature can to a large extent be explained by the fact that for some experiments the samples were deliberately contaminated which may result into extremely high concentration levels. Also the electrostatic stability of the formed pairs directly influences the gettering capability as pointed out by Cerofolini *et al* (1987) and Polignano *et al* (1988). This can explain the higher gettering efficiency for P and B compared to Sb and As in the case of Au atoms.

Gettering sinks can also be formed by extended defects such as e.g. dislocations. The interaction between the dislocation and the metal impurity can be either elastic due to the interaction with the strain field associated with the impurity or electrical by the acceptor type dislocation behaviour related to the dangling bonds. The elastic interaction energy is caused by the size effect of the impurity and can be calculated for different impurities and dislocation types. Although dislocations can getter impurities it is surely not a prerequisite for gettering action, as explained in the paragraphs above. Extended defect gettering is also playing an important role in heteroepitaxial systems as the strain field of misfit dislocations in combination with solubility effects leads to gettering. Case studies for Au and Ni gettering in Si-Ge systems have been discussed by respectively Salih *et al* (1984) and Lee *et al* (1988).

2.6 Impact on device properties

The impact of dislocations on the electrical properties of devices have been studied since the early days when Shockley (1953) put forward the fundamental idea that edge dislocations can act as acceptors in semiconductors with a diamond structure. Already in 1956 Kurtz *et al* reported on the correlation between the dislocation density and the minority carrier lifetime of germanium and silicon crystals showing that the lifetime is inversely proportional to the dislocation density. A few years later, Chynoweth and Pearson (1958) found experimental evidence that dislocations have a direct impact on the electrical behaviour of pn-junctions, i.e. avalanche breakdown microplasmas occur preferentially on sites where dislocations pass through junctions. The strong degrading impact of metal precipitates has first been reported in 1960 by Goetzberger and Shockley. During the following decades a huge amount of papers have been published on the possible influence of crystallographic defects on the electrical performance of MOS transistors, power devices (John 1967) high voltage devices (Kato *et al* 1976), bipolar devices (Ashburn *et al* 1977), Charge Coupled Devices (Claeys *et al* (1977) and Jastrzebski *et al* (1981)), memory devices (Strack *et al* 1979), ... As it is not intended to give a review of this topic, the given references are only selected to illustrate the type of work that has been performed. In-depth reviews concerning the impact of process-induced defects on the electrical performance of integrated circuits have been published by Kolbesen and Strunk (1985) and Kolbesen *et al* (1991).

Defect Control

The defect generation caused by standard (Claeys and Vanhellemont 1990) and more advanced processing steps (Claeys *et al* 1991, Tamura *et al* 1991) have been reported in the literature. For each of the different processing steps such as e.g. thermal processing, ion implantation, isolation techniques, diffusion processes, rapid thermal processing, silicidation ... a large number of papers have been published during the last decades. Although originally, the experimental results have stignated many debates, for most of the processing steps the fundamentals of the defect generation are nowadays well understood and can therefore be described by generally accepted theoretical models. This implies that defect control has more become a *science* instead of a collection of experimental results based on trial and error engineering. Of course, as the driving force towards further miniaturisation in order to achieve higher packing densities and improved speed performances, inevitably will necessitate the introduction of new or even still exotic processing steps, defect control engineering will remain a hot topic of key importance to the microelectronic fabrication community.

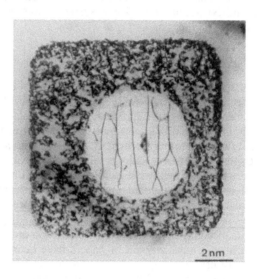

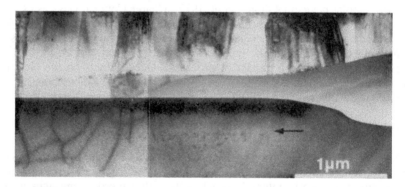

Figure 2.26 Plan view (a) and cross-section (b) TEM images of harmless defects in shallow As implanted junctions.

It should be kept in mind that a good insight into the fundamentals of the process-induced defect generation is only a part of the defect control engineering activities. In many cases it will not be possible to eliminate the crystallographic defects or contamination completely, so that the only solution is relying on the neutralisation of the electrical activity of the defects by the implementation of an appropriate gettering strategy. Figure 2.26 illustrates the presence of secondary defects in shallow As junctions. As the defects are not decorated, no electrical influence is seen on the leakage current performance. Over the last decades a large number of extrinsic and intrinsic gettering techniques have been reported. Some of these techniques are schematically illustrated in figure 2.27, as suggested by Shimura (1989). Within the scope of this chapter it is not possible to review this topic in detail. However, some relevant aspects of gettering strategies have briefly been discussed in section 2.5.

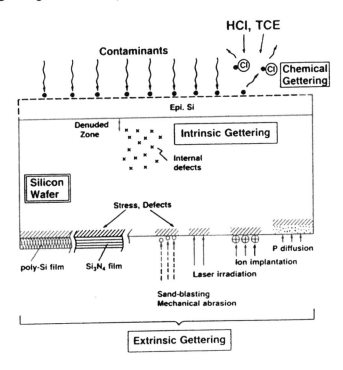

Figure 2.27. Schematical representation of different gettering approaches (Shimura 1989).

Lifetime engineering

Whereas gettering is aiming at removing as much as possible the crystallographic defects or at least rendering them electrical inactive, it is also possible to rely on the presence of crystallographic defects and their impact on the electrical performance of the devices. One of the oldest applications can be found in the bipolar technology where metallic impurities such as Au have been used to control the minority carrier lifetime to a defined low level. This is done by intentionally introducing in a well-controlled manner defects with well-defined numbers and types and in well-defined areas of the devices. In some case one is also interested in having a well-defined depth distribution of the defects. A very powerful technique to achieve these goals is the beneficial use of irradiations with high energy particles (electrons, protons) for lifetime engineering. For this purpose it is common use to define a so-called damage coefficient K_g, relating post-irradiation τ_g and the pre-irradiation lifetime τ_{g0}

and given by

$$\tau_g^{-1} = \tau_{g0}^{-1} + K_g \Phi \tag{2.45}$$

with Φ the irradiation fluence and K_g the damage coefficient which has to be determined for the different experimental conditions. The damage coefficient depends on the parameters of the silicon (doping type and density) and on the parameters of the irradiation (type of irradiation, energy, temperature). In general, for the same irradiation energy the damage coefficient is lower for electrons than for neutrons and is the highest for protons (Ma and Dressendorfer 1988). However, in order to allow a good control the variation of the lifetime as function of the energy should be a slow varying function, which is obtained for proton irradiations.

It is also feasible to develop advanced or new type of devices based on the controlled generation of misfit dislocations at strained interfaces such as in heteroepitaxial structures. Rozgonyi and Kola (1990) have demonstrated the use of Si-Ge layers for controlling in well-defined regions the minority carrier lifetime and the surface recombination velocity, so that CMOS devices with enhanced latch-up immunity and improved radiation hardness can be fabricated. By confining the misfit dislocations and controlling their electrical properties, it even becomes possible to contact the ends of the dislocation and to build new types of electronic devices such "buried wires" and so-called dislocations MOSFETs where the conduction occurs along a dislocation branch and remains controlled by the gate electrode. Such a transistor structure is schematically shown in figure 2.28. The expansion of the space charge layer is blocked by the space charge regions around the dislocation and near the dislocation the minority carrier lifetime is strongly reduced. The conductance of the channel can be controlled by selective decoration of the misfit dislocations by metal precipitates (Kola and Rozgonyi 1988). The speed of the devices can be increased by doping with noble metals, which results in a degradation of the carrier lifetime. Such device structures show a great potential for deep submicron transistors (e.g. 0.1 μm) due to the absence of doping requirements.

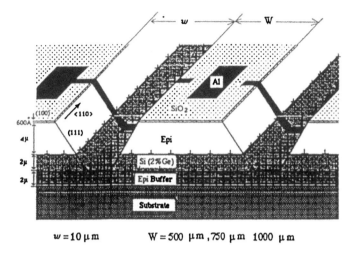

$w = 10\,\mu m$ $W = 500\;\mu m\,,750\;\mu m\;\;1000\;\mu m$

Figure 2.28 A schematic "dislocation MOSFET", after Rozgonyi and Kola (1990).

2.7 REFERENCES

Abe T, Harada H and Chikawa J 1983 Physica 116B 139

Abe T and Kimura 1990 Semiconductor Silicon eds H R Huff, K Barraclough and J Chikawa (Pennington: Electrochem. Soc.) 105

Alexander H 1986 Dislocations in Covalent Crystals ed N F R Nabarro (Amsterdam: North Holland) 7 114

Alpern P, Bergholz W and Kakoschke R 1989 J. Electrochem. Soc. 136 3841

Amelinckx S 1979 Dislocations in Particular Structures ed N F R Nabarro (Amsterdam: North Holland) 2 67

Ashburn P, Bull C, Nicolas K H and Booker G R 1977 Solid-State Electron. 20 731

Bean A R, Newman R C and Smith R S 1970 J. Phys. Chem. Solids 31 739

Bender H and Vanhellemont J 1988 Phys. Stat. Sol. (a) 107 455

Bender H and Vanhellemont J 1992 Mat. Res. Soc. Symp. Proc. 262.

Bender H and Vanhellemont J 1993 Handbook of Semiconductors, Vol. 3 (2nd ed), ed. S. Mahajan, (North Holland, Amsterdam), in press.

Bentini G, Correra L and Donolato C 1984 J. Appl. Phys. 56 2922

Benton J L 1990 J. Cryst. Growth 106 116

Borisenko V E and Yudin S G 1987 Phys. Stat. Sol.(a) 101 123

Bronner G B and Plummer J D 1987a J. Appl. Phys. 61 5286

Bronner G B and Plummer J D 1987b Proc. 17th ESSDERC Conf eds P Calzolari and G Soncini (Technoprint) 537

Bube R H 1960 Photoconductivity of Solids (New York: Wiley)

Celotti G, Nobili D and Ostoja P 1974 J. Mater. Sci 9 821

Cerofolini G F, Polignano M L, Bender H and Claeys C 1987 Phys. Stat. Sol A 103 654

Chidambarrao D, Peng J P and Srinivasan G R 1991 J. Appl. Phys. 70 4816

Chikawa J and Shirai S 1978 Jpn. J. Appl. Phys. 18 153

Chynoweth A G and Pearson G L 1958 J. Appl. Phys. 29 1103

Claeys C, Laes E, Declerck G and Van Overstraeten R 1977 Semiconductor Silicon eds H R Huff and E Sirtl (Princeton: Electrochem. Soc.) 773

Claeys C and Vanhellemont J 1987 Proc. Gettering and Defect Engineering in the Semiconductor Technology (GADEST) ed. H. Richter (Frankfurt-Oder: Inst. Phys. of Semiconductors) 3

Claeys C and Vanhellemont J 1989 Solid State Phenomena 6-7 21

Claeys C and Vanhellemont J 1990 Defect Control in Semiconductors ed K Sumino 565

Claeys C and Vanhellemont J 1993 Rad. Effect & Defects in Solids (in press)

Claeys C, Vanhellemont J and Simoen E 1991 Solid State Phenomena 19-20 95

Coene W, Bender H and Amelinckx 1985 S Phil. Mag. A 52 1509

Conwell E and Weisskopf 1950 Phys. Rev. 77 388

Corbett J W, Watkins G D, Chrenko R M and McDonald R S 1961 Phys. Rev. 121 1015

Corbett J W, Watkins G D and McDonald R S 1964 Phys. Rev. 135 A1381

Corbett J W and Bourgoin J C 1975 Point Defects in Solids eds J H Crawford Jr and L M Slifkin (New York:Plenum Press) ch 1

de Kock A J R 1973 Philips Res. Rept Suppl. 1

de Kock A J R, Roksnoer P J and Boonen P G T 1974 J. Cryst. Growth 22 311

de Kock A J R, Stacy W T and van den Wijgert W M 1979 Appl. Phys. Lett. 34 611

de Kock A J R and Wijgert W M 1980 J. Cryst. Growth 49 718

De Wolf I, Vanhellemont J, Romano-Rodríguez, Norström H and Maes H E 1992 J.Appl.Phys. 71 898

Fahey P M, Griffin P B and Plummer J D 1989 Reviews of Modern Physics 61 289

Fahey P M, Mader S R, Stiffer S R, Mahler R L, Mis J D and Slinkman J A 1992 IBM J. Res. Dev. 36 158

Föll H and Kolbesen B O 1975 Appl. Phys. Lett. 8 319

Föll H and Carter C B 1979 Phil. Mag. A 40 497

Frank W 1975 Inst. Phys. Conf. Ser. 23 23

Fukuda T and Ohsawa A 1991 Defect in Silicon II eds W M Bullis, U Gösele and F Shimura (Pennington, NJ: Electrochem. Soc.) 173

Gilles D, Weber E R and Hahn S 1990 Phys. Rev. Lett. 64 196

Goetzberger A and Shockley W 1960 J. Appl. Phys. 31 1821

Gösele U, Frank W and Seeger A 1982 Solid Stat. Commum. 45 31

Gösele U and Tan T Y 1991 MRS Bulletin XVI 42

Griffin J A, Hartung J and Weber J 1989 Mat. Sci Forum 38-41 619

Graff K 1983 Aggregation Phenomena of Point Defects in Silicon eds E Sirtl and J. Goorissen (Pennington: Electrochem. Soc.) 5286

Graff K, Heffner H A and Hennerici W 1988 J. Electrochem. Soc. 135 952

Gupta M C and Ruoff A L 1980, J. Appl. Phys. 51 1072

Ham F S 1958 J. Phys. Chem. 6 335

Heijmink Liesert B J, Gregorkiewicz T and Ammerlaan C A J 1992 Mat. Sci. Forum 83-87 407

Hölzlein K, Pensl G and Schulz M1984 Appl. Phys. A 34 155

Hornstra J 1958 J. Phys. & Chem. Solids 5 129

Hull D 1965 Introduction to Dislocations (Oxford: Pergamon Press)

Hu S M 1975 J. Appl. Phys. 46 1465

Hu S M 1980 J. Appl. Phys. 51 5945

Hu S M 1985 J. Appl. Phys. 57 1069

Hu S M 1989 J. Appl. Phys. 66 2741

Hu S M 1990 J. Appl. Phys. 67 1092

Hu S M 1991a " Process Physics and Modelling in Semiconductor Technology eds. G R Srinivasan, J Plummer and S Pantelides (Pennington: Electrochem. Soc.) 548

Hu S M 1991b J. Appl. Phys. 69 7901

Hu S M, Klepner S P, Schwenker R O and Seto D K 1976 J. Appl. Phys. 47 4098

Inoue N, Wada K and Osaka J 1981 Semiconductor Silicon eds. H R Huff, R J Kriegler and Y Takeishi (Pennington: Electrochem. Soc.) 81-5 282

Jain SC, Gosling TJ, Willis JR, Totterdell DHJ and Bullough R 1992 Phil.Mag. A 65 1151

Jastrzebski L, Levine P A, Fisher W A, Cope A D, Savoye E D and Henry W N 1981 J. Electrochem. Soc. 128 885

John H F 1967 Proc IEEE 55 1249

Kamiura Y, Hashimoto F and Yoneta M 1989 J. Appl. Phys. 65 600

Kanamori A and Kanamori M 1979 J. Appl. Phys. 50 8095

Kang J S and Schroder D K 1989 J. Appl. Phys. 65 2974

Kasper E and Daembkes H 1987 Inst. Phys. Conf. Ser. 82 93

Kato T, Koyama H, Matsukawa T and Fujikawa K 1976 Solid-State Electron. 19 955

Katsura J, Nakayama H, Nishino T and Hamakawa Y 1982 Jpn. J. Appl. Phys. 21 712

Kimerling L C and Benton J L 1983 Physica B 116 297

Kimerling L C, Asom M T, Benton J L, Drevinsky P J and Caefer C E 1989 Materials Science Forum Vols. 38-41 141

Kola R R and Rozgonyi G A 1988 Special Topics in Electronic Materials eds B R Appleton et al (MRS Ext. Abstr.) EA-18 15

Kolbesen B O and Strunk H P 1985 VLSI Electronics Microstructure Science ed N G Einspruch (New York: Academic Press) 12 ch. 4

Kolbesen B O, Cerva H, Gelsdorf F, Zoth G and Bergholz W 1991 Defects in Silicon II eds W M Bullis, U Gösele and F Shimura (Pennington: Electrochem. Soc.) 371

Kosevich A M 1979 Dislocations in Solids ed N F R Nabarro (Amsterdam: North Holland) 1 33

Kovacs I and Zsoldos L 1973 Dislocations and Plastic Deformation ed D Ter Haar (Oxford: Pergamon)

Lambert J L and Reese M 1968 Solid-State Electron. 11 1055

Lelikov Yu S, Rebane Yu T and Shreter Yu G 1989 Inst. Phys. Conf. Ser. 104 113

Leroy B and Plougonven 1980 J. Electrochem. Soc. 127 961

Ma T P and Dressendorfer P V 1988 Ionizing Radiation Effects in MOS Devices and Circuits (New York: Wiley Intersci.)

Mainwood A, Larkins F P and Stoneham A M 1978 Solid-State Electron. 21 1431

Matthews J W and Blakeslee A E 1974 J. Cryst. Growth 27 118

Milnes A G 1973 Deep Impurities in Semiconductor (New York: Wiley Intersc)

Murray D C, Evans G R and Carter J C 1991 IEEE Trans. Electr. Dev. 38 407

Nakayama H, Nishino T and Hamakawa Y 1980 Jpn. J. Appl. Phys. 19 501

Nakashima H and Shiraki Y 1978 Appl. Phys. Lett. 33 257

Newman R C 1988 Mat. Res. Soc. Symp. 104

Ning T H 1987 Properties of Silicon EMIS Data reviews

Pensl G, Schulz M, Hölzlein K, Bergholz W and Hutchison J L 1989 Appl. Phys. A 48 1989

People R and Bean J C 1985 Appl. Phys. Lett. 47 322

People R and Bean J C 1986 Appl. Phys. Lett. 49 229

Polignano M L, Cerofolini G F, Bender H and Claeys C 1988 J. Appl. Phys. 64 689

Prussin S 1961 J. Appl. Phys. 32 1876

Queisser H J 1961 J. Appl. Phys. 32 1776

Queisser H J 1982 Defects in Semiconductors II eds S Mahajan and J W Corbett (Pittsburg: Mat. Res. Soc.) 323

Ravi K V 1981 Imperfections and Impurities in Semiconductor Silicon (New York: John Wiley and Sons)

Read W T 1953 Dislocations in Crystals (New York: Mc Graw-Hill)

Roksnoer P J and van den Boom M M B 1981 J. Cryst. Growth 53 563

Romano-Rodriguez A and Vanhellemont J 1992 Mat. Sci Forum 83-87 303

Rotsaert E, Clauws P, Simoen E and Vennik J 1991 Solid-State Comm. 77 415

Rozgonyi G A and Kola RR 1990 Defect Control in Semiconductors ed K Sumino (Amsterdam: Elsevier) 589

Salih A S M, Kim H J, Davis R F and Rozgonyi G A 1984 Semiconductor Processing ed D C Gupta (Philadelphia: Am. Soc. Test. Mat.) 272

Shockley W 1953 Phys. Rev. 91 228

Scholz F J and Roach J W 1992 Solid State Electron. 35 447

Schröter W, Seibt M and Gillis D 1991 Materials Science and Technology ed W Schröter 4

Seeger A and Chik K P 1968 Phys. Stat. Sol. 29 455

Shimura F 1989 Semiconductor Silicon Crystal Technology (New York: Academic Press)

Simoen E and Claeys C 1993 IEEE Trans. Electron. Dev. (in press)

Simoen E, Vanhellemont J and Claeys C 1991 Proc. Conf. on Defect Control and Related Yield Management (Brussels: Semi Europe) 74

Spry R J and Compton W D 1968 Radiation Effects in Semiconductors (New York: Plenum) 421

Steele A G, Lencgyshun L C and Thewalt M L A 1990 Appl. Phys. Lett. 56 148

Stiffler S R, Lasky J B, Koburger C W and Berry W S 1990 IEEE Trans. Electron Dev. 37 1253

Strack H, Mayer K R and Kolbesen B O 1979 Solid-State Electron. 22 135

Sumino K 1981 Semiconductor Silicon eds H R Huff, R J Kriegler and Y Takeishi (Pennington: Electrochem. Soc.) 220

Sumino K 1985 Proc. Gettering and Defect Engineering in the Semiconductor Technology (GADEST) ed. H. Richter (Frankfurt-Oder : Inst. Phys. of Semiconductors) 41

Sumino K and Imae 1983 M Phil. Mag. A 47 753

Sun Q, Yao K H, Lagowski J and Gatos H C 1990 J. Appl. Phys. 67 4313

Tajima M, Kishino S, Kanamori M and Iizuka T 1980 J. Appl. Phys. 51 2247

Tajima M, Stallhofer P and Huber D 1983 Jpn. J. Appl. Phys. 22 L586

Tamura M, Isomae S, Ando T, Ohyi K, Yamagishi H and Hashimoto A 1991 Defects in Silicon II eds W M Bullis, U Gösele and F Shimura (Pennington: Electrochem. Soc.) 3

Tan T Y 1986 Proc. Mat. Res. Soc. 59 269

Tan T Y and Gösele Y 1985 Appl. Phys. A 37 1

Teodosiu 1982 Elastic Models of Crystal Defects (Berlin: Springer Verlag)

Van den hove L, Vanhellemont J, Wolters R, Claassen W, De Keersmaecker R and Declerck G 1988 Advanced Materials for ULSI eds M Scott, Y Akasaka and R Reif (Pennington: Electrochem. Soc.) 165

van der Merwe J H 1963 J. Appl. Phys. 34 117

Vanhellemont J and Claeys C 1987 J. Appl. Phys. 62 3960

Vanhellemont J and Claeys C 1988a J. Electrochem. Soc 135 1509

Vanhellemont J and Claeys C 1988b J. Appl. Phys. 63 5703

Vanhellemont J and Claeys C 1989 Mat. Sci. Forum 38-41 171

Vanhellemont J and Claeys C 1991a Process Physics and Modelling in Semiconductor Technology eds. G R Srinivasan, J Plummer and S Pantelides (Pennington: Electrochem. Soc.) 583

Vanhellemont J and Claeys C 1991b Defects in Silicon II ed W M Bullis, U Gösele and F Shimura (Pennington: Electrochem. Soc.) 263

Vanhellemont J and Claeys C 1992 J. Appl. Phys. 71 1073

Vanhellemont J and Claeys C 1993 Proc. RTP '93 eds R.B. Fair and B. Lojek (RTP '93, Scottsdale, AZ) 62

Vanhellemont J, Amelinckx S and Claeys C 1987a J. Appl. Phys. 61 2170

Vanhellemont J, Amelinckx S and Claeys C 1987b J. Appl. Phys. 61 2176

Vanhellemont J, Bender H and Claeys C 1989 Inst.Phys.Conf.Ser. 104 461

Vanhellemont J, Vanstraelen G, Simoen E and Claeys C 1990 Ext. Abstr. Electrochem. Soc. Meeting (Pennington: Electrochem. Soc.) 90-2 591

Vanhellemont J, De Wolf I, Janssens KGF, Frabboni S, Balboni R and Armigliato A 1993 Appl. Surf. Sci. 63 119

Voronkov V V 1982 J. Cryst. Growth 59 625

Watkins G D 1975 Point Defects in Solids eds J H Crawford Jr and L M Slifkin (New York: Plenum Press) ch 4

Watkins G D, Troxel J R and Chatterjee A P 1978 Inst. Phys. Conf. Ser. 46 16

Weber E R 1983 Appl. Phys. A 30 1

Weertman J and Weertman J R 1965 Elementary Dislocation Theory eds M E Fine, J Weertman and J R Weertman (New York : Mc Millan)

Wijaranakula W 1991 J. Appl. Phys. 70 3018

Wijaranakula W, Mollenkopf H and Matlock J H 1990 Appl. Phys. Lett. 56 764

Willis J R, Jain S C and Bullough R 1990 Phil. Mag. A 62 115

Yang D, Lu J, Li L, Yao H and Que D 1991 Appl. Phys. Lett. 59 1227

Yeh T H and Joshi M L 1969 J. Electrochem. Soc. 116 73

Zeldovich J 1942 J. Exp. Theor. Phys. 12 525

Zhong L, Wang Z, Wan S, Zhu J and Shimura F 1992 Appl. Phys. A 55 313

3

Molecular Beam Epitaxy of Silicon, Silicon Alloys, and Metals

E. Kasper and C.M. Falco

3.1 INTRODUCTION

The traditional source of molecular beams is the well-known Knudsen cell. However, more recently evaporation cells with larger openings (effusion cells), gaseous sources, and electron beam evaporators also have been developed for the generation of molecular beams. These newer sources have been developed to the point where molecular beam epitaxy of metals and silicon-based materials is now mainly conducted by the usage of electron beam evaporator sources. Since many of the technical requirements for the growth of silicon, silicides, and metals by MBE are similar, this chapter treats these classes of materials together. In this chapter we describe the reasons for the choice of electron beam evaporators, the technical solutions needed to use such sources, and the implications for the growth process. Especially with the growth of silicon based materials, industrial aspects have to be strongly considered. Future electronics production may benefit from MBE if new equipment routes succeed.

3.2 ELECTRON BEAM EVAPORATORS AND METALS MBE

In MBE, the deposition beam flux is created through thermal evaporation of the desired material under ultra-high vacuum conditions. Because of the wide variety of materials used in thin film production, a broad range of evaporation temperatures must be accessible. The earliest MBE systems were designed for low-temperature/high vapor pressure materials that could be easily evaporated with radiatively heated furnaces (Knudsen cells). Through precise temperature control of these effusion cells, extremely steady evaporation rates of less than 0.01 Å/sec could be attained allowing for sub-monolayer control of the deposition. The ultra-high vacuum conditions present in an MBE system place the evaporating atoms in their mean-free-path limit and hence allow line-of-sight travel from the furnace to the substrate. Deposition is then easily controlled through the use of mechanical shutters to block the beam flux thereby allowing the effusion cells to be maintained at their required temperatures.

The usefulness of effusion cells is degraded as the required temperature for evaporation of a given material increases. The main disadvantage is the intimate contact the deposition material must make with the containing crucible. Because the heating process is indirect, the entire cell must be brought to a high temperature. For many materials, reactions can then take place liberating unwanted contaminants into the beam flux. The most popular crucible material is pyrolitic boron nitride (PBN) developed mainly for III-V MBE systems. At temperatures above 1200 °C, significant amounts of boron and nitrogen are released causing potential problems in sample preparation. Current interest in the production of high temperature superconductors in MBE environments has motivated several companies to offer higher temperature effusion cells (Kasper 1994a). These furnaces use pyrolitic graphite crucibles capable of temperatures in excess of 1800 °C. However, carbon reactivity and contamination problems still exist for many materials.

Though implementation of effusion cells is quite straight forward, many low-vapor-pressure materials cannot be successfully evaporated with them. Silicon is a prime example. For useful evaporation rates, Si must be heated to 1700 °C where it becomes molten, and hence reactive with the crucible containers. This problem can be avoided through the use of electron-beam (e-beam) evaporators. These devices rely on a focussed high-energy electron beam impinging on the target material directly thereby causing heating and evaporation. The material is contained in a water-cooled high thermal conductivity copper hearth. In the case of Si with its lower thermal conductivity, the target only locally heats. A pool of molten Si forms in the center of the bulk Si target and therefore never comes into contact with the walls of the hearth. Only the deposition material is heated significantly, and hence impurity levels are much reduced.

There are several commercially available e-beam evaporators suitable for a vacuum environment (Kasper 1994b). These units are generally designed for the high evaporation rates (2,500 nm/min) used in commercial optical coating production and difficulty can arise when attempting the small rates (< 1 nm/sec) required for MBE. Small modifications must be performed on some models (Kasper 1994c) to make them totally UHV compatible. These include the removal of all lubricants, replacement of rubber o-ring seals by ceramics, gold plating of fasteners, etc. Generally, e-beam evaporators consist of a filament for electron production located beneath the sample charge. The accelerated electrons (from 6 to 12 keV) are deflected via a magnetic field 270 degrees into the sample material. This deflection keeps the evaporated atoms away from the electron filament, prolonging its life and allowing normal incidence of the e-beam onto the target. For some target materials (Si is again a good example) the localized electron spot simply burrows a narrow hole into the sample. This is due to poor thermal conductivity of the Si, and hence strong local heating. To avoid this condition, the e-guns have deflection capabilities built in to allow the beam to be swept into a larger pattern to produce more even heating. There are multiple pocket e-beam evaporators commercially available (Models) that allow several deposition materials to be selected in the vacuum. Care must be taken to insure that the bearing mechanisms that allow the pockets to be moved into the e-beam are truly UHV compatible. At present the first of these devices are being installed and tested in true MBE machines (Chow).

Unlike effusion cells, temperature is no longer a good parameter for rate control for e-beam evaporators. Instead, the evaporation rate must be monitored independently and used to control the e-beam power in a feed-back loop. Here lies one of the main disadvantages of e-beams for very small evaporation rates. With effusion cells, the rate at a given temperature can be determined from the time average of an accumulated film's thickness. To reproduce this rate, the system need merely be heated to the same temperature and regulated. Because the e-gun power supply relies on feed-back from the real-time measured rate, a lower limit is set for reliable rate control depending on the sensitivity of the flux monitor. Currently, *in situ* rate monitors in common use consist of three types; quartz crystal oscillators, electron impact emission spectroscopy (EIES), and quadrupole mass spectrometers. The quartz crystal is not actually a rate monitor, but measures total deposited film thickness. A time derivative therefore must be calculated to obtain an evaporation rate, leading to increased noise. The quartz crystal also becomes loaded as the deposited material increases, leading to a finite lifetime and requiring periodic breaking of the vacuum for replacement of the crystal. The EIES, on the other hand, utilizes the intensity of characteristic optical radiation emitted from transitions caused by bombarding electrons into the evaporant beam as a real-time measurement of mass transfer. This has the desirable feature of being sensitive to different chemical species simultaneously, however, with varying degrees of sensitivity. For most metals the signal to noise is such to allow for rate control to well below 0.1 Å/sec. However, for silicon with its reduced radiated intensity, this is an effective lower limit. In commercial EIES units (Leybold) the radiation intensity is converted to an evaporation rate using a calibration from an *in situ* quartz crystal monitor. As in the case of effusion cells, the evaporating atoms from an e-beam source are allowed to deposit on the substrate by opening a mechanical shutter. Because the e-gun power is regulated by the measured evaporation rate, the flux monitor must see the beam flux at all times. In this way any transients are avoided when the shutters are opened.

Another difficulty in using e-beam evaporators arises from the amount of stray electrons released into the growth environment. Typically as much as 30% of the incident electrons are scattered into the surroundings of the target material. These electrons can cause the evaporation of small amounts of the hearth material. In instances where trace elements in the prepared sample could have detrimental effects, the surrounding hearth walls should be lined with plates of the same material as that being deposited. The stray electrons also have deleterious effects on the ion pressure gauges in the system making their readings during growth unreliable. At the substrate, the electron flux has been measured as high as $10^{13}/cm^2$ during growth. The resultant effects on sample integrity are as yet not well known. Also, a portion of the evaporated beam (typically $\approx 0.1\%$) is ionized as well.

One measurement that is highly desireable during growth is reflection high energy electron diffraction (RHEED). This surface probe is very sensitive to layering characteristics and can be used to determine a samples growth mode. The use of e-beam evaporators makes the use of RHEED in real-time difficult at present. There are three inherent problems that must be overcome. For low vapor-pressure/high-temperature materials, the intensity of visible radiation released when they are at their operating temperatures is so great that it bathes the entire growth chamber in bright white light. This makes observation of RHEED patterns on the phosphor screen very difficult. Care could be taken to reduce this radiation when the flux shutter is closed, however, it must be remembered that the EIES rate monitor must see the evaporant flux at all times making light tight shielding difficult. If the deposition material does not radiate so intensely such as for Si, another difficulty arises when the e-beam sweep is utilized. The stray fields produced from the sweep generator tend to sweep the RHEED pattern as well. Again, it is possible with careful shielding that this effect can be reduced or eliminated. Finally, the stray electrons from the e-guns impinge directly on the phosphor screen causing a background glow.

3.2.1 Recent Developments in Metals MBE

Exploitation of modern vacuum technology and vapor deposition techniques for the synthesis of nearly flawless semiconductor superlattices has been highly successful. Recently, interest in such compositionally modulated materials has expanded to include metals (Chang 1985, Schuller 1987). Actually, as Fig. 3.1

shows, research on metallic multilayers and superlattices dates back over fifteen
years. Although the two curves in the figure include different types of data (e.g.

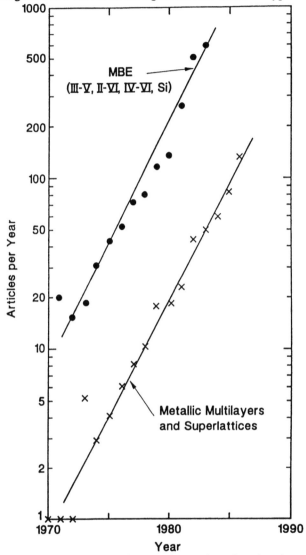

Figure 3.1 Data assembled by the authors on the number of semiconductor and metal
MBE-related papers published each year since 1970.

the top curve includes all MBE-grown materials, not just heterostructures and
superlattices), it is interesting to note how closely the fields have paralleled each
other in growth.   The exponential growth in research on MBE-grown

semiconductors can partially be attributed to the large number of practical devices arising from this research. Various applications of metallic superlattices are now appearing. A few examples are optics for soft x-rays (Ceglio 1986, Koch 1990) and magneto-optic data recording (Zeper 1990, Bennett 1990, Tanaka). Although it is dangerous to extrapolate exponential growth curves, the emergence of such applications should continue to stimulate research in this field, and should mean a strong future for both basic and applied research on metallic superlattices and other thin film structures.

In the case of artificially layered materials made partially or completely of metallic components, the absence of strong covalent bonding makes the growth of high-quality crystals more difficult than for semiconductors. However, while highly perfect layers are required for semiconductors in order to exhibit phenomena of interest, it is important to realize that metals can tolerate greater levels of structural and chemical imperfections and still exhibit useful properties. In many ways metals are more forgiving of such imperfections than are semiconductors. Many physical properties of metals are not strongly affected by a less than perfect material, whereas semiconductor properties are seriously degraded by defects, impurities, etc. While it is more difficult to grow a "perfect" metal crystal than it is a perfect semiconductor crystal, at the same time it is usually less important. However, for certain materials and physical properties great care must be taken in the fabrication of the metallic film. For example, the superconducting properties of Nb-containing superlattices are degraded by

oxygen contamination. Since Nb is a strong oxygen getter, UHV and/or high deposition rates are required to deep oxygen contamination negligible in the growing film. Similarly, very high quality Dy-Y superlattices are grown in UHV on a buffer layer of Nb to prevent reaction with the sapphire substrate (Salamon 1986).

Due to the high cost, we estimate fewer than a dozen MBE machines are currently in use or on order for the fabrication of metallic thin films or superlattices (Kasper 1994d). However, as for semiconductor superlattices, MBE is a powerful technique, and should find increasing use as further applications of these materials are identified.

The requirements for an MBE machine capable of producing such metallic multilayers are different than for a GaAs machine. Since many of the materials of interest have high melting points, reasonable vapor pressures often cannot be achieved using Knudsen cells. Consequently, electron beam evaporation sources must be incorporated in the growth chamber. Provision should also be made for a higher temperature substrate platform for epitaxy of refractory metals. These, and other, requirements are very similar to those needed for the epitaxy of silicon

so that MBE systems based on the new generation of Si MBE machines can also be used for MBE of any metal. A schematic diagram of such a machine is shown in Figure 3.2.

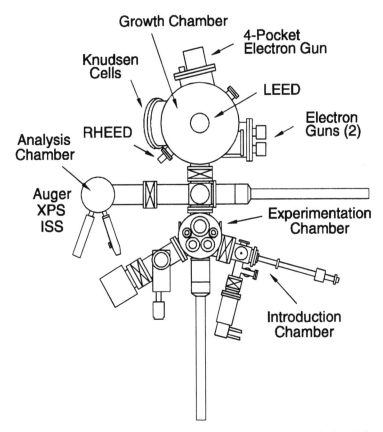

Figure 3.2 Schematic diagram based on a commercial MBE system designed for silicon epitaxy and also suitable for epitaxial growth of refractory metals (Perkin-Elmer model 433-S).

RHEED Oscillations

Oscillations in RHEED patterns during MBE growth of semiconductor materials have been commonly observed for several years (Cohen 1987, Dobson 1988) but have only recently been observed during growth of metallic layers (Kozoil 1987). Such oscillations are extremely valuable, both as an indicator that two-dimensional (2D) layer-by-layer growth is occurring and as a tool for determining in real time precisely the number of monolayers deposited. It should be noted, however, that absence of RHEED oscillations does not necessarily indicate lack of layer-by-layer growth, since mechanisms other than 2D nucleation may be occurring in a given situation (e.g. 2D growth via dislocation steps or misorientation steps).

In situations where RHEED oscillations do occur, they can be used to define what one means by a monolayer of material during the growth of the overlayer. Although similar information can be gleaned from the detailed analysis of Auger line intensities as a function of coverage, the time required for such measurements makes it impossible to use the latter method to give the real-time feedback which makes RHEED oscillations so useful. The highest quality semiconductor superlattices have been grown by using RHEED oscillations to determine when each new layer should start.

The simplest picture for understanding RHEED intensity oscillations is that when ideal 2D-nucleated layer-by-layer growth occurs. For this case, each monolayer almost completely covers the surface before the next monolayer begins to grow. At the moment when a layer is completed, the intensity of the diffraction pattern is at it's maximum. This pattern is characteristic of the 2-dimensional reciprocal lattice representing the surface. When the next layer is partially completed, the intensity will be lower due to the (destructive) interference between it and the previous layer which is still partially exposed. As the layer reaches completion the intensity is again maximized. If more than a few layers are forming simultaneously the oscillations will not appear. If the substrate temperature is too low (or if impurity adatoms are on the surface), the arriving atoms will not have enough energy to move about the surface and find a suitable lattice position. This will limit the growth perfection. On the other hand, if the temperature is too high, excessive diffusion into the bulk of the previously formed layers may occur, often destroying the desired layered structure.

Continuing investigations have shown that RHEED oscillations can be observed in the growth of many metallic overlayers under the proper growth conditions. Purcell et al. have demonstrated RHEED oscillations for Fe, Ni, Au, and Mn overlayers epitaxially grown in various combinations on Fe, Ni, Au, Ag, and Ru substrates (Purcell 1988). The oscillations are observed for substrate temperatures between room temperature and 450 °C, with the least damped oscillations occurring at the lower temperatures. This result is in contrast to the case of semiconductors where much higher temperatures are typically required to observe the oscillations.

Metallic Superlattices

"Metallic superlattices" are multilayered materials with several characteristics. Among the most important are sharp boundaries (< 5% of the thinnest layer) between the two components. In addition, there is long range structural coherence maintained across many layers of the material. Thus, one has the roughly natural lattice spacing within the layer modulated by the imposed superlattice layer periodicity. Two examples of superlattices are Mo/Ta (grown by sputtering (Makous 1987)), and Gd/Y (grown by MBE (Kwo 1985)). The Gd/Y superlattices were shown to be single crystals with coherence lengths of several thousand Ångstroms and interfaces spanning only two atomic layers.

Dy/Y is another single-crystal rare-earth superlattice which has been produced by MBE (Salamon 1986). Neutron diffraction measurements show that the helical periodicity of the Dy spins remains coherent across several bilayers. Since bulk Dy has an incommensurate helical magnetic order, the authors concluded that this magnetic structure is "modulated, but not interrupted, by the intervening layers of 'nonmagnetic' Y."

These rare-earth superlattices have proven to be interesting model systems for the study of magnetism in part because of their well known and nearly perfect crystal structure. Although very useful magnetic multilayer films have been produced by other methods, such as sputtering, MBE-grown structures offer the best chance to study the physical properties of such structures in a controlled setting on an atomic level.

Metal Silicides

The physical properties and growth of metal silicides have been studied intensively for more than a decade. Much of the interest stems from the desire to use these silicides for interconnections and gates in silicon-based integrated circuits. In addition to chemical stability and low resistivity, many of these materials have the added attraction of having a small lattice mismatch with Si, thus raising the possibility of producing epitaxial silicide films. Although these films have been produced by many methods, the recent application of MBE techniques has produced an explosion of activity and progress (Tung 1988). The ability to produce single-crystal silicon-silicide structures is interesting in the short term since such a structure can provide a compact high-conductivity path for device applications. The long-term outlook for such technology is also bright since such simple structures are the first step toward more complicated three-dimensional structures with novel properties which can be tailored to a particular application.

Most of the recent research has been directed toward epitaxial growth of the transition metal silicides; $NiSi_2$, $CoSi_2$, $PtSi$, Pd_2Si and the refractory metal disilicides; $CrSi_2$, WSi_2, $MoSi_2$. Typically such experiments involve deposition of the metal, or co-deposition of Si and the metal, onto a clean Si substrate under UHV conditions. The material may be deposited onto a room temperature substrate and subsequently annealed, or it may be deposited directly onto a heated substrate. In all cases the structure of the final silicide film is extremely sensitive to the details of the fabrication process. It is possible to form anything from a randomly oriented polycrystalline film to a single-crystal epitaxial one.

As an example, we will consider the case of $MoSi_2$. Perio et al. deposited Mo films onto Si (100) by electron-gun evaporation under UHV conditions (Perio 1984). When the Mo was deposited onto 650 °C substrates, $MoSi_2$ films were formed with preferential orientation. If the deposition is performed at room temperature with subsequent annealing at 650 °C, a randomly oriented polycrystalline film is formed. The physical picture used to explain this result is the following. When the substrate is hot, the silicide growth rate is greater than the Mo deposition rate so that initially oriented silicide nuclei form. But when the Mo is deposited onto room temperature Si, the as-deposited film is unoriented. Post annealing causes the Si to diffuse into the Mo grains forming $MoSi_2$ but the large density of grain boundaries and their associated stresses interfere with reorientation of the silicide grains. Very similar behavior was observed for W deposited on Si (100) (Torres 1985).

A few single-crystal Si/silicide multilayer structures have been successfully fabricated and described in detail elsewhere (Tung 1988, Mandl 1992). $Si/NiSi_2/Si$ and $Si/CoSi_2/Si$ three-layer structures were grown using a "template" method whereby sharp interfaces are formed by first creating a thin high-quality template layer and then growing, under different conditions, a thicker epilayer of that material. X-ray diffraction and ion channeling showed these structures to be single crystals with sharp interfaces.

3.2.2 Some Prospects for the Near Future

Multilayer Optics for Soft X-Rays

Advances in microfabrication technology have made possible the development of multilayer optical coatings with near-atomic perfection. Coupled with the increasing availability of synchrotron light sources, this has generated a revival of interest in reflective x-ray optics, in particular for the soft x-ray regime. The soft x-ray spectral region can be considered to lie between the "hard" x-rays (λ < 10 Å), conventionally used for materials analysis and medical radiology, and the vacuum ultraviolet (λ > 500 Å).

Multilayer-based soft x-ray imaging devices can be extremely useful in many areas, such as astronomy, plasma diagnostics, and synchrotron optics. The high spatial resolution which can be expected is also of much interest in microelectronics fabrication. It has been estimated that by 1997 there will be 175 commercial synchrotron radiation sources dedicated to producing integrated circuits by x-ray lithography (Yanoff 1988). The high resolution is also important for microscopy with the additional advantage that live cells are observable with soft x-rays without the need for staining or processing. Mirrors and beam splitters are needed for soft x-ray laser resonator cavities. Development of such lasers would generate a large number of additional applications requiring optical elements. Unfortunately, at x-ray wavelengths the refractive index approaches unity for every material. As a consequence, not much reflection or refraction occurs at any surface, as Röntgen himself found. In addition, absorption of x-rays is not negligible. Thus, conventional lenses or mirrors are not possible at these wavelengths.

Because of the small normal-incidence reflectivity of all materials in the soft x-ray and X-UV regions of the electromagnetic spectrum, optical systems must be either grazing-incidence or based on interference or diffraction phenomena (multilayers or Fresnel zone plates (Baez 1952 1961, Kearney 1991)). Unlike grazing-incidence optics, these later two phenomena allow for normal-incidence use. This geometry makes possible larger apertures for a given size device plus the avoidance of the severe aberrations, astigmatism in particular, associated with grazing incidence optics. Crystals have been used for a long time as optical elements for hard x-rays. Many attempts to fabricate artificial crystals as "Bragg reflectors" for soft x-rays have been made throughout this century (Underwood 1981). More recently, and from a different perspective, similar structures were suggested by Spiller (1972) with the specific aim of producing near-normal-incidence optical elements. Positive results were obtained soon thereafter.

Multilayers for this spectral region consist of layers of high-absorption materials separated by "spacer" layers of low-absorption materials. Because of the very small thicknesses of the layers required (< 100 Å), very good deposition control is necessary. Also, information about the microstructure of the films has to be taken into account, since the optical constants depend strongly on it in the X-UV region.

Much better performance in a periodic structure consisting of two materials can be obtained by reducing the relative thickness of the more highly absorbing material. Some additional improvement is possible with slightly aperiodic designs. If the nodes of the standing wave which result from interference of the incident and reflected waves are located within the high-absorption material, reducing the relative thickness of the high absorptance material causes the overall absorption per unit length to be reduced so that a greater number of interfaces can then contribute to the reflection process. For a many-period structure of the type described, and at a given incidence angle, a particular wavelength will be selectively reflected and transmitted through the structure, in accordance with an analog of Bragg's condition.

The choice of the pair of materials in a reflecting multilayer is primarily a compromise between pairs with the greatest difference in refractive indices and materials with the lowest absorptances (Spiller 1976). However, other considerations must be made. For example, the difficulty with which the materials can be deposited in the very thin layers required, and the interfacial stability and microroughness which will result for the given pair of materials. The latter is usually not known in advance.

Very high reflectances can be predicted for ideal multilayer structures in the soft x-ray region (Rosenbluth 1983, 1988). However, actual structures can deviate strongly from the ideal model. Film materials are not generally homogeneous but can contain impurities and microvoids. The microstructure of the films depends on growth conditions. Oxidation and compound formation can also occur. The most important deviation from ideal multilayers is the fact that most interfaces are not atomically sharp, but show diffusion and microroughness. All of these film characteristics will affect the optical properties of the materials.

Diffusion enhances transmission at the interfaces, reducing reflection, while (due to absorption) there is no compensating increase in the effective number of layers contributing to reflection.

Microroughness at the interfaces causes scattering losses. Such losses in multilayers for soft x-rays are usually included in an *ad hoc* manner during the recursive calculation of reflectance. This is accomplished by multiplying the reflectance amplitude coefficient for a given interface by the square root of the exponential decay factor

$$\exp\left[\frac{4\pi\sigma\sin\theta}{\lambda}\right]^2$$

where σ is the rms roughness value, and θ is the glancing angle. This factor appears in diffraction theory of x-rays or neutrons by crystals when accounting for the effects of thermal motion of atoms in the lattice (James 1948). The same factor appears in the scalar treatment of electromagnetic wave scattering by single rough surfaces (Beckmann 1963). The computational scheme used to account for scattering losses in multilayers also can be modified to account for reflectance losses due to interdiffusion (Spiller 1986).

With state-of-the-art MBE deposition methods, it is possible to reduce contamination and oxidation to negligible levels. Interfacial roughness, diffusion, and microstructural properties are material or process-dependent and are difficult to control in thin-film fabrication.

Any fabrication method for soft x-ray multilayers must meet rather demanding criteria, since layer thicknesses range from approximately 200 Å to a few angstroms. Ångstrom-level constant deviations from the design thicknesses will displace the wavelength of peak reflectance, which then will not occur at the desired wavelength. Random thickness errors of each layer of the order of one percent will produce non-negligible reduction in reflectance. The effect of such errors in the multilayer period increases linearly with decreasing period.

In general, it is desirable to keep contamination and oxidation of the layers as low as possible. Damage to the sample surface during fabrication must be avoided, since it will increase microroughness. To date, multilayer reflectors for soft x-rays have been fabricated by electron-beam evaporation (Spiller 1981, Slaughter 1989), sputtering (Barbee 1985), and laser evaporation (Gaponov 1981). These techniques can produce amorphous or polycrystalline layers with precisely controlled thicknesses and very low impurity levels. The major problem, particularly for small layer thicknesses, is the lack of control over the interface roughness and microstructural properties. MBE is a promising fabrication technique which has not yet been applied to multilayer x-ray optical coatings. This technique could allow the fabrication of essentially roughness-free interfaces in soft x-ray multilayers, substantially improving reflectances. In addition, the possibility of fabricating thousands of layers with the required tolerances would allow structures designed for the most difficult part of the soft x-ray spectrum, $\lambda < 40$ Å.

MBE may make possible fabrication of fully crystalline multilayers. These could withstand the effects of high intensity radiation much better than polycrystalline or amorphous ones, since re-crystallization could not occur. Recrystallization can greatly increase surface roughness or even destroy the layered structure in a multilayer, as has recently been shown by Ziegler et al. (1986) for carbon/metal multilayers.

The requirements for a MBE machine capable of producing such multilayers are different than for a GaAs machine. The materials most often used for x-ray optics are: C or Si as the spacer and W, Mo or another refractory metal as the high-absorption material. Since the materials of interest have high melting points, MBE systems based on the new generation of Si MBE machines, such as the machine shown in Fig.3.3 can also be used to produce soft x-ray optics.

Magnetic Structures

The growth and study of magnetic thin films have become of major interest in recent years, driven by both their fundamental scientific interest and technological usefulness. Sub-monolayer control of the production of these structures provides, for the first time, an experimental realization of theoretically tractable ideas in magnetism. These well characterized, low-dimensional systems can eliminate many of the complicating factors that made first-principles theoretical progress in magnetic materials so difficult. From a technological viewpoint, improving materials for magnetic recording and devices is currently a highly competitive and lucrative pursuit. There are strong possibilities of producing controlled epitaxial layers of magnetic materials to enhance their magnetic moments and exploit interface induced anisotropies to increase their magnetic domain density. Another approach in the creation of new materials is to lattice match the magnetic elements with substrates that produce meta-stable systems with properties very different from that of the bulk material. The use of MBE techniques in magnetic film growth allows for a fundamental understanding of these useful effects, which can then direct the progress of technological development.

Through the use of epitaxial growth, the resulting lattice of the deposited material can differ significantly from that of its bulk crystalline structure. Many materials possess a different structure at high temperatures or in alloys. By choosing a suitable substrate material to closely match the desired structure, meta-stable crystals can be formed and maintained. This has been demonstrated with bcc Co on GaAs (Prinz 1985). By extrapolating the lattice constant versus composition of the Fe-Co alloy system to 100% Co, the author was able to predict the lattice spacing for a bcc Co crystal that did not exist in nature. This lattice constant of 2.819 Å was only 0.2% of that for 2 x 2 construction on a GaAs lattice, and so could be grown with a small mismatch. The resulting bcc crystal persisted to thicknesses up to 357 Å where it began to revert to the stable bulk hcp phase.

It is important to note the difference between a metastable phase and an unstable, forced structure. If the total energy of the systems is calculated, local minima in phase space represent metastable states that ordinarily have a higher energy than the bulk stable phase. These phases can be stabilized as in the Co on GaAs. Forced structures, conversely, rely on imbedment in a foreign matrix and are non equilibrium states. Theoretical energy calculations can, therefore, predict structures that can now be realized experimentally. Technologically, this opens up a whole new class of materials that can be created to have various desired properties.

An interesting and potentially useful effect recently discovered in thin magnetic films is that of interface anisotropy. For most thin films, the shape anisotropy tends to force the magnetic easy-axis to lie in the plane of the film. For a few monolayers of Fe on Ag, it was demonstrated (Jonker 1986, Heinrich 1987, Koon 1987) that the polarization was preferentially perpendicular to the plane. Similar results have been observed in very thin layers of Co on Au (Velu 1988) and Pd. (Engel 1991, Draaisma 1987, Purcell, den Broeder 1987). Interestingly, no such effects is seen in monolayers of Co on Cu (Dariel 1987). A fundamental understanding of the influence of the interface on magnetic anisotropy is one of the current challenges of the theory of magnetic materials. Proper control of this phenomenon in the development of layered materials (superlattices) can lead to such things as higher density media for magnetic recording and enhanced magneto-optical devices.

3.3 SILICON MBE MOLECULAR BEAM EPITAXY (SI-MBE)

Molecular Beam Epitaxy realizes the simplest principle of synthesizing a crystal. Molecular beams of that material of which the crystal is composed are directed onto the heated surface of the single-crystalline substrate in the clean environment of a UHV-chamber (Fig. 3.3). If dopants are required, flux densities of the dopant beams are several orders of magnitude smaller than the flux densities of the matrix element or elements. We mean explicitly when using the terminology Si-MBE (Kasper, 1988a) not only the growth of doped/undoped silicon but the syntheses of silicon-based epitaxial structures. Frequently used matrix elements include Si, Ge, $CoSi_2$, $NiSi_2$, CaF_2. Usual dopant elements are Sb, Ga, B and As (ion).

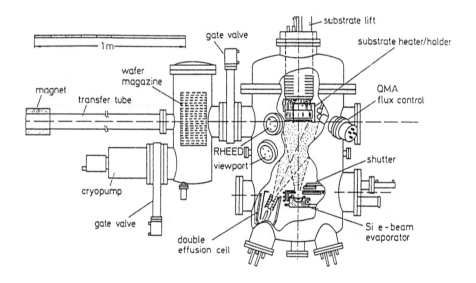

Fig. 3.3: Si-MBE apparatus (basic unit) following the design principle of maximum simplicity. Growth chamber (right) and storage chamber (left) with a 25 wafer magazine (for wafers up to 150 mm diameter). An automated wafer transfer from wafer magazine to substrate heater is provided (Kasper, 1989).

Particularly complex equipment, often with unnecessary or questionable subsystems, has sometimes been described in the past for Si-MBE. It is our opinion that Si-MBE equipment should clearly reflect a straight-forward MBE scheme. As an example, Fig. 3.3 shows a Si-MBE basic unit constructed following the design principle of maximum simplicity (Kasper, 1989). This design principle requires replacement of all subsystems which are not proven to be necessary. The basic unit can be connected with further processing chambers (e.g. metalization, ion implantation) or analytical chambers. The growth chamber contains only the three sub-systems: molecular beam sources (e-beam evaporators and effusion cells), substrate holder/heater with substrate voltage supply for doping by secondary implantation (Jorke, 1986) and in situ monitoring (e.g. quadruple mass spectrometry for flux density) for process control. Excellent process vacuum is obtained by proper combination of pumping systems, careful choice of materials, and a reproducible baking procedure. In the given example, pumping combines the continuous gas outlet of a non gettering turbomolecular pump, with the high pumping speed of a gettering titanium sublimation pump. Typical construction materials for sources, heaters and shieldings are pyrolytic boron nitride (PBN), high purity graphite and ultrapure silicon itself (Fig.3.4). Reproducible baking is provided by a water pipe winding (overnight heat up to 80 ∞C), and an electrical heater (up to 350 ∞C) welded to the outside thermally isolated chamber wall.

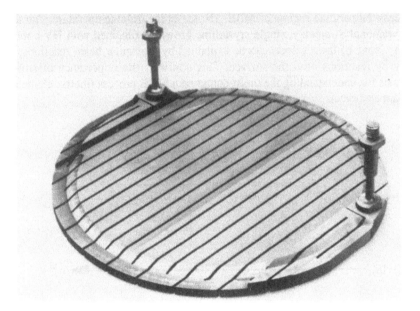

Fig. 3.4: High purity materials used for construction of Si-MBE components. Graphite, silicon and boron nitride for substrate heater, electron beam evaporator crucible, and effusion cell, respectively. Shown is the large area (150 mm diam.) graphite meander used as wafer heater.

3.3.1 Surface Cleanliness as Key for High Quality Material

Surface physics plays an essential role in the epitaxy process. Adsorption, desorption, adatom motion, and in some cases chemical surface reactions determine the incorporation of impinging atoms into the growing crystal. Chemical surface reactions include dissociation of molecular beam species (e.g. Sb_4, B_2O_3 or HBO_2), desorption of surface oxide ($Si + SiO_2 \rightarrow 2\ SiO$), or silicide formation from thin metal films on silicon ($Ni + 2\ Si \rightarrow NiSi_2$).

A very fundamental result of early MBE investigations concerns the influence of surface cleanliness on crystal quality (Kasper, 1988 b). Vacuum conditions much more stringent than are simply required for molecular beam propagation are necessary for good crystal quality. Under excellent process vacuum, growth of crystals with high perfection is obtained (Fig. 3.5). Under improper UHV conditions effects such as dopant-contamination reactions, defect generation (dislocations and stacking faults), and surface etching (during growth interruptions) degrade the crystal perfection. Poorer vacuum conditions, in the High Vacuum (HV) range, result in growth of polycrystalline material in the low process temperature regime of MBE. (Note: In the usual temperature regime of conventional Si-epitaxy, single crystalline growth is obtained with HV-conditions). None of these effects can be explained by molecular beam reactions, but only by reactions onto the surface. This confirms the importance of surface physics for understanding the multi-component MBE process (matrix elements - dopant elements).

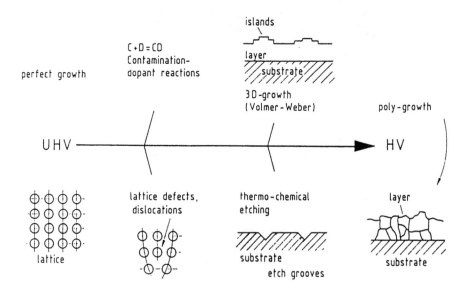

Fig.3.5: General scheme showing influence of vacuum conditions on Si-MBE material quality. Crystal quality degrades from perfect material to polycrystalline material by varying vacuum from ultrahigh vacuum (UHV) to high vacuum (HV).

Since the flux density of dopant elements is much lower than those of the matrix elements, let us first consider the incorporation behavior of the matrix elements. The growth rate. R of a single component system (e.g. homoepitaxy Si on Si substrate) is given by

$$R = j_M \, \eta \, \Omega_0 \tag{3.1}$$

where j_M is the matrix element flux density, the incorporation coefficient, and Ω_0 the atomic volume. The atomic volume can be calculated from macroscopic data (N_A Avogadro's number, M molecular weight, specific weight), or from the lattice constant a_0 as

$$\Omega_0 = M / (N_A \rho) = a_0^3 / 8 \text{ (diamond lattice)} \tag{3.2}$$

For a multicomponent system the growth rate is given by the sum of all individual contributions from Eq. (3.1). Usually Si-MBE is performed under process conditions which allow nearly complete incorporation of the matrix elements ($\eta \approx 1$). For silicon itself, the nearly complete incorporation of the molecular beam was confirmed with MBE experiments using process temperatures up to 950 °C (Kasper, 1982). This was explained by a unity sticking coefficient and by the high mobility of surface adatoms with a low barrier (< 1.1 eV) for surface motion which effectively reduces desorption because adatoms are earlier fixed at steps.

Strained layer epitaxy (heteroepitaxy of lattice mismatched materials) with biaxial strain ε in the layer plane and opposite vertical strain $\varepsilon_{perp.}$ changes the atomic volume Ω to

$$\Omega/\Omega_0 = 1 + 2\varepsilon - \varepsilon_{perp} = 1 + 2\varepsilon \, (1\text{-}2v) / (1\text{-}v) \tag{3.3}$$

where v is Poisson's ratio. In thermodynamic equilibrium the number of adsorption events equals that of desorption events, independent of how long the mean times for both processes are. On a growing surface the freshly created surface is covered by the next monolayer with thickness d_{ML} after a mean time t_{ML}.

$$t_{ML} \, R = d_{ML} \tag{3.4}$$

On the (100) surface a monolayer (ML) thickness is given by $a_0/4$. On the (111) surface with its AaBbCc sequence, a ML thickness (d_{ML}) is given by $a_0/\sqrt{3}$ (distance A - B). Typical times for ML growth under MBE conditions range from 0.1 sec. to 10 sec. All processes with characteristic time constants far below t_{ML} are not influenced by the advancing surface. For all other processes

the boundary conditions vary discontinuously at times t = 0, t_{ML}, 2 t_{ML}, ...,
because the growing surface discontinuously advances due to its atomistic nature.

The mean rate equation of a particle species for times t larger than t_{ML} is given
by

$$j = -\frac{\delta n_s}{\delta t} + j_{des} + Rn \qquad (3.5)$$

where n_s is the adatom density, n the bulk concentration, j the atomic beam flux
density, and j_{des} the desorbing flux density. A solid surface is in equilibrium
with its vapor if $j_{des} = j - j_0$. Growth needs supersaturation σ which can be
measured by

$$\sigma = j / j_0 - 1 \qquad (3.6)$$

where j_0 is the equilibrium flux density.

Because of its low process temperature MBE is a kinetically driven process with
high supersaturation. MBE of silicon is performed under supersaturations which
probably are the highest ever applied for single crystal growth (Vogtländer, 1986)
of high quality materials (Table 3.1). Single crystal growth is possible even to
very low temperatures (Kasper, 1988 a), more than 1150 K (R = 0.4 nm/s)
below the melting point (Note: conventional Si-epitaxy runs typically 300 K
below the melting point). As the growth temperature is lowered further to near
room temperature, within a small transition region the growth switches to an
amorphous regime (Fig. 3.6), as was proven by temperature ramping experiments (Jorke, 1989). However, there is no polycrystalline growth regime,
except on contaminated surfaces.

Table 3.1: Supersaturation σ for MBE of silicon (growth rate R = 0.4 nm/s)
(Vogtländer, 1986)

Growth temperature (°C)	450	550	650	750	850	950
Supersatura-tion	3×10^{16}	4.5×10^{12}	4.7×10^{9}	1.8×10^{7}	2×10^{5}	4.4×10^{3}

The macroscopically vertical growth is microscopically caused by the lateral motion of atomic surface steps. The motion of the monatomic steps is based on catching moving adatoms as described by the Burton-Cabrera-Frank theory (Burton, 1951). However, deviations from this so called two-dimensional (2D), or Frank-van der Merwe, growth mode are possible at heterointerfaces or on contaminated surfaces.

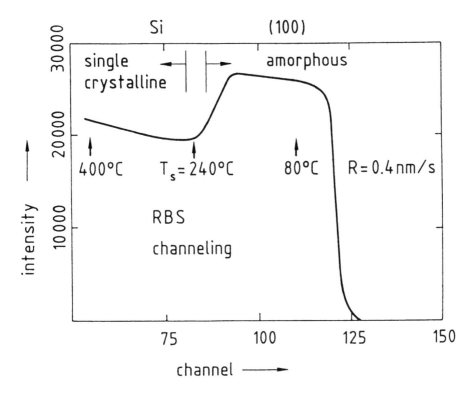

Fig. 3.6: Temperature ramping experiment for Si-growth (Jorke, 1989). MBE starts at 750 ∞C, and decreases to room temperature. Rutherford backscattering (RBS) proves rather sharp transition from single crystalline growth to amorphous growth at 240 ∞C growth temperature (growth rate R = 0.4 nm/s).

Possible sources of surface steps are dislocations, local misorientation toward a low index plane, and nucleation from adatom clusters. However, dislocations are not very important in these days good quality substrates, whereas commercial substrates are oriented toward the nominal low index plane within a few tenths of a degree. This unintentional misorientation (angle i) causes a step sequence with a mean step distance L given by

$$\tan(i) \approx i = d_{ML} / L. \qquad (3.7)$$

According to Eq. (3.7), a typical misorientation of a quarter of a degree should cause a mean step distance of 35 nm (assuming monatomic steps on a 100 surface). There are several indications that the polished substrate surface is locally more misoriented (wavy surface) leading to an even higher step density as growth starts, which is reduced to the above mentioned value during MBE-growth via atomic step flow (Kasper, 1982).

At low temperatures stable nuclei are formed from small adatom clusters (possibly by two joining atoms) under high supersaturation conditions. Such nuclei grow by capturing adatoms. These 2D-nuclei annihilate each other after completion of each ML growth stage, and nucleation has to be restarted.

At low growth temperatures, when all adatoms do not have sufficient mobility to reach the existing misorientation steps, this nucleation process helps to create the additional steps necessary for continuation of 2D-growth mode. Adatoms can directly stick to the ledges of steps, or move along the ledge to a sticking place at a kink of the ledge. Direct sticking on the ledge is favored for low MBE temperatures, whereas sticking at kinks is favored for high MBE temperatures. Ishizaka (1982) assumes that transition from ledge to kink sticking occurs at $T_m/2$ (T_m melting point). At roughly the same temperature, transition from misorientation step movement to 2D-nucleation occurs. Following a modified form of Ishizaka's rule, 2D-growth mode via misorientation steps dominates roughly in the temperature range

$$3T_m / 4 > T > T_m / 2.$$

2D-growth mode via nucleation steps dominates from $T_m/2 > T > T_m/4$ (Fig. 3.7). Above $3T_m/4$ and below $T_m/4$ the 2-D growth mode breaks down because of surface roughening and amorphous growth, respectively. For Si with melting point $T_m = 1680$ K, the 2D-growth regime in Ishizaka's rule should be confined to temperatures from 420 K to 1260 K, with misorientation steps dominating in the range above 840 K, and nucleation steps below that $T_m/2$ temperature.

It is our experience that Ishizaka's rule gives a good description of the dominant growth behavior, although the detailed values naturally are dependant on the material system, the growth rate the surface orientation, and the given misorientation. As a rule, higher misorientation reduces the transition temperature from misorientation steps to nucleation, and can eventually decrease the transition to amorphous growth. This has been experimentally confirmed for 111 Si surfaces. In all cases the minimum number of surface steps is given by the misorientation of the substrate; nucleation will periodically increase the step density (Fig.3.7) with a maximum for deposition of (k + 1/2) ML (k = 1,2,3, ...).

The experimental investigation of step structure in Si-MBE was pioneered by observation of oscillations in LEED (Gronwald, 1982), (Phaneuf, 1987) and RHEED (Sakamoto, 1986) intensities, and by surface microscopy with high resolution (UHV - REM (Kahata, 1989), LEEM (Telieps, 1987) STM (Schwarzentruber, 1990). Unfortunately, the experimental results were someti-mes misinterpreted, in that oscillations were taken as necessary proof of 2D-growth. It is now known that oscillations can only be observed for the 2D-nucleation dominated regime of 2D-growth mode (Neave, 1990). Oscillations are missed for both rough surfaces and the smoothest surface of 2D-growth mode via misorientation steps. In agreement with Ishizaka's rule the most intense oscillations were observed in the lower temperature regime of MBE-growth. Furthermore, to increase the dominance of 2D-nucleation special surface prepa-ration techniques were used in these investigations.

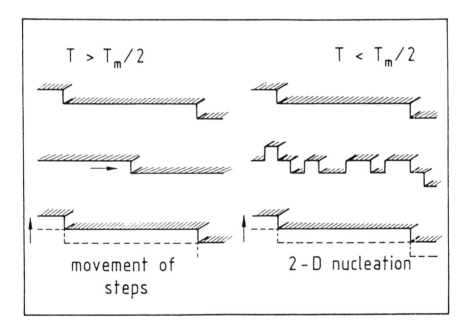

Fig. 3.7: After Ishizaka's rule (Ishizaka, 1988) 2D-growth mode is dominated by motion of misorientation steps at higher temperatures (>T_m/2), and by 2D-nucleation at low temperatures.

3.3.2 The Doping Problem

Potential dopant elements for n-type silicon are P, As, and Sb. For p-type silicon dopants are B, Al, Ga, and In. For physical and technical reasons only Sb (n-type) and Ga (p-type)-beams are easily produced by effusion cells in a UHV-surrounding. One of the hardest challenges in the development of Si-MBE techniques is that both of these elements failed to behave in a manner expected from a simple picture of the MBE-process. Beams of Sb or Ga coevaporated with Si will not be incorporated in the growing crystal, but instead will segregate on the surface. A measure of this behavior is the segregation ratio r_s, defined as

$$r_s = n_s / (n \ d_{ML}) \tag{3.8}$$

where n_s is the surface adatom concentration, n the bulk concentration, and d_{ML} the ML thickness.

In a simple picture, the adatoms arriving at the surface would be buried by each ML growth. The expected segregation ratio in this case would be unity ($r_s = 1$). However, experimentally measured segregation ratios are up to seven orders of magnitude higher, indicating severe segregation (Fig. 3.8). This segregation phenomenon, which also is observed with some dopant elements in III/V MBE, is not yet well understood. A possible explanation (Jorke, 1988) assumes an energetically unfavorable subsurface state (by energy E_I) which is separated from the surface state (with desorption energy E_D) by an activation barrier E_A. This model can also explain the strange dependance of the segregation ratio on the growth temperature (Fig.3.8). To overcome this severe problem a number of different technical solutions were developed, many of which include help from ionizing the dopant species. Amongst them we describe briefly the four solutions believed to be the most successful.

In I^2 MBE only growth of the matrix is performed by MBE, whereas doping is achieved by low energy ion implantation (I^2) from an implanter connected to the MBE chamber. Doping elements are As and B.

Doping by secondary implantation (DSI) utilizes knock-on of the segregating Sb by Si ions accelerated toward the substrate by an applied substrate voltage. The required amount of dopant adlayer is created by pre build up (PBU) and flash off (FO) techniques. The recently developed high temperature effusion cells allow evaporation of B, which is far less segregating than Ga.

A process modification (solid phase-MBE), with deposition of the amorpheous film at room temperature followed by solid phase recrystallization at around 600 ∞C immediately afterwards, allows complete incorporation of dopant materials at levels up to several times $10^{20}/cm^3$, or into thin layers (planar doping, or atomic layer doping ALD or doping).

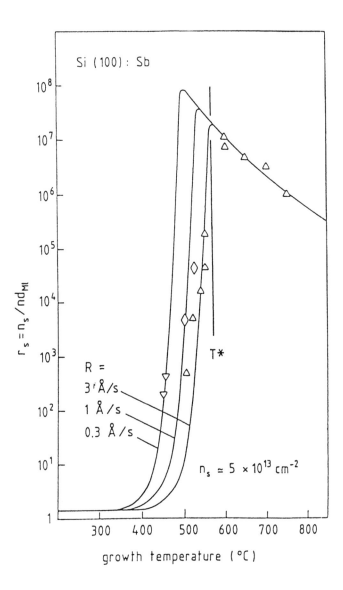

Fig. 3.8: Sb-doping by coevaporation on Si (100). Segregation ratio r_s is plotted as a function of growth temperature T_s for different growth rates (Jorke, 1988). A sharp decrease of the segregation ratio is seen below a growth rate dependant transition temperature T^*.

At rather low growth temperatures (type 300 °C - 450 °C) the segregation of coevaporated dopant atoms is kinetically suppressed leading to rather sharp profiles (LTD low temperature doping).

3.3.3 Differences from III/V MBE

The general lines of modern MBE techniques were developed for III/V materials. Many of the procedures successfully used for these III/V materials were then taken directly for Si-MBE, without analyzing or remembering the roots of the solutions. Some of these procedures are invalid, or at least questionable, in Si-MBE and metal-MBE, with their completely different surroundings (re-fractory materials, electrons and ions). Examples include the usage of liquid nitrogen cooled shrouds, the wafer pretreatment in separate preparation chambers, and the handling of temperature measurements and optical monitoring of the surface.

In III/V MBE, with its highly volatile group V components, liquid nitrogen (LN) cooled shrouds considerably improve material quality. However, in Si- and metal-MBE, with its components which are completely nonvolatile at room temperature, LN-cooled shrouds are not only meaningless, but also dangerous for material quality because of electron stimulated desorption of gettered gases, and because of brittleness of the cold deposited film.

In the early days of III/V MBE the samples were attached to a Mo block by In solder. Thorough outgassing of this sample holder, not really of the sample itself, was necessary in a special preparation chamber. With silicon, naked wafer handling is necessary for technological acceptance of the method. The only *in situ* preparation of the clean wafer required is removal of the very thin surface oxide, which is best done immediately before epitaxy.

MBE of a certain III/V system is usually performed in a relatively small temperature window. Within this window many other parameters (e.g. molecular species and flux ratio) influence the growth, the result of which can be monitored in a qualitative manner by observation of the RHEED pattern. Si-MBE is performed over a wide temperature range for example, from 150 °C to 950 °C. Because of this wide range, rather extensive temperature calibration procedures are required. In our case we measure heater temperature by means of a thermocouple, and wafer temperature by pyrometer (for IR-transparent semiconductors, this is easily applicable only above 550 °C). Both of these readings are then calibrated with a special wafer which contains several thermocouples cemented with Al_2O_3 into small holes ultrasonically drilled into the wafer.

RHEED analysis of Si samples gives very useful information, especially when quantitatively done (spot profile, separation, intensity, and oscillations) and properly interpreted. However, to analyze the proper surface cleaning, the expected submonolayer dopant coverage, or the ionization degree of the beam additional information from optical and electrical monitoring is necessary. Such measurements include light scattering, second harmonic light generation, spectroscopic ellipsometry, electron and ion flux spectroscopy.

3.3.4 Heteroepitaxy

Interfaces between different materials (heterointerfaces) are an exciting topic, both for physics and technology. MBE is a well suited method to overcome some of the general problems connected with achieving heteroepitaxy, namely:

- lattice mismatch
- three-dimensional growth mode
- differences in deposition conditions

The main questions were discussed, and details of metal/ semiconductor, insulator/semiconductor, and semiconductor/silicon interfaces were given, in the review book (Kasper, 1988 a). Here we shall mainly point toward the unique potential of the SiGe/Si system, which follows from the large metastable pseudomorphic (commen-surate) growth regime typically for brittle materials grown at low temperatures. The metastability allows realization of many device structures (table 3.2) which would otherwise not be grown from largely mismatched material couples. The lattice mismatch in this system (4.2% between Si and Ge) can be accommodated by strain (strained layer epitaxy SLE) or by misfit dislocations (strain relaxation by plastic deformation). Epitaxy in strained layers only exists up to a critical thickness t_c; above which thickness misfit dislocations are created and move toward the interface. For large lattice mismatch, the theoretically predicted critical thickness (Fig.3.9) is too small for many useful device applications of SLE. However, MBE experiments have shown a large metastable regime of strained SiGe growth on Si substrates (pseudomorphic or commensurat interface), thereby increasing the experimentally obtained critical thickness considerably by reducing the growth temperature to 550 °C or below. Together with strain-adjustment methods, this metastability allows tailoring of SiGe/Si-based structures and devices not possible in unstrained materials.

The in plane stress of lattice mismatched materials modifies the valence - and conduction band offsets by destroying the degeneracy of light and heavy holes (at momentum $\kappa = 0$) and by splitting the sixfold degeneracy of electrons Δ point,

Brillouin zone) into fourfold degenerated (in plane) and twofold degenerated (perpendicular to the heterointerface) energy levels (see, e.g., the review (Pearsall, 1991). The main consequences for the SiGe/Si system are: (i) Strained SiGe on unstrained Si exhibits only a very small conduction band offset, which means a flat conduction band for this unsymmetrical strain conduction. (ii) The band gap of strained SiGe is decreased compared to relaxed SiGe (infrared shift). (iii) Strain adjustment (tensile strain for Si, reduction of compressivestrain for SiGe) results in a type II band ordering where the conduction band edge of the smaller band gap material SiGe is higher in energy than that of the wider band gap Si.

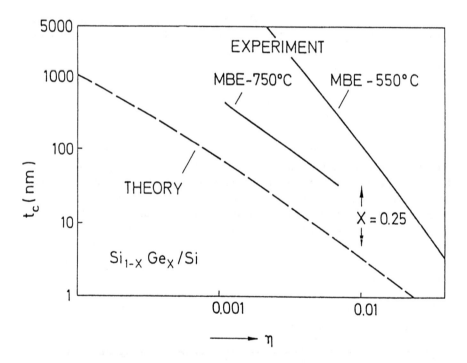

Fig.3.9: Critical thickness t_c versus lattice mismatch η_o for the SiGe/Si heterosystem. A large metastable growth regime extends between the theoretical equilibrium curve and the experimentally found curves (MBE at 750 °C and 550 °C, respectively).

Band gap engineering by strain adjustment is appreciated as additional design freedom, but the limits of stability/metastability of strained layer structures have to be considered. The basic information needed contains lattice mismatch (Ge content), thickness, growth temperature, and strain distribution. Table 3.2 gives a selection (Kasper, 1992) of strained SiGe/Si structures for electronic or optoelectronic applications, which shows the typical submicron or nanometer range of stable/metastable thickness.

Table 3.2:   Selection of strained SiGe/Si structures characterized by their Ge content x, thickness t, and growth temperature T, and their strain status (strain adjustment by a substrate other than Si or by a relaxed buffer layer on Si).

Structure	Growth T (∞C)	x	t (nm)	Strain adjustm.	Remarks
HBT [a]	550	<0.07	45	-	Stable
HBT	640	0,21	60	-	Metastable
p MODMOS [b]	500	<0.25	20	-	Metastable
n MODFET [c]	550	0/0,5	7.5/30	+	Metastable
p RT [d]	550	0,21	25+4+25	-	Metastable
LED [e]	610	0,2/0	30/60	-	Metastable
Waveguide APD [f]	500	0,6/0	3/29	-	Metastable
Superlattice	300	0/1.0	0.8/0.6	+	Metastable

[a] Heterobipolar transistor.

[b] Modulation doped metal oxide silicon transistor.

[c] Modulation doped field effect transistor.

[d] Resonant tunneling device.

[e] Light emitter diode.

[f] Avalanche photodiode.

The physical phenomena utilized include carrier injection across one heterointerface, carrier confinement by at least two heterointerfaces (quantum well) or material modifications by a periodic stacking of heterointerfaces (multiquantum well, superlattice). The phenomena can be modified and varied by superposition of dopant structures. The superposition of a p-n junction onto a SiGe/Si heterojunction delivers a simple, but very conclusive example. Within an iso junction (same doping type on both sides) of strained SiGe/ unstrained Si the motion of electrons is nearly unaffected (flat conduction band), whereas the holes are forced to jump to the SiGe side. Within a p-SiGe/n-Si junction the injection probability of holes is the same as in a Si *p-n* junction, but the injection of electrons is strongly increased compared to Si.

A rapid transfer from research to development and production can be expected for the exciting results now obtained for the SiGe/ Si-heterobipolartransistor (HBT). Other device projects are at an earlier stage but also with rapidly improving performance data comparable to the development at the HBT area a few years ago (see the respective chapters in this book).

3.4   APPLICATIONS OF SILICON MBE

Some of the applications of metals MBE were described in the previous section. Here we want to stress the importance of MBE for experimental surface physics, and discuss the controversial opinions about MBE as a production method. MBE is already established as a tool for creating submicron or nanometer structures, and for developing advanced classes of devices.

3.4.1 MBE for Clean Surface Studies

Surface physics studies must be performed in a UHV environment which is sufficiently good to maintain cleanliness for the entire length of the investigation. The substrate, usually chemically pretreated, often is then *in situ* heated to flash off the surface oxide. Many different cleaning schemes have been developed for this purpose, but a very representative one was published by Ishizaka (1982).

Analysis of the substrate surface by RHEED, LEED, XPS or Auger is commonly believed to prove the surface cleanliness. However, contamination levels of less than 1/1000 ML are difficult to detect by these methods, with the principle contamination species usually carbon, oxygen, boron (Casel, 1987), and aluminum. With the much higher sensitivity of SIMS, investigations usually find one or a combination of the above impurities with concentration levels up to several times $10^{12}/cm^2$. Also, transmission electron microscopy studies of the substrate/overlayer interface often show it marked by strain fields caused by contamination. It is unrealistic to expect the step structure and cleanliness of a cleaned substrate to be as perfect as one wants. Effects such as step bunching, 3D-nucleation and enhanced dopant incorporation rule out use of such cleaned surfaces for many very sensitive surface physics experiments. However, growth of a thin MBE-layer (typically 100 nm) on the "clean" substrate would result in a more reliable and defined surface. An excellent example of this technique is the TEM study (Ourmazd, 1987) of the oxide interface on (100) Si.

3.4.2 Future Electronic Systems: Will MBE be the Key Technology for Realization?

At the present time several thousand tons of semiconductor material are annually produced as chips for electronic systems. Except for a small percentage, all of this semiconductor material is silicon. Within a decade the semiconductor chip weight probably will increase to several ten thousand tons per year. For physical, technological, economic and ecologic reasons (wafer size, thermal conductivity, crystal perfection, price, resources and harmlessness) there is strong reason to believe that silicon will be the main material base for these future electronic systems. On the other hand, III/V technology has demonstrated very excellent device performance using heterostructure concepts. Therefore, very likely these future electronic systems will benefit from a monolithic integration of conventional IC's (which define the complexity level) with heterojunction devices (which define the ultimate performance) on a silicon substrate (Fig. 3.10). Prime candidates for the heterojunction couple are GaAs/Si and Si-Ge/Si, with some chances for silicide/Si and CaF_2/Si.

Let us now speculate how such a future electronic system can be realized. Our proposal is based on a 3-step sequence of processing. In a first processing sequence, the conventional IC without the metalization level is defined. In a second step by MBE and possibly using X-ray lithography, the heterojunction devices are defined in a low temperature process. Then the metalization is defined. MBE would be a key technology necessary for realization of this, if one assumes that all other techniques would develop from existing "state of the art" technologies, rather than from some as yet unrecognized technology.

Concerning MBE, this concept is criticized from the viewpoint of the low production capacity of MBE process. Let us assume an epitaxial thickness of 0.2 μm for future heterodevices on a 500 μm thick Si substrate. This would mean the annual MBE-production of 4 tons of heteroepitaxial layers per 10 thousand tons of silicon substrates. Assuming 400 production systems were used for this (capital investment below one billion dollars), it would require the annual production of 10 kg of heteroepitaxy per system.

Indeed, existing equipment, which is mainly university type research oriented rather than industrial production oriented, fails to fulfill the above sketched production capacity because of low uptime, insufficient condensation area, and small source volume. However, realistic concepts have been proposed to overcome each of these problems, including proposals for high uptime systems (e.g our design of maximum simplicity), for large area processing (J.C.Bean's concept of batch processing (Bean, 1988), and for source volume enhancement (H.Kibbel's concept of load lock refill station (Kibbel, 1991). In this context gaseous beam sources should also be considered. The potential of chemical vapour deposition (CVD) is described in the next chapter.

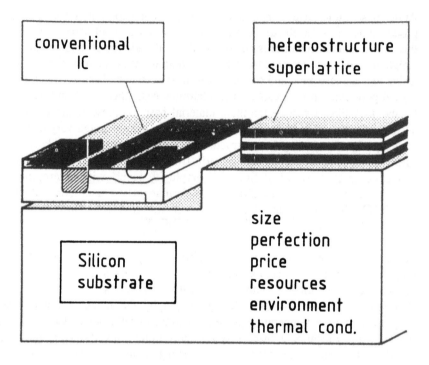

Fig.3.10: Scheme of monolithic integration of conventional IC's with heterojunction (or superlattice) devices on a Si substrate.

It is clear that the realization of production equipment of the type required needs close cooperation between the electronics and the equipment industries. The groups which succeed would benefit from an attractive equipment market, and from electronic systems superior in performance and multifunctionality.

3.5   REFERENCES

Bean J, Kasper E 1988 *Industrial Applications: Possible Approaches in Silicon Molecular Beam Epitaxy* (Boca Raton: CRC)

Baez A V 1952 *H. Opt. Soc.* Am. **42** 756-762

Baez A V 1961 *J. Opt. Soc.* Am **51** 405

Barbee Jr. T W 1985 *Proc. SPIE* **563** 3

Beckmann P and Spitzichino A 1963 *The Scattering of Electromagnetic Waves from Rough Surfaces* (Oxford) Pergamon)

Bennett W R, Person D.C, and Falco C M. 1990 *Proc. 4th Topical Meeting on Optical Data Storage* (Washington DC: Optical Society of America)

den Broeder F J A, Donkersloot H C. Draaisma H J G, and de Jonge W J M 1987 *J. Appl. Phys.* **61** 4317

Burton W K, Cabrera N, Frank F C 1951 *Phil. Trans. Ray. Soc.* **243** A 299

Casel A et al. 1987 *J.Vac. Sci. Technol.* *B5 1650*

Ceglio N M and Dhez P eds 1986 *Multilayer Structures and Laboratory X. Ray Laser Research* (Bellingham, W.A: SPIE) various articles

Chang L L and Giesssen B C eds 1985 *Synthetic Modulated Structures* (New York: Academic Press) various chapters

Chow P, Superioir Vacuum Corporation, private communication

Cohen P 1 and Pukite P R. 1987 *J. Vac. Sci. Technol.* **A 5** 2027 and references within

Dariel M. Bennet L H. Lashmore D S, Lubitz P, Rubenstein M. Lecchter W L, and Harford M 1987 *J. Appl. Phys.* **61** 4067

Dobson P J, Joyce B A. Neave J H, and Zhang J 1988 *Surface and Interface Characterization by Electron Optical Methods eds* A Howie and U Valdrè (New York: Plenum Press) p 185

Draaisma H J G, de Jonge W M, and den Broeder F J S 1987 *J. Magn. Magn. Mat* **6** 351

Engel B N, England C D, van Leeuwen R A, Wiedemann M H, and Falco C M 1991 *Phys. Rev. Lett.* **67** 1910

Gaponov, Gluskin E S, Guser S A, Luskin B M, and Salashchenko N.N.
1981 *Opt. Commun.* **38** 7

Gronwald K D and Henzler M 1982 *Surf. Sci* **117** 180

Heinrich B, Urquhart K B, Arrott A S, Cochran J F, Myrtle K, and Purcell S T
1987 *Phys. Rev. Lett* **59** 1756

Ishizaka A et al. 1982 *Coll. Pap. MBE CST II* P. 183 (Tokyo)

Ishizaka A 1988 *Jap. J. Appl. Phys.* **27** 883

James R W 1948 *The Optical Principles of the Diffraction of X-Rays* (London: Bell)

Jonker B T, Walker K H, Kisker E, Prinz G A, and Carbone C 1986
Phys. Rev. Lett **57** 142

Jorke H, and Kibbel H 1986 *J. Electrochem. Soc.* **133** 774

Jorke H 1988 *Surf. Sci.* **193** 569

Jorke H et al. 1989 *Phys. Rev.* **B 340** 2005

Kahata H and Yagi K 1989 *Jap. J. Appl. Phys.* **28** L 1042

Kasper E 1982 *Appl. Phys.* **A28** 129

Kasper E and Bean J C 1988a *Silicon Molecular Beam Epitaxy* (Boca Raton:
CRC Press)

Kasper E and Jorke H 1988b *Growth Kinetics of Si-MBE in Springer Series in
Surface Sciences* **Vol. 10** (ed. Vanselow, New York: Springer)

Kasper E, Kibbel H, Schäffler F 1989 *Electrochem. Soc.* **136** 1154

Kasper E and Jorke H 1992 *J. Vac. Sci. Technol.* **B5** 1650

Kasper E and Falco C 1994a *To our knowledge Riber was the first company to
introduce commercial high temperature cells. They are now available from several
companies.*

Kasper E and Falco C 1994b UHV *compatible electron guns are available from
various suppliers, including Balzers, Temescal, and Leybold-Heraeus*

Kasper E and Falco C 1994c *For example, several manufacturers offer UHV electron guns based on modified versions of Temescal models.*

Kasper E and Falco C 1994d *We are aware of MBE machines sed for growing metallic multilayers at the University of Arizona, AT&T Bell Laboratories, University of California at San Diego, IBM Almaden and Yorktown, University of Illinois, Johns Hopkins University, University of Michigan, and the Naval Research Laboratory.*

Kearney P A, Slaughter J M, and Falco C M 1991 *Opt. Eng.* **30** 1076

Kibbel H and Bägl A 1991 *German Patent Application*

Koch E E and Schmahl G eds 1990 *Soft X-Ray Optics and Technology* (Bellingham, WA: SPIE) various articles

Koon N C, Jonker B T. Volkenig F A, Krebs J J, and Prinz G A 1987 *Phys. Rev. Lett* **59** 2463

Koziol C. Lilienkamp G, and Bauer E 1987 *Appl. Phys. Lett* **51** 901

Kwo J, Gyorgy E M, McWhan D B, Hong M, DiSalvo F J, Vettier C, and Bower J E 1985 *Phys. Rev. Lett* **55** 1402

Leybold-Heraeus Sentinell III

Makous J, Falco C M, Cucolo A M, and Vaglio R 1987 *Japanese J. Appl. Phys.* **26** 1467

Mantl S 1992 Science Reports **8** 1

Models are available from Balzers, Leybold-Hereaus and Temescal, among others.

Neave J H et al. 1985 *Appl. Phys. Lett.* **47** 100

Niskinage T and Cho K J 1988 *Jap. J. Appl. Phys.* **27** L12

Ourmazd A et al. 1987 *Phys. Rev. Lett.* **59** 213

Pearsall T P 1991 *Strained Layer Superlattices, Science and Technology* (New York: Academic Press)

Perio A, Torres J, Bomchil G, Arnoud d'Avitaya F, and Pantel R 1984 *Appl. Phys. Lett* **45** 857

Phaneuf R J and Williams E D 1987 *Phys. Rev. Lett.* **58** 2563

Prinz G A 1985 *Phys. Rev. Lett* **54** 1051

Purchell S. T, Arrott A S, and Heinrich B 1988 J. *Vac. Sci. Technol.* **B6** 794

Purchell S T, Johnson M T, Mcgee N W E, Zeper W B, Hoving W *J. Mag. Mag. Matt.* (submitted)

Rosenbluth A E 1983 *Ph.D. Dissertation* Univ. of Rochester

 Rosenbluth A E 1988 *Revue Phys. Appl.* **23** 1599

Sakamoto K et al. 1989 *J. Electrochem. Soc.* **136** 2705

Salamon M B, Sinha S, Rhynne J J, Cunningham J E, Erwin R W, Borchers J, and Flynn C P *1986 Phys. Rev. Lett* **56** 259

Schuller I K and Dow J D eds 1987 *Interfaces, Superlattices and Thin Films* (Pittsburgh: Materials Research Society)

Schwartzentruber B S et al. 1990 *Phys. Rev. Lett.* **65** 1913

Slaughter J M, Burkland M K, Kearney P A, Lampis A R, Milanovic Z, Schulze D W, and Falco C M 1989 *Proc. SPIE* **1160** 235

Spiller E 1972 *Appl. Phys. Lett.* **20** 365

Spiller E 1976 *Appl. Opt.* **15** 2333

Spiller E 1981 *AIP Proc.* **75** 125

Spiller E and Rosenbluth A E 1986 *Opt. Eng.* **25** 954

Tanaka M, Yuzurihara H, and Tokita T, *IEEE Trans. on Magnetics* (in press)

Telieps W 1987 *Appl. Phys.* **A44,** 55

Torres J, Perio A, Pantel R, Campidelli Y, and Arnaud d'Avitaya F 1985 *Thin Solid Films* **126** 233

Tung R T 1988 *Silicon-Molecular Beam Epitaxy vol. II* eds E Kasper and J C Bean (Florida: CRC Press) example

Underwood J H and Barbee T W 1981 *AIP Conf. Proc.* **75** 170 (for a discussion of early work)

Velu E, Dupas C, Renard D, Renard J P, and Seiden J 1988 *Phys. Rev.* B 37 668

Vogtländer K et al. 1986 *Appl. Phys.* **A39** 31

Yanoff A as quoted in: Williams G P 1988 *Synchrontron Radiation News* **1** pp 21-27

Zeper W B, Greidanus F J A M, van Kesteren H W, Jacobs B A J, Spruit J H M, and Carcia P F 1990 *Proc. SPIE* **1274**, 282

Ziegler E et al. 1986 *Appl. Phys. Lett* **48** 1354

ACKNOWLEDGEMENTS

One of us (E.K.) acknowledges fruitful discussions with many members of the Si-MBE and the mm-wave devices groups of Daimler-Benz Research Center, Ulm. Special thanks are due to Horst Kibbel for his intuitive contributions to many areas of MBE. Equipment was designed and constructed in cooperation with companies Atomika, Balzers, Leybold, VG and Riber. It is a pleasure for one of us (C.M.F.) to thank Dr. Brad Engel and Dr. Jon Slaughter for their contributions to this chapter, and to acknowledge the work of the other postdocs and students currently working on metallic superlattices. Many discussions over the years with R.C. Dynes, G. Güntherodt, J.B. Ketterson, I.K. Schuller, and R. Vaglio contributed to this work. Also, without the technical expertise of Dr. Peter Chow of Perkin Elmer in designing our 433-S MBE system, non of this research would have been possible.

4

Low Thermal Budget Chemical Vapour Deposition Techniques for Si and SiGe

Matty R. CAYMAX
W.Y. LEONG

4.1   INTRODUCTION

Whereas growth of strained $Si_{1-x}Ge_x$ layers in MBE systems has been studied since the mid seventies, in CVD systems the first results were only obtained about fifteen years later. Although epitaxial growth by CVD is being done already since the sixties and in the mean time has matured, the standard growth temperatures which were higher than 1000 °C prevented the growth of strained SiGe layers as well as the realisation of the extremely sharp dopant transitions which are required for the device applications associated with this topic. The lowest temperatures reported were above 850 °C, and then the layer quality was still questionable. It is only with the advent of the UHV CVD and RTCVD/LRP techniques, that high quality and very accurately controlled, very thin layers of epitaxial Si (and, more recently, SiGe) could be deposited in CVD machines. This was mainly the result of the insight, that the most important parameters in CVD were the purity of the substrate surface, of the growth environment and of the gases, and, in addition, the availability since that time of wafers, gases, wet chemicals and reactor techniques which offer the required purity. From then on, the further development went very rapidly, and a host of advanced device structures fabricated with low temperature CVD epi layers were reported. It is just fair to recognize here the booster role which has been played in this whole development by the results and realisations which have been and still are being obtained in the Si MBE world.

In this chapter, we will concentrate on the epitaxial growth of Si and SiGe in CVDsystems at low temperatures, i.e. between 500 and 750 °C, and review associated realisations and remaining problems.

4.2   PREREQUISITES FOR LOW-TEMPERATURE GROWTH

Already in the early sixties (Nakanuma, 1968 ; Batsford, 1963) one was aware of the importance of obtaining a perfectly clean substrate for the growth of good quality epitaxial layers at temperatures below 900 °C. By lack of appropriate cleaning techniques, 'there (was) is no surface better than the surface itself freshly deposited from the vapor phase' (Nakanuma, 1968). This fresh surface was obtained by coating the surface with Si from the hydrogen reaction of $SiCl_4$ at 1200°C immediately prior to the low temperature deposition. Although very effective, this very high temperature coating technique is not compatible with the low thermal budget processing currently required. The problem with obtaining a clean surface has to be dealt with in an other way.

We can broadly distinguish four kinds of contamination on the wafer surface : particles, metals, carbon and oxygen. Particles are of major concern for the yield of VLSI and ULSI processes because of their increasing impact on the ever smaller line widths used. They result in easily recognisable and localised defects such as stacking faults, pit holes or dislocations (e.g. in strained $Si_{1-x}Ge_x$ growth, see par 1.4a), but do not represent really fundamental problems from the chemical or physical point of view for successful epitaxial growth. Nevertheless, in real production applications they are unwanted for yield reasons and wafer cleaning schemes should take the removal (or addition !) of particles into account. The impact of metallic contamination on crystalline versus electrical characteristics of the deposited layers has to be distinguished. Whereas ppb's of certain metals can be deleterious for gate oxide integrity (Verhaverbeke, 1991), much higher levels can probably be tolerated at the wafer epi interface before they start influencing the crystalline quality e.g. by formation of silicide-like precipitates. With electronic device performance in mind, the tolerable level will depend on the application (bipolar, MOS, optical sensors, ...). In any case, as will be discussed further on, the improved wet cleaning technology and nowadays chemical purity perform well enough in removing trace metal impurities from wafer surfaces for these applications.

The situation is more complicated with carbon and oxygen. C contamination can result from organics such as trace solvent contamination in the chemicals used, polymeric materials outgassed from plastic wafer boxes, packaging materials, floor coverings etc., or that are washed out from containers, bottles, tubings etc., light airborne hydrocarbons such as methane, and others (Budde, 1992). Oxygen contamination on the Si surface is primarily a relict from the incomplete removal of an oxide layer, whether it be native or grown in a wet chemical cleaning cycle. In the following paragraphs, we will describe the actual state-of-the-art of wafer cleaning with respect to subsequent low-temperature epitaxial layer growth.

Principally, preparing wafers for low temperature epi means : removal of virtually all contamination from the wafer to expose the bare Si surface. This can be done before or after introduction of the wafer into the reactor (i.e. ex-situ or in-situ). In order to accomplish this completely in-situ, one could consider the use of advanced gas phase chemical cleanings, but to date no complete cleaning cycle that removes all previously mentioned impurities has been designed. Sometimes, especially for growth in MBE reactors, one simply loads the as-received wafers into the preparation chamber, thermally evaporates the native oxide, and a buffer layer of pure Si is grown, burying all remaining contaminants. This normally results in unacceptably high numbers of point defects ($\geq$ 100 cm^{-2}) due to remaining particles and other imperfections. By using the ex-situ cleaning route, one can take advantage of the considerable know-how that has been gathered in recent years. After cleaning the wafers, the highly reactive bare Si surface has to be protected against recontamination during transport to and within the reactor prior to deposition. A first such approach was reported in 1982 by Shiraki and coworkers (Ishizaka, 1982 ; Ishizaka, 1986), in which a thin (< 8 Å) non stoechiometric oxide layer is formed during a modified RCA cleaning sequence. The idea is that this oxide is much less reactive, so that less organic contaminants will adsorb during transport, and what is more, it is more volatile and easily desorbed during heating of the wafer in the preparation station. This heating to temperatures around 750 °C for 15 minutes or so serves to desorb the protective oxide layer thermally. Actually, it appears that higher temperatures (over 800 °C) are required to remove the oxide layer completely, as is our (Caymax, 1990) and other's (Xie, 1986 ; Hull, 1991) experience. Even at this high temperature, although, a certain amount of carbon remains at the interface. So, again the cleaning method involves quite a high thermal budget. In case one wants to remove the C completely, a temperature of even up to 1250 °C is required (at which the C diffuses into the substrate).

A more convenient agent to protect the surface after a complete cleaning cycle is a passivating H layer, resulting from an HF dip, proposed by Meyerson (Meyerson, 1986). This layer too will decrease the susceptibility for contamination by hydrocarbons, but also by oxygen and water. Morita et al. (Morita, 1989) have shown this passivation to be very stable at room temperature as well as in DI water. From temperature programmed desorption studies (see Par. 1.3.b.5) , several determinations of the kinetic parameters for the breaking up of this passivation layer at elevated temperatue show activation energies for desorption around 49 kcal/mole for the Si-H, and around 43 kcal/mole for the Si-H$_2$ species, resp. From Meyerson (1990a) we find that at 400 °C only 0.5 % of a monolayer of H is lost per minute.

The complete cleaning cycle then could look as follows : first, the wafers are subjected to the classic RCA cleaning (Kern, 1970). Recently, it has been found that the Standard Clean Step 1 (SC1) of this scheme, which consists of a 10 min treatment in a boiling solution of 1 H$_2$O$_2$: 1 NH$_3$: 5 H$_2$O , roughens the surface of state-of-the-art polished wafers due to a continuous oxidation etching action. This can be avoided by lowering the amount of ammonia (Meuris, 1992). Depending on the purity of the chemicals used, some metal contamination may be left on the surface. It has been shown by Ohmi (1992) that for impurity levels below the ppb level, SC1 does not leave any measurable trace of metal contamination. In any case, if any metallic contamination would be left left after this step it is easily removed by SC2, the 1 H_2O_2 : 1 HCl : 5 H_2O solution. In between SC1 and SC2 and after SC2 the wafers are rinsed in overflowing DI water. The self-passivating oxide film left is stripped by dipping in HF/H_2O solutions, typically for 30 s in 2 % HF, which simultaneously removes the very last traces of the metallic contamination left by SC1 (Verhaverbeke 1992). We have shown (Caymax, 1992) that rinsing in DI water after this HF dip increases the O concentration at the epi wafer interface later on, and has to be omitted. Drying of the wafer by means of N_2 blowing immediately after the HF dip is preferred. After this, the wafers have to be loaded into the reactor within half an hour or so. A particular feature is the presence of interfacial B contamination, well-documented in the MBE literature (Xie, 1986). It has been suggested by Robbins et al. (Robbins, 1990) that this B is caused by air borne contaminants ;

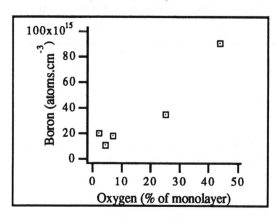

Caymax et al. (1992) showed a correlation between the presence of B and that of O (Figure 4.1), indicating that both these species are remnants from the pre-epi cleaning treatment. By applying an appropriate cleaning scheme as pointed out, the B contamination can be kept in the 10^{10}/cm^2 range.

Figure 4.1 Relationship between interfacial O an B contamination

Upon passing a (moderately heated) loadlock, the wafers are put into the reactor where they start heating up to the growth temperature. At wafer temperatures above, say, 500 °C, thermal H_2 desorption becomes increasingly significant, thereby exposing the bare Si surface to the reactor ambience. During the deposition, hydrogen is formed as a reaction product from the chemisorbed Si source gas but is also desorbed. Therefore, also during steady state growth one may expect a non neglibible fraction of the surface to be prone to recontamination. [#] In order to avoid this, one has to provide a sufficiently clean growth environment, which includes both reactor background conditions and gas cleanliness. It is interesting to quantify the requirements that have to be fullfilled in order to grow epitaxial layers of acceptable quality. It is difficult to say anything meaningful about C because C

[#] As stated by Meyerson (1990a), 'the Si-surface is no longer passive' ; probably the situation is not that bad, because it is known from kinetical studies that even at 600 - 650 °C a considerable fraction of the Si-surface is coverd by H (Greve, 1991).

contamination can result from a variety of sources. Therefore, we restrict ourselves to O.

We will try to calculate the film O content to be expected as a function of water vapour pressure. It is reasonable to suppose that this content will be given by the ratio of the rates with which Si and O species arrive at the surface and are built into the lattice. Therefore, the O concentration, C_O will be determined by the ratio of the net resulting fluxes towards the growing surface of O and Si, F_O and F_{Si} (with net resulting flux, we mean the net number of atoms that per second and per cm^2 definitely is grown-in on the surface, i.e. the difference between adsorbed and desorbed precursor molecules):

$$C_O = N_V \, (F_O/F_{Si})$$

where N_V = the volume density of the Si crystal (6.6×10^{22} atoms/cm^3). The net flux of Si is defined as

$$F_{Si} = R_{Si} \, (N_S \, / \, d) = R_{Si} \cdot 8.3 \times 10^{12}$$

with F_{Si} in atoms/cm^2, R_{Si} the growth rate in Å/min, N_S the surface density (6.78×10^{14}/cm^2) and d the thickness of a one-atomic Si layer (1.36 Å). Further on, we may assume that all incorporated O originates from H_2O vapour, because normally the H_2O content exceeds the O_2 concentration by orders of magnitude, whereas the efficiency of building in is comparable. Therefore, we may write

$$F_O = F_{H_2O} \, s_{H_2O} \vartheta$$

with F_{H_2O} the flux (in atoms/cm^2.s) of H_2O molecules towards the surface, which at 650 °C is given by kinetical gas theory as p_{H_2O}(torr) . 2×10^{20}, s_{H_2O} the sticking coefficient of H_2O and ϑ the fraction of free surface sites. At temperatures above 650 °C, we may assume $\vartheta = 1$ (Greve, 1991), i.e. the surface is essentially H free. For sake of simplicity, we restrict ourselves here to growth from SiH_4, such that no Cl from SiH_2Cl_2 can be present on the surface. From Ghidini and Smith (1984) we know that $s_{H_2O} \cong 0.003$, almost independent of temperature [#] . Taking all together we obtain

---

[#] Actually, Ghidini and Smith experimentally determined the critical water vapour pressure at which Si becomes oxidised as a function of temperature. They found two boundaries, an upper one (at which a firm oxide-layer is formed) and a lower one (where only a non-passivating superficial Si-O-H layer is formed). The sticking coefficient of H_2O to the Si-surface can be defined as the ratio of the SiO-formation rate via the global reaction

$$H_2O + Si* \rightarrow SiO* + H_2 \uparrow$$

and the arrival rate of H_2O-molecules R_{H_2O}. For both boundaries they found different values ; here, we use the value appropriate for the lower boundary, because actual reactor

$$C_O \text{ (atoms/cm}^3) = 5.4 \ 10^{27} \ p_{H_2O}/R_{Si}$$

When growing epitaxial Si from SiH_4 in a very low pressure system at 600 °C, a growth rate of about 5 Å/min is typical. Let us assume the concentration of O in the epi layer is acceptable if it is lower then or equal to this in Cz-Si, i.e. about 2 x 10^{18}/cm^3. Taking this into account, we find for the total allowable water partial pressure in the growth ambient a critical value of 10^{-9} mbar. For a typical water contamination of state-of-the-art SiH_4 gases (i.e., without point-of-use gas-purification) of 0.5 ppm, and when working at typically 2 x 10^{-3} mbar, from the gas we may expect about 10^{-9} mbar of water. This means the additional contribution from the reactor background should be below 10^{-10} mbar. Our (Caymax, 1992) and others' (Racanelli, 1991) experience shows a very good correlation between these calculations and experimental results.

It is worthwile to refer here once more to the work performed by Ghidini and Smith (Smith, 1982 ; Ghidini, 1984), who determined critical pressures of O_2 and H_2O (as a function of temperature) above which oxidation of Si occurs with formation of stable oxide layers, whereas below these pressures, no stable oxide is formed. This can be explained by means of a very simple model, based on the following reaction equations :

$O_2(g) \rightarrow O_2(ads) \rightarrow 2 \ O(ads)$	r.1
$Si + 2 \ O \rightarrow SiO_2$	r. 2a
$Si + O \rightarrow SiO \uparrow$	r. 2b

When a lot of O_2 is present, reaction r.2a is favoured over r.2b (giving rise to a stable oxide layer) and vice versa, and high temperatures enhance SiO evaporation.

Therefore, to maintain an oxide free surface at the lower the temperature, a lower O_2 (or H_2O) pressure must be achieved. It has been proposed (Meyerson, 1986 ; Greve, 1990) to use these data for setting up specifications for background pressures in low-temperature epitaxial CVD reactors in order to evaporate the native oxide of Si wafers or to prevent reoxidation of oxide stripped wafers. Two remarks are here in order . First, at the pressures derived from the Ghidini and Smith expressions, e.g. 10^{-10} mbar at 600 °C, it takes roughly 3 hours in order to form 1 monolayer of oxide assuming O_2 has a sticking coefficient of one (which actually is rather in the order of 10^{-2}.) Therefore, the time for formation of one monolayer is even in the order of 300 hours, so the argument of preventing oxide layer formation does not make really sense. A second remark pertains on the kinetics of the removal of a thin oxide

---

conditions will probably be in in this region (low p_{H_2O}, low temperature). Although this boundary is strongly temperature-activated, s_{H_2O} apparently is almost temperature-independent. The reason is that the critical water vapour pressure is proportional to the SiO equilibrium partial pressure, which is itself a strong function of temperature.

layer (or some oxide contamination) under low O_2 and H_2O partial pressures, which could be suggested by Ghidini and Smith's data. We have found that even at temperatures as high as 800 °C in a UHV environment, a so-called "Shiraki" oxide (Ishizaka, 1986) is removed only very partially (Caymax, 1990) by nucleation of small holes in the oxide layer. This could have to do with the fact that SiO has to penetrate through a full oxide layer, which is not readily accomplished. In addition, Racanelli (Racanelli, 1991) has calculated the time constant for reaching an equilibrium coverage of the surface with O atoms by H_2O adsorption. It is clear from this that the time needed to obtain a very clean surface is much longer than is practically useful. Because of all this, we prefer to stick to Meyerson's point of view in (Meyerson, 1990a), according to which the H passivated wafers are not really prone to reoxidation in the reactor irrespective of the reactor atmosphere as long as the H passivation is not disrupted. As pointed out before, the *determination of the sticking coefficient of H_2O (and O_2)* at the temperature of interest by Ghidini and Smith, rather than their kinetic measurements, can then be used to set limits to the tolerable O_2 and H_2O pressures in the reactor and in the gases during growth, based on tolerable O concentrations in the film bulk.

4.3   GROWTH SYSTEMS

In the mid-eighties, two techniques designed for low temperature (< 850 °C) epitaxial growth emerged more or less simultaneously, i.e. UHVCVD by Meyerson at IBM Yorktown Heights (Meyerson, 1986) and a plasma enhanced CVD technique designed by Reif at MIT (Donahue, 1986), relying on a low energy Ar ion bombardment cleaning step. At the same time, CVD epi growth in a Rapid Thermal Processing reactor was announced by the group of Gibbons (Gibbons, 1985). Although initially the process temperature aimed at was higher (~ 950 ° C), thereby keeping the total thermal budget as low as possible by reducing the dwell time at really high temperature, later on process temperatures were considerably lowered. In this section, we will describe the two most widespread techniques in somewhat more detail, whereas a number of other techniques will be covered more briefly.

4.3.1 Ultra High Vacuum Chemical Vapour Deposition (UHVCVD)

By applying rigorously the concept of creating/maintaining a very pure Si surface in an ultra clean ambience, the first UHV CVD reactor was designed and built by Meyerson (1986), followed later on by the groups of Greve (1990) at Carnegie Mellon University, Pittsburgh, and Caymax (1990) at IMEC, Leuven, Belgium. Although differing in details, the principal characteristic of these systems is to offer an environment for growth with water partial pressures in the 10^{-10} mbar range and oxygen and hydrocarbon levels even lower by orders of magnitude. The ultimate goal of this kind of growth techniques being its application in production environments, two important aspects are high throughput and good layer uniformity and reproducibility. Although the low process temperature causes a relatively low growth rate, this is more than compensated by the batch type processing : there are no fundamental problems in scaling up this kind of furnace to large diameter wafers (200 mm or more), in batches of hundreds of wafers. In principle, combination of such a multi wafer reactor with other reaction chambers into larger clusters is feasible. The required homogeneity and reproducibility is assured by the physical and chemical characteristics of the growth process. On the one hand, the deposition process is essentially an equilibrium process of reactant (SiH_4) adsorption and reaction product (H_2) desorption. The rate of this process will, depending on temperature and partial reactant pressure, be determined by surface kinetics (in case of high H coverage where H_2 desorption is rate limiting) or by reactant supply (in case of low H coverage, where SiH_4 adsorption controls the growth rate), or, which is often the case, by a combination of these. In the reactant supply case, we know the conversion efficiency or sticking coefficient of SiH_4 to be small (10^{-4} - 10^{-3} ; see section 4.4). Because of the very low deposition pressure (10^{-3} mbar range), the main gas flow pattern in the cylindrical furnace tube is intermediate between the viscous and molecular regime (so called Knudsen flow). In among the wafers, mass transport will be purely molecular though, and therefore very efficient. As a consequence of these two facts (low sticking coefficient and quick mass transport), gas fluxes will be very uniform and, as a consequence, deposition rate as well. A very uniform temperature distribution throughout the full wafer batch and across the wafers is easily obtained with the hot wall design. Temperature control and measurement is also easier compared to RTCVD systems (see 4.3.2). A disadvantage is the very slow temperature response in comparison with single wafer reactors. The "closed" character of these systems hampers the use of in-situ measurement techniques for studying various aspects of the growth process as contrasted with e.g. MBE chambers or cold wall CVD systems (see 4.3.3).

In Figure 4.2, a schematical drawing of the UHV VLPCVD

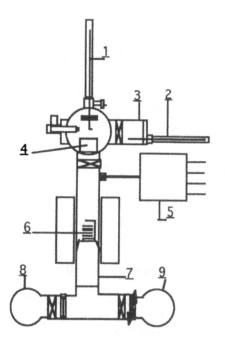

Figure 4.2 UHV VLPCVD epi reactor : 3 - loadlock ; 1,2 - transfer sticks ; 4 - UHV furnace ; 7 - VLPCVD-furnace ; 6 - wafers ; 5 - gassystem ; 8, 9 - pumps

reactor at IMEC gives an outline of the most important parts. Essentially, it consists of three turbomolecular pumped chambers : i) a loadlock chamber with a stainless steel, magnetically coupled transfer rod ; ii) a stainless steel Ultra High Vacuum chamber equipped with an internal resistively heated furnace capable of heating a full batch of 5" wafers to temperatures of 1000 °C ; iii) a 200 mm inner diameter quartz tube surrounded by a three zone resistivelyheated furnace. In the Very Low Pressure Chemical Vapour Deposition tube (VLPCVD tube), a background pressure in the 10^{-9} mbar range (with the furnace at a nominal temperature of 650 °C) is maintained by means of a 3000 l/s cryopump ; gases are pumped by means of a 510 l/s plasma turbomolecular pump. The standard pressure during deposition is 2.6 x 10^{-3} mbar. The quartz tube is sealed at both ends by means of removable, all metal sealings so facilatating etching and cleaning. Further, all other parts are connected by means of standard Conflat® flanges according to UHV standards. Gas lines are welded where possible or VCR fitted. All gases are normally required to contain less than 1 ppm of H_2O, O_2, CO, CO_2 and hydrocarbons. Up to 7 wafers of 5" diameter can be loaded simultaneously by means of a quartz wafer holder. After introducing the holder into the loadlock, this is pumped down in 15 minutes to the 10^{-6} mbar range.The holder with the wafers is then transferred into the UHV chamber. The wafers are thoroughly outgassed at 400 °C in the internal furnace. During this treatment, the base pressure rises from 2 - 3 x 10^{-10} mbar to 2 - 5 x 10^{-9} and then falls back to approximately 10^{-9} mbar. After 10 min, the wafers are transported into the vertical VLPCVD reactor by means of a quartz transfer stick ; the reactor background pressure does not rise higher then about 5 x 10^{-8} mbar. This way of loading has two advantages : the total loading action takes only half an hour, and the wafers, once at a temperature higher than 400 °C, do not see water vapour pressures higher than 10^{-10} mbar. Other ways of introduction differ somewhat in details, such as loading under H_2 flow (Meyerson, 1986) or baking overnight in the furnace tube at deposition temperature (Racanelli, 1991). Whereas an overnight bake at a temperature of 600 °C will reveal the bare Si surface, which can collect a lot of O contamination, this is completely removed by baking the wafers in situ for a short time (10 min) at 800 °C, leaving no SIMS detectable trace of oxygen at the interface above the Cz Si background level. On the other hand, comparable to MBE growth (Casel, 1987), higher interfacial B levels seem to result (Racanelli, 1991), whereas with our UHV degassing treatment at 400 °C or Meyerson's loading under H_2, these levels are limited to 5 x 10^{10}/cm^2 or less (Caymax, 1992 ; Meyerson, 1990b). No difference in C contamination for the several experimental approaches is detected, resulting in interfacial C doses of 0.01 monolayer or less. So, the exact way of preparing and loading wafers is the result of a number of factors and considerations : reactor design, whether preference is given to lower O or lower B interfacial contamination (in some cases), required throughput, personal preferences, and possibly others as well.

Si and SiGe growth is obtained from the heterogeneous thermal decomposition of the hydride source gases SiH_4 and GeH_4, while doping is obtained by adding B_2H_6 for p-type and PH_3 (or AsH_3) for n-type.

An important issue is the uniformity of deposition across large diameter wafers in a closely stacked multi wafer system such as these. Generally speaking, one may expect that at these very low pressures (and temperatures), the deposition is fully controlled by heterogeneous reactions at the wafer surface since the homogeneous gas phase decomposition of SiH_4 giving rise to the gas phase reactive SiH_2 radical here is in the fall off regime, resulting in negligibly slow reaction rates (Meyerson, 1987). Of course, the rate of this heterogeneous process and therefore its uniformity is governed by the temperature distribution on the wafers, which is assured by the hot-wall CVD concept. Therefore, if the diffusional transport of source gas molecules is fast enough, such that no gas phase depletion occurs, deposition rate and doping will be very uniform. Again, the very low pressure enhances the diffusivity of gas species vastly, such that, as can be verified experimentally, the non uniformity is limited to a few percents or lower (Caymax, 1990). Greve and Racanelli have modelled this more quantitatively by representing the mass transport process as the result of multiple collisions of a gas species between two wafers and assuming a random walk for this species (Greve, 1991). They arrived at essentially the same conclusion. This reasoning can be extrapolated to the wafer-to-wafer uniformity for which values of lower than a few percent are obtained.

Doping of epitaxial CVD Si and SiGe layers will be discussed in more detail in section 4.5. Important to note here is that p-type doping is simply achieved by adding B_2H_6 to the growth ambience. Very sharp dopant profiles are possible, as well as a very broad range of dopant concentrations between less than $10^{14}/cm^3$ and up to $10^{21}/cm^3$. N-type doping on the other hand introduces a number of difficulties, but which are not insurmountable.

The kinetics of the growth of Si and SiGe layers will be described in section 4.4. For the temperature range between 550 and 650 °C, very "workable" growth rates are obtained (0.5 - 10 nm/min) for the applications envisaged, combining a very low thermal budget, and very accurate control. Because of the batch type reactor and a relatively short processing time, throughput is no problem.

A few examples of applications of UHV/CVD epi material to be mentioned here are base layers in Heterojunction Bipolar Transistors with layer thicknesses in the order of 100 nm, (see, e.g., Patton, 1989 ; Poortmans, 1992) and quantum effect devices such as Double Barrier Resonant Tunneling Diodes and superlattice systems, requiring individual layer thicknesses around 5 nm for which control on the nm level becomes a stringent requirement (Wang, 89). At the same time, such devices demonstrated the abruptness of heterojunctions achievable with UHVCVD as well as of doping transitions which are reported to be sharper than 2 nm/decade, i.e. not resolvable anymore by SIMS. An extensive review of such devices is given in Chapters 6 and 7.

4.3.2 Limited Reaction Processing - Rapid Thermal CVD

The epitaxial growth of Si by Limited Reaction Processing (LRP) has been pioneered by the group of Gibbons at Stanford in the mid-eighties (Gibbons, 1985), lateron followed by a number of other groups. The LRP technique finds its origin in the Rapid Thermal Annealing systems, developed in the early eighties. The reaction chamber consists of a mostly horizontal quartz tube (Figure 4.3) and can contain one wafer, resting on quartz pins. The wafer is heated up by one or two banks of IR quartz halogen lamps, with an output power of tens of kW.

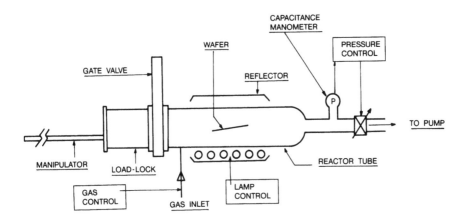

Figure 4.3 Typical outline of an LRP/RTCVD reactor (after Gibbons, 1985 and Sturm, 1991b)

The tube can be air or water cooled. The gases used for Si epi are typically SiH_2Cl_2, GeH_4, B_2H_6, AsH_3 or PH_3, together with a carrier gas such as H_2, Ar or N_2, all controlled by Mass Flow Controllers. The reactor is operated at low pressures (in the mbar range) by means of a simple pumping system. Depending on this, the background pressure can vary between 10^{-3} and 10^{-7} mbar ; it is mostly limited due to the use of elastomeric seals. Thanks to the high available powerand low thermal mass, the heating and cooling rate of the wafer is in the order of 30 - 1000 °C/s, enabling very fast switching in temperature.

Temperature control and homogeneity is one of the major issues in RT processing. Originally, open loop systems were used controlling the lamp power only. The temperature was known from calibration tests with thermocouples welded to a wafer or melting point standards. A drawback was the poor reproducibility (only within less than 20 °C (Gibbons, 1985)). This reproducibility becomes worse when the wafer surface changes because absorption/reflection of radiation is very susceptible to the optical characteristics of this surface. An important improvement was the use of closed loop temperature control by means of optical sensors. A first technique was pyrometry, which works quite well for "standardized", non changing wafer surfaces. A problem is the dependence of the emissivity at the pyrometer's wavelength on the material (Si <-> SiGe, oxide...) and on the surface finish. A solution to overcome this problem has been found by Sturm (Sturm, 1990). He makes use of a property of Si that its transmission of IR light with wavelengths near the bandgap (1.3 - 1.5 µm) decreases with increasing temperatute, due to bandgap reduction and an increased phonon population. Experimentally, it was verified that the growth of a SiGe layer did not appreciably influence the optical characteristics, such that successive growth of Si and SiGe is possible without influencing the temperature controller (Sturm, 1991b). Advanced industrial designs such as the Centura (Applied Materials) and the Integra Three (AG Associates) have improved considerably these thermal characteristics ; temperature uniformities of ± 6 % and repeatabilities of ± 4 % (both within 3 σ) over 200 mm wafers are claimed (Sinha 1993 ; Gat, 1993). A very tight closed loop power control (within 0.1 %) seems to perform even better than closed loop temperature controlling, regardless of possible emissivity variations of the wafer surfaces. Moslehi et al (1991) have shown a temperature difference of less than 2 °C over a 150 mm wafer in an Texas Instrument TI-AVP reactor with multi-zone lamp heating and multi-point wafer temperature sensors. The importance of a very uniform temperature is illustrated by the fact that, with an activation energy of 47 Kcal/mole, a deviation in temperature of only 1 °C (which is only 0.1 % !) results in a growth rate variation of 3 %, which is the maximum tolerable for VLSI requirements.

The difference between Rapid Thermal CVD (RTCVD) and LRP is related to the "switching mechanism". In the original LRP technique, the gas is continuously flowing, and the deposition process is controlled by ramping up and down the temperature. Typically, growth takes 30 - 200 s at 900 - 1000 °C. Although this seems to result in very low thermal budgets, one must take into account that dopant impurity diffusion is a very strongly thermally acitvated process, with activation energies in the order of 80 kcal/mole (Fair, 1981). With a growth rate of about 500 nm/min at 900 °C (King, 1988) and of 5 nm/min at 700 °C (Sturm, 1991b), the growth rate ratio is 100 ; the diffusivity ratio on the other hand in this range is approximately 1000. This implies that dopant diffusion for equally thick layers at 700 °C will be only 10 % of this at 900 °C. This characteristic allows to apply a "conventional" CVD approach in an RTP reactor : with the wafers continuously at the growth temperature, the growth itself is controlled by switching the gas flows on and off. This technique is called RTCVD (Green, 1989). Sturm has refined this RTCVD technique further by optimising the individual layer temperature in order to obtain acceptable growth rates : 625 °C for SiGe (~ 100 Å/min) in order to avoid islanding and to achieve metastable layers, and 700 °C for Si (~ 30 Å/min) in order to accelerate the deposition rate (Sturm, 1991b).

An important issue in LRP/RTCVD is the incorporation of O (and C) at the interface(s) and in the film bulk. Contamination at the substrate epi interface has not been studied in detail. Most LRP or RTCVD techniques use a high-temperature bake (1000 - 1200 °C) for 20 - 120 s in flowing H_2. HCl, which was originally used in low concentrations (Gibbons, 1985 ; Green, 1989) has been abandoned lateron (King, 1989 ; Green, 1990. Only Green et al. give some information on the effectiveness of this cleaning cycle (Green, 1989), which results in 2 % of a monolayer of O and 20 % of a monolayer of C at the interface. Normally, before starting the growth of the device layers, a thick (1 μm) buffer layer of pure Si is grown at high temperature (1000 °C) after the H_2 prebake in order to bury all possible contamination and surface roughness which could have been left by the high temperature prebake. Jung et al. have studied the effect of Si buffer layers as compared to the direct growth of SiGe on the original surface and found a reduction in defect densities with two to three orders of magnitude (Jung, 1991). In conclusion, LRP/RTCVD requires the use of a high temperature H_2 prebake and a thick Si buffer grown at high temperature in order to create a good starting surface for low-temperature epi growth.

A second point of concern is the O and C contamination in the bulk of the films. Distinction must be made between Si and SiGe films. For pure Si, it has been shown by Green et al. (1990) that the O concentration in layers grown from SiH_2Cl_2 at T < 800 °C is above $10^{20}/cm^3$ and depends on the source gas purity. At higher T, this concentration drops quickly to below $10^{18}/cm^3$. For SiH_4 as source gas, which contains less H_2O as impurity, this transition temperature is about 700 °C. This difference between the low and high temperature regime is mainly due to volatilisation of SiO at high T, removing much of the O contamination regardless of the source gas. For very pure source gases, the reactor's background water pressure will be the limiting factor. For SiGe on the other hand, O appears to incorporate more readily at a given growth temperature : according to Hoyt et al. (1990), a Si layer grown at 700 °C contains typically $10^{18}/cm^3$ of O, whereas a SiGe layer grown in the same conditions exhibits more than $10^{20}/cm^3$ of O. The reason for this seems to be a cooperative effect of Si and Ge in promoting O incorporation. This problem has been overcome by lowering the reactor's background impurity levels by adding a loadlock to the process chamber. After a number of consecutive runs without exposing the inner of the reactor chamber to atmosphere, the film impurity level decreased by more than two orders of magnitude for 625 °C films. C incorporation comes out to be of lesser concern, with typical levels of $10^{18}/cm^3$ over the entire T range (Green, 1990).

An important advantage of LRP/RTCVD machines when performing n-type doping is their temperature flexibility. As discussed in Section 4.5, at lower temperatures problems of extended memory effects and sluggish dopant concentration transitions are encountered. This can be solved by working at higher temperatures, above 800 °C (King, 1989 ; Matutinovic-Krstelj 1991), at which the PH_3 decomposition products (P or P_2) desorb much more readily. Extensive reports on n-type doping features in LRP/RTCVD have not been reported in literature, but apparently it is at least useful for growth of in-situ doped collector and emitter layers in npn HBT's (King, 1989 ; Sturm, 1989).

With respect to p-type doping transitions, RTCVD performs at the same level as UHVCVD, when low temperatures (625 - 700 °C) are chosen for layer growth. Sturm et al. report on devices containing double, symmetric two dimensional hole gases at Si/SiGe heterojunctions (Venkataraman, 1991). The 2D confinement of the hole gas at this modulation doped junction is very sensitive to the heteroepitaxial interface region. If dopant segregation occurs, the symmetry of the normal and inverted interfaces which are grown successively in one run, is destroyed. This would result in different carrier mobilities. Low temperature magnetoresistance experiments on this double heterostructure clearly point to symmetrical interfaces. The results suggest that the B content decays with three orders of magnitude in about 20 Å, which is comparable to earlier UHVCVD results (Wang, 1989).

An interesting feature of epitaxially grown layers in advanced VLSI applications is the possibility to grow selectively epi Si (or SiGe) in windows in oxide layers. It has been shown by several authors (e.g. Regolini, 1989) that the use of SiH_2Cl_2 permits almost perfect selective epitaxial growth, although it influences the introduction of dopants (Zaslawski, 1992). This topic will be discussed further in Section 4.6.

A last topic in this overview of RTCVD and LRP reactors concerns single wafer atmospheric pressure CVD (APCVD). The main difference with LRP/RTCVD is the operational pressure which is slightly above 1 atm, and the presence of a SiC coated graphite susceptor which slows down the "Rapid" character of this reactor due to its high thermal mass (typically 3 °C/s - de Boer and Meyer 1991). This kind of reactor was more or less simultaneously pioneered on a laboratory scale by the group of Sedgwick at IBM (Sedgwick, 1989) and at full production level by ASM Epitaxy, resulting in the commercially available Epsilon one system (originally designed for high T CMOS epi applications). They are load locked an have point-of-use purifiers in the main gas line (usually H_2). As in RTCVD/LRP, the low T epi growth step is preceded by a H_2 prebake step at 850 °C (de Boer, 1991) or an HCl etch step at 1190 °C for several minutes (Kamins, 1991). A surprising fact is that, in spite of the relatively high partial pressures of O_2 and H_2O to be expected (about 10^{-5} mbar as can be calculated from the known impurity level of the carrier gas), high quality epi is grown at T = 650 - 850 °C with O levels below $10^{18}/cm^3$. Agnello and Sedgwick (Agnello, 1992b) attribute this to a depletion of O_2/H_2O in the boundary gas layer due to diffusional mass transport limitations (contrary to the much higher diffusivities in low pressure systems such as RTCVD/LRP), resulting in much lower gas-phase oxidiser concentrations in the direct neighbourhood of the wafer surface.

The use of SiH_2Cl_2 at atmospheric pressure brings an other interesting point with it concerning n-type doping. Agnello et al. have shown that n-type doping using PH_3 as well as AsH_3 in APCVD at temperatures below 800 °C is very well controllable and results in sharp doping transitions (Sedgwick, 1992 ; Agnello, 1992a). The memory effect is much less pronounced than in UHV/CVD : residual doping levels in doping superlattices is $5 \times 10^{16}/cm^3$. The films are of excellent crystalline quality, and diodes, made by depositing B doped poycrystalline Si on a P doped epi layer exhibit ideal behaviour. For dopant contents below about $5 \times 10^{19}/cm^3$, 100 % activation is achieved when grown at temperatures of 750 °C or higher. Another surprising observation was the growth rate enhancement at lower temperature. These phenomena are further discussed in Section 4.5.

Because the use of SiH_2Cl_2 always leads to selective growth, Agnello et al. have also tested n-type doped growth from SiH_4 in APCVD (Agnello, 1992b) in order to deposit blanket layers of Si. The growth rate dropped in this case considerably, and dopant transitions are as bad as in UHVCVD, with very high residual doping levels after closing the dopant gas line. Even worse is the crystalline quality which is heavily deteriorated. For non doped growth from SiH_4, the crystalline quality was generally acceptable but a heavy interfacial O contamination caused twinning defects when oxide precipitates larger than 300 Å resulting from this contamination, were formed.

In conclusion, we can state that RTCVD or LRP is a mature technology, that, generally speaking, can compete with UHVCVD. Definite advantages are its temperature flexibility and the ability to do in-situ n-type doping. Some disadvantages are the required high temperature prebake, and the difficulties when using SiH_4 for growing non selective layers. Another difficult-to-solve problem could be the extreme temperature uniformity (< 1 °C !) over large diameter wafers, which is required at low temperatures. This, together with indispensable parts such as loadlocks, wafer handling mechanisms and sophisticated gas handling systems make RTCVD, LRP or single wafer APCVD certainly not simpler machines than UHVCVD.

4.3.3. Other techniques

Plasma-Enhanced CVD

Plasma enhanced CVD is widely used for the deposition of dielectric films such as silicon nitride and silicon oxide. The main distinguishing feature of plasma-enhanced CVD is the dissociation of the source gas molecules in a plasma glow discharge into their respective precursors for the subsequent deposition. The production of precursors in a plasma leads to more energetic species impinging on the substrate surface compared to conventional thermal CVD. This has the benefit of lowering the deposition temperature (Reif, 1984).

The operating pressures in a PECVD system vary over the range of 10^{-3}- 10^{-1} torr. The plasma glow discharge can be arranged to occur further upstream (i.e. remotely) such that the substrate is located outside the plasma region. This arrangement minimises direct radiation damage to the substrate surface and offers additional control over the precursor production pathway (Rudder et al, 1986). The substrate may also be biased dc with respective to the rf electrodes to enhance the removal of native oxide by in-situ plasma sputtering (Donahue et al, 1984).

Pure or hydrogen-diluted SiH_4 and GeH_4 are typically used as the main source gases in PECVD. In remote PECVD, however, a noble gas such as argon or He is used as the rf plasma species, whilst SiH_4 and GeH_4 diluted in He are introduced downstream. For pre-epitaxy surface treatment, a Ar/H_2 plasma was reported to be more effective (Fukuda et al, 1991).

Early work of Towsend and Uddin (1973) showed that Si epitaxy could be achieved as low as 800°C by PECVD. Subsequently, the epitaxial temperature was further reduced to 700-750°C by Suzuki and Itoh (1983), and to 650-800°C by Donahue and Reif(1985). Using a UHV reactor chamber and in-situ remote H plasma cleaning prior to epitaxy, Breaux et al (1989) have achieved Si homoepitaxy at 150°C. Fukuda et al (1991) recently reported Si epitaxy without external substrate heating in a ultraclean electron-cyclotron-resonance (ECR) PECVD system.

Comfort and Reif (1987) observed that PECVD was beneficial to in-situ heavy arsenic doping in Si epitaxy at 800°C. In LPCVD, the surface morphology of the epitaxial layers degraded significantly as the As doping concentration exceeds 10^{18} cm^{-3} due to As saturation coverage of the bare Si surface (Meyerson and Yu, 1984). This is accompanied by a drastic reduction in the growth rate. It is believed that the more energetic radicals in PECVD altered the saturation coverage leading to improved epitaxy at high As doping concentrations.

Gas Source Si-MBE

Solid source Si-MBE (Gravesteijn et al, 1991) has pioneered many of the advances in low temperature epitaxy of Si and SiGe alloys. To date, the best performance in Si based bipolar transistors were obtained in MBE grown Si/SiGe material (Gruhle et al, 1992). Si-MBE however suffers a number of serious disadvantages related largely to the use of the e-beam evaporator to generate the Si flux. The deposited layers have particulate defects due to 'spitting' when uncontrolled Si deposits on the chamber walls of the chamber flake off and fall back into the molten Si pool. Particulates falling near the e-

beam can become charged and deflected towards the substrate. Furthermore, the solid Si charge used in the evaporator has a limited life leading to low throughout.

In gas source Si-MBE (Hirayama et al, 1987), a source gas such as SiH_4, Si_2H_6 or GeH_4 is used instead. Si deposition is now confined only to the heated substrate (and the hot heater) hence eliminating the problem of unwanted Si deposition on the cold chamber wall. Most workers have used Si_2H_6 as the Si source in preference to SiH_4 since the Si growth rates achieved with Si_2H_6 is much higher and the liquid nitrogen cold shroud is more effective in trapping disilane (Hirayama et al, 1988).

Doping can be readily achieved using gaseous B_2H_6, AsH_3 and PH_3. The operating pressure in gas source MBE is typically $<10^{-5}$ torr compared to 10^{-9} torr in Si-MBE. Due to the higher operating pressures, turbomolecular pumping is generally chosen. In practical terms, the main difference between gas source MBE and LP-CVD is its much lower operating pressure during growth and the attainment of ultra-high vacuum background by high temperature system bakeout. The lower pressure permits the use of high vacuum diagnostic techniques such as Reflection High Energy Electron Diffraction (RHEED) to monitor the growth (Liu et al, 1992), and low energy ion implantation and Knudsen cell for in-situ doping.

Photochemical Vapour Deposition (Photo-CVD)

In the photo-CVD process, ultraviolet radiation in used to excite and dissociate the reactant gases non-thermally. This leads to enhanced surface reaction and migration of the precursors on the substrate with the potential of lowering the deposition temperature. To increase the ultraviolet radiation absorption, the reactant molecules can be photo-sensitized with a small amount of mercury vapour (e.g. by passing the source gases through a thermally controlled Hg vaporizer). Disilane (Si_2H_6) is usually preferred to silane (SiH_4) due to its higher probability of photochemical decomposition (Pollock et al, 1973).

Nishida et al (1986) have demonstrated epitaxial growth of Si by photo-CVD using a low-pressure mercury lamp (1849 and 2537Å) and a gas mixture of Hg photosensitized Si_2H_6 and SiH_2F_2 diluted in H_2 at 200°C . The growth rate reported at 200°C was about 20Å/min. They found that the radicals from SiH_2F_2 played an important role in achieving the very low temperature epitaxy. They suggested its presence might assist the breaking of the hydrogen bond which has been reported to hinder the micro-crystallite growth in a-Si. Using the same growth system, Yamada et al (1989) reported heavily phosphorus-doped layers grown at 250°C with electron concentrations above 2×10^{21} cm^{-3}.

Lian et al (1991) reported a different approach in which an ArF excimer laser (193 nm) was used to excite the Si_2H_6 molecules above and parallel to the substrate surface in order to minimise any pyrolytic effect from the irradiation. The deposition was performed in an UHV system. Use of SiH_2F_2 and the mercury photosensitizer were avoided as they are considered to be non UHV-compatible. Growth rates between 0.5-4Å/min were obtained for substrate temperatures ranging 280-450°C. Epitaxy was achieved at 330°C under 0.5W laser power.

4.4 KINETICS OF CVD GROWTH OF Si AND SiGe in SiH_4/GeH_4 SYSTEMS

4.4.1 Introduction

The deposition of amorphous, polycrystalline and epitaxial Si films by CVD methods is a very important process in the micro-electronics industry for over 25 years already. An extremely broad range of deposition conditions and reactor designs has been tried out in an attempt to fullfil the needs of this industry. Temperatures between 500 and 1200 °C and pressures between 10^{-6} and 1 atmosphere have been used in variety of different reactor geometries ranging from the single 1" wafer reactor to systems with hundreds of 6" wafers or even bigger. At the same time, a huge number of publications on this subject has emerged, describing the deposition kinetics primarily from a phenomenological point of view. The result was considerable confusion : for deposition from SiH_4, the activation energies found range between 9 and 55 kcal/mole, the reaction order for the SiH_4 concentration ranges beween 0 and 1.8 and for the H_2 concentration beween -1 and 0. In order to put the specific study of CVD epi growth kinetics at low temperature into a more general perspective, some topics will very briefly be touched here. Whereas, according to an authoritative review of the state-of-the-art of CVD modelling (Gokoglu, 1990), the deposition rate can be accurately modeled as long as mass transport is rate governing, many questions remain about surface kinetics. Besides that, for an extended overview of the literature about results of kinetical studies, we refer to a paper by Comfort and Reif (1989).

For deposition at temperatures above 1000 °C and at atmospheric pressure, the homogeneous gas phase decomposition of silane as well as the heterogeneous reactions of adsorption, incorporation into the lattice and reaction product desorption are very rapid. Therefore, the growth rate is determined mainly by mass transport through the gasphase, a process which is not very sensitive to temperature and shows an Arrhenius activation energy of only about 5 kcal/mole. Modelling of these systems primarily implied a solution of the relevant balanced mass and energy transport equations (Bloem, 1978). At lower temperatures (~ 750 - 1100 °C) and pressures (10^{-3} - 1 bar), the chemical reactions become slower and the transport properties of the gas phase species change. The arrival rate of Si bearing species at the surface (and the removal rate of reaction byproducts away from the surface) becomes a function of complicated diffusion phenomena as well as of chemical reaction rates. Here, the homogeneous silane pyrolysis, which occurs via a very complex system of hundreds of elementary gas phase reactions with tens of intermediate products, has to be taken into account. Furtheron, the reactions of physical/chemical adsorption, diffusion, incorporation and concurrent desorption have to be added to the model. A quite detailed model, describing epitaxial Si growth rates in the near-atmospheric range (down to 5 torr), including a sophisticated hydrodynamical treatment together with an elaborated gas phase reaction scheme was presented by Coltrin, Kee and Miller (Coltrin, 1986).

In the lower pressure regime (mtorr range), Coltrins' model becomes inappropriate mainly because the primary gas phase reaction

$$SiH_4 + M \rightarrow SiH_2 + H_2 + M$$

slows down in proportion to the decreasing pressure. This is because the rate of this unimolecular reaction is determined by collisional activation, which is proportional to the total pressure. Meyerson and Jasinsky (1987) have calculated the rate constant of this reaction according to the RRKM formalism : whereas at 1 atm and 650 °C SiH_2 still contributes about 30 % to the deposition according to Coltrin, at 5 torr this decreases to 2 % and at a typical VLPCVD pressure of 2 mtorr to 10^{-3} % which is completely negligible. So, in conclusion, at lower temperatures (500 - 750 °C) and pressures (mtorr - torr), the rate controlling step is to be found in the heterogeneous reactions occurring at the Si surface. Only for cases where the supply of Si bearing species through the gas phase would be too slow, some transport phenomena need to be considered, too.

4.4.2 Kinetics of the silane system

The study of Si film growth kinetics at low temperatures and especially at very low pressures can become quite complicated due to the varying hydrogen surface coverage. This explains partly the large discrepancy in the vast amount of results from kinetic studies (for a thorough review, see Comfort, 1989 and Buss, 1988). Almost all these studies, though, relied on one or two phenomenological observations (H_2 evolution and/or growth rate), which then were interpreted in terms of a hypothetical reaction sequence. Mostly this sequence consists of some adsorption reaction :

$$SiH_4 + * \rightarrow SiH_4*$$

which is followed by a "deposition" step resulting in solid Si and molecular H_2 which has to be evaporated :

$$SiH_4 * -> film + 2 H_2\uparrow.$$

This sequence can be further refined by adding various intermediate reaction steps (dissociation of adsorbed silane, reevaporation of Si adatoms). Nevertheless, most authors admit that other reaction schemes could equally well be fitted to their experimental results. In order to elucidate unambiguously the underlying fundamental reaction steps, one needs to identify and quantify the intermediate reaction products upon adsorption of silane onto the Si surface, preferably in conditions of controllable surface coverages i.e. under ultra high vacuum. With respect to this, a particularly interesting series of papers has been published by Gates and Greenlief, a good overview of which is presented in Gates, 1990, providing further references. The authors used Static Secondary Ion Mass Spectroscopy to identify the SiH_x species formed upon adsorption of silane on Si (100)-(2x1) as well as on Si(111)-(7x7) surfaces for temperatures between -163 and 400°C. The reactive sticking probability of silane to the surface was measured by means of Temperature Programmed Desorption of H_2 as a function of H coverage. According to these findings, silane adsorbs dissociatively, consuming two active sites :

$$2 SiH_4(g) + 2 * -> 2 H* + 2 SiH_3*$$

This is followed by a number of steps in which SiH_x fragment are consecutively dehydrogenated untill the Si adatom is incorporated into the growing film, releasing the active site and desorbing H_2 :

$$2 SiH* - > H_2(g) + 2 * + film$$

The elementary mechanism as proposed by Gates, though, is contested in a recent publication by Hirose et al. (Hirose, 1991). From thermal desorption measurements on SiH_4 saturated surfaces, the authors conclude that silane adsorbs dissociatively, not into $H* + SiH_3*$, but into 4 monohydride units (one of which is a SiH adatom).

An important conclusion is that the rate controlling steps are dissociative silane adsorption and H_2 desorption. Incorporation of Si adatoms in the lattice does not seem to play an important kinetical role, nor does the desorption of Si adatoms as has been shown by Robbins (Robbins, 1987). In order to put now this knowledge into a quantitative model, it is assumed that every SiH_4 molecule that hits the surface is eventually converted into solid Si (i.e. film growth rates equal silane adsorption rates). (This is in contrast to Buss' work (Buss, 1988) who takes also a SiH_2* recombination reaction, resulting in SiH_4 desorption, into account.) In this way, the rate of deposition is given by

$$R = k_1 F N^2 \vartheta^2$$

with k_1 the rate constant of the adsorption reaction (dimensions : cm^4), F the flux of SiH_4 towards the surface, N the surface site density (which is for Si (100) equal to $6.78 \times 10^{14}/cm^2$) and ϑ the vacant site fraction. Buss et al. (1988) have shown the adsorption step to be essentially not temperature dependent, a statement which has been taken over by others. For the activation energy of the H_2 desorption reaction, we take 47-48 kcal/mole [#]. This is also the value of the activation energy for Si growth found by many workers (Gates, 1991, ; Buss, 1988 ; Liehr, 1990a) ; somewhat lower values, although probably identical within experimental uncertainty limits, are 45 kcal/mole (Poortmans, 1993) and 40 kcal/mole (Jang, 1991). Based on these two assumptions, Racanelli and Greve have fitted a model as described to their growth rate data (Greve, 1991). When we change their expression for the growth rate (which is given as $R = k_1 F N \vartheta^2$) according to the equation given previously, and recalculate their data, we find for k_1 a value of $1.2 \times 10^{-32} cm^4$ (at T = 600 °C). For the H_2 desorption step, the value of the first order rate constant comes out to be $1.6 s^{-1}$. This compares very well to Sinniah's value at 600 °C of $1.23 s^{-1}$. From an analogous fit, Gates found for k_1 an expression $\varepsilon = 2 k_1 N^2 \vartheta^2$ where he calls ε the reaction efficiency, which is the growth rate divided by the silane flux. Due to a wrong calculation of this flux, the authors found an ε value of $2 - 5 \times 10^{-4}$; with a corrected flux, this becomes at 400 °C $1 - 2.5 \times 10^{-3}$ and so, assuming with Gates an activation energy of 3 kcal/mole, k_1 takes on a value of $2 - 5 \times 10^{-33} cm^4$ at 600 °C. In Buss' work (Buss, 1988) the rate constant for SiH_4 adsorption is defined taking the gas phase silane concentration rather than the silane flux into account ; recalculating, we find then $k_1 = 1.5 \times 10^{-33} cm^4$. This agrees well with Gates' findings, but disagrees to an order of magnitude with Greve's results. It has to be pointed out, however, that it is difficult to compare these figures directly : on the one hand, Greve and Buss use CVD growth systems, whereas Gates works with a Chemical Beam machine. This can introduce differences into the SiH_4 adsorption efficiency because of thermal activation of SiH_4 which occurs in a CVD system due to gas phase or wall collisions. This does not happen in beams. On the other hand, the differences in model are not negligible.

In conclusion, it can be stated that the insight in the detailed growth mechanism is rapidly progressing since a few years now. Nevertheless, some questions and contradictions still remain to be solved. For this, growth rate data in the low temperature regime as a function of pressure over a broader pressure range (10^{-3} - 100 torr) are needed. Additionally, fundamental surface studies preferentially in real CVD conditions in view of the very important role played by H would be very welcome.

[#] Various authors have found different values for this energy ; if we re-interpret their results according to the rule of thumb (Rhodin, '76) that E_a(kcal/mole) $= 0.06 \times T_p$ with E_a the activation energy and T_p the peak temperature for thermal desorption, then we find in all cases values between 47 and 48 kcal/mole : Greenlief, '89 (*) ; Meyerson, '90 (*) ; Liehr, '90 (*); Hirose, '91 (*) ; Sinniah, '89 ; an * indicates re-interpretation.

4.4.3 Kinetics of the silane germane system

In order to grow $Si_{1-x}Ge_x$ alloy layers, GeH_4 is added to the normally used growth mixture of SiH_4, dopant gas and possibly some carrier gas. Important questions concerning growth of SiGe are : i) the incorporation ratio of Ge ; ii) influence of GeH_4 on the growth rate and the temperature dependence of the growth rate ; iii) influence of GeH_4 on dopant incorporation.

The incorporation ratio of Ge can be defined as $\chi_{Ge}/[F(GeH_4)/(F(GeH_4) + F(SiH_4))]$ with χ_{Ge} the atomic fraction of Ge in the layer, $F(GeH_4)$ and $F(SiH_4)$

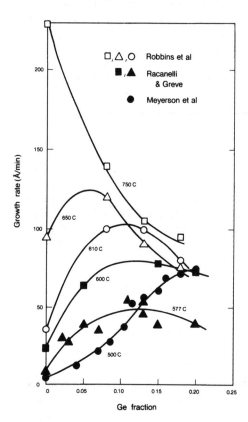

the resp. gas flows. According to most authors, this ratio is about 3, and independent of temperature between 580 and 700 °C (Racanelli, 1990 ; Robbins, 1991 ; Poortmans, 1993). Meyerson finds about 2 at 520 °C (Meyerson, 1988), and Reif reports a ratio of 3.3 - 4.2, varying with temperature between 570 and 700 °C in a random fashion (Jang, 1991).The influence of GeH_4 on the growth rate appears to be rather complicated. Figure 4.4, which has been reproduced from Racanelli, 1991, summarises results obtained by Meyerson (Meyerson, 1988), Greve and Racanelli, and Robbins (Robbins, 1991). At a lower Ge fraction, the growth rate is enhanced considerably, whereas at higher Ge contents it decreases again. The peak of highest growth rate shifts from x = 0.2 at 500 °C towards lower x with increasing temperature, vanishing completely at 750 °C. Reif has reported similar results. Figure 4. 5, reproduced from (Jang, 1991) is an Arrhenius plot of growth rates with the Ge content as parameter. For pure Si (x = 0), it shows a straight line between 570 and 700 °C with an activation energy of about 20 kcal/mole.

Figure 4.4 Influence of Ge-fraction on the epitaxial growth rate of $Si_{1-x}Ge_x$ layers (reproduced from Racanelli, 1991by permission of the Journal of Metals)

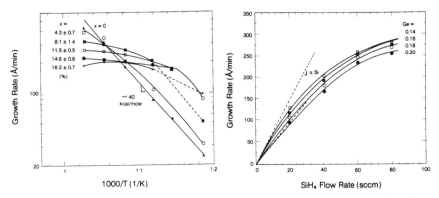

Figure 4.5 Temperature dependence of growth rate for $Si_{1-x}Ge_x$ (reproduced from Jang, 1991, with permission from Applied Physics Letters)

Figure 4.6 Growth rate as a function of SiH_4 flow rate (reproduced from Jang, 1992, with permission from Applied Physics Letters)

For small x, the temperature dependence starts to deviate, esp. at higher T ; the higher x, the stronger the deviation, with ever maller activation energy, and this extends to ever lower temperatures.

An explanation for this behaviour at low x has been suggested for the first time by Meyerson (Meyerson, 1988), whereas the phenomenon at high x has been explained by Robbins (Robbins, 1991). We refer here to the growth model presented for pure Si in previous section. The rate limiting step again is H_2 desorption, which is assumed to proceed easier from a Ge site at the surface than from Si, reducing the activation energy at low x. At higher x, though, the surface contains more and more Ge, and therefore, less H. So, the rate control shifts to dissociative MH_4 adsorption (where M = Si or Ge). Robbins suggested that the sticking coefficient for MH_4 is lower at Ge sites, thus slowing down the effective adsorption rate k_1 :

$$k_{1,M}(x) = k_{1,M}(0)/(1 + \rho\, x)$$

with ρ a coefficient which, according to a fit to the experimental data, takes on the value of 12. The ratio of sticking probabilities for silane and germane is found to be approximately constant in the temperature range 610 - 750 °C for $0 \le x \le 0.19$, with a value of about 4.7. In a certain sense, Ge content and temperature play an identical role in the growth rate control mechanism : both low Ge content and low temperature favour a high H coverage, whereas increasing the Ge contents and process temperature decrease this coverage. As a result, the rate control shifts from the H_2 desorption reaction to the MH_4 adsorption reaction, possibly dependent on mass transport limitations.

The formal model proposed by Robbins is also applicable to results of Reif (Jang, 1992) : these authors studied the influence of SiH_4 pressure on the growth rate with Ge content as parameter (Figure 4. 6). The reaction efficiency, which in fact is k_1, can be found as the slope α of these curves at zero pressure (where H coverage is zero). It is clear, that this efficiency decreases for increasing x, which confirms the applicability of Robbins' model.

In conclusion, the experimentally observed growth rate dependencies on various deposition parameters such as temperature, pressure (in a rather narrow range) and GeH_4 content have been combined in quantitative models, although the underlying physical phenomena are not yet completely unravelled. More insight can be gained by further surface science studies of $Si_{1-x}Ge_x$ material, preferentially under CVD conditions as noted previously.

4.5 DOPING OF LOW TEMPERATURE Si and $Si_{1-x}Ge_x$ LAYERS

In sharp contrast to Si MBE (Kubiak, 1991), doping in CVD Si is achieved simply by adding the appropriate dopant gaseous species, e.g. AsH_3, PH_3 and B_2H_6, to the source gas stream during growth. The transition abruptness of the doping profile is determined by the gas switching system used in the reactor and the incorporation kinetics of the dopant species at the growth front

van Opdorp and Leys (1987) have examined the various factors which govern the design of a fast gas switching system. A conclusion which can be drawn from this is that fast gas switching is rather easily accomplished in UHV/CVD and VLPCVD, thanks to the typical high pumping speeds and high molecular rates obtained at the very low pressures used. In higher pressure systems on the other hand, the design becomes more involved and factors such as flow patterns and the like have to be taken into account.

We will take now a closer look at the incorporation of the dopant species aspect.

4.5.1 p-type doping

P-type doping by adding B_2H_6 to the source gas is a rather straigthforward matter in low temperature CVD. The B content in the grown films follows very closely the gas phase B_2H_6 concentration over several orders of magnitude (Meyerson, 1992). An arbitrary B doping profile grown at 700 °C is shown in Figure 4.7 showing high-low B doping transitions (i.e. in the upward direction) which are on the order of 20 nm/decade (Poortmans, 1993). The low-high transitions are even sharper (in the order of 10 nm/decade), and cannot be resolved by SIMS anymore. Any desired concentration value between the reactor background doping level (less than $10^{14}/cm^3$) and up to about $10^{21}/cm^3$ can easily be obtained. Typically, full electrical activity is obtained except for the highest levels, where carrier concentrations are in the order of 1- 2 x

$10^{20}/cm^3$. Still, this exceeds the solid solubility at the growth temperature (Meyerson, 1987) by about 2 orders of magnitude. The growth rate of Si and

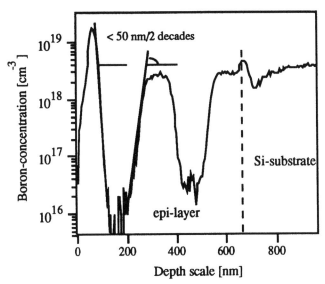

Figure 4.7 Boron profile in a UHV/CVD epi layer, grown at 700 °C (Poortmans, 1993)

SiGe layers is only very slightly influenced by the addition of B_2H_6 : it is increased by about 10 % when going to $5 \times 10^{19}/cm^3$ of B (Poortmans, 1993). From this and other studies (Meyerson, 1987 ; Greve, 1992) we can conclude that the sticking coefficient of B_2H_6 is slightly larger than this of SiH_4. LRP/RTCVD, UHV/CVD and other forms of low temperature CVD (Venkataraman, 1991 ; Wang, 1989 ; Roksnoer, 1991) appear to perform equally well concerning sharpness of dopant transition and B incorporation.

4.5.2 n-type doping

Typically, n-type doping is obtained by adding AsH_3 or PH_3 to the source gas. In contrast to p-type doping, n-type doping behaves rather differently in the various CVD systems. In Figure 4.8, SIMS measurements of B, P and Geprofiles in a multiple epi layer grown at 600 °C in a UHV/CVD system are represented (Caymax, 1991). The layer is composed of three sublayers : I, highly B doped and without P ; II, intermediate P doping ; III, highest P doping. Some features to be noted are : i) P doped layers show an increasing P concentration in the upper few tens of nm of the layer ; ii) at the substrate layer interface, (except for sample #97-4, which was the first growth run to introduce PH_3 in into the reactor), a P spike is found and the first, intentionally P free sublayer I shows a background P content of about $2 \times 10^{17}/cm^3$; iii) the highest

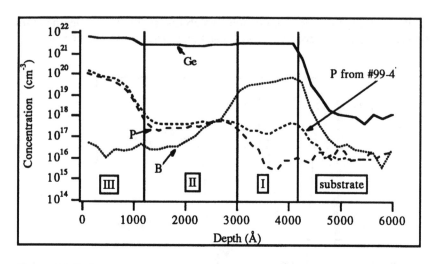

Figure 4.8 B, P and Ge profiles in a UHV/CVD multilayer grown in the first growth run with PH_3 at 600 °C ; in overlay : P-profile of sample #99-4, grown in the same conditions. a few days later

P level (sublayer III, below the superficial P peak) is in the order of 5 - 7 x $10^{19}/cm^3$; iv) the low-high transition is still reasonably sharp (about 30 nm/decade, but compare to B !), but a sharp high-low transition is impossible. Simultaneously, a reduced growth rate is noted (about a factor of two lower under these conditions). SRP measurements show about 40 - 50 % of the P to be electrically active.

Leong et al. (1988) observed that the incorported As and P concentrations in layers, grown in a cold (stainless steel) wall VLPCVD system were limited to an upper level around $10^{18}/cm^3$. At these concentrations, the surface morphology and electrical quality of the layers were severely degraded.

A better n-type doping control is obtained in LRP/RTCVD systems. As noted previously, extensive literature reports on n-type doping in LRP/RTCVD are not known to us, but apparently, the use of PH_3 in a SiH_2Cl_2 ambience at temperatures above 800 °C allows at least the growth of the (less demanding) emitter and collector layers in npn HBT's (King, 1989, Sturm; 1989). The best results (see section 4.3.2) have been obtained in APCVD for AsH_3 as well as for PH_3, using SiH_2Cl_2 in a H_2 main gas stream at temperatures below 800 °C.

The explanation for the observed phenomena in low temperature CVD growth from SiH_4 is two-fold. On the one hand, as is known from Si MBE work (Leong, 1988), As and P tend to segregate towards the surface in the solid state, producing an enriched adlayer on the growing surface and rather sluggish concentration profiles. On the other hand, PH_3 and AsH_3 are known to adsorb to the Si surface while dissociating into P/As and Si-H units (Meyerson, 1984), with a sticking coefficient near unity. At temperatures above 400 °C, H desorbs and the surface becomes fully covered with P/As. This results in a blocking of all active sites, reducing the growth rate considerably. All heated reactor parts become covered likewise by P/As, and, in this way, form a continuous source of dopants upon release into the gas phase. This explains the pronounced memory effect. At higher temperatures (> 750 °C), P (or P_2) and As will desorb more quickly, so releasing again the active surface sites. Because of the difficulties with temperature changes in hot wall CVD systems, it is cumbersome to thermally desorb the poisoning P from the reactor parts by such a temperature increase ; a possible - but far from ideal - solution to this problem would be to provide a second tube to the reactor, exclusively devoted to n-type doping.

The remarkable results in SiH_2Cl_2 based APCVD at temperatures below 800 °C have been explained as follows (Agnello, 1992) : during undoped growth, the most abundant species at the Si surface is Cl, a reaction by-product that has to desorb as HCl (or $SiCl_2$, but this does not contribute to the deposition). The authors argue that Cl is stronger bound to the surface than As or P, so that a poisoning effect such as in a pure Si-H system cannot occur. Therefore, the growth rate does not drop as in SiH_4 systems. Further, because Cl desorption can become accelerated due to formation of volatile AsCl or PCl components, the growth rate increases. This reaction assists also in desorption of As, or P, so making much sharper dopant transitions possible.

Interesting to note here are the experiments of Roksnoer et al. (1991), who have grown extremely sharp P spikes in an atmospheric pressure Fast Gas Switching CVD reactor, using PH_3 and Si_2H_6 in a H_2 carrier stream at temperature between 800 and 850 °C. Leading and trailing edges of 2 and 7 nm/dec., resp, have been reached. After turning off the PH_3 flow, growth was interrupted for 15 min in order to let P desorb from the surface before continuing the growth.

Also Hirayama and Tatsumi (1989) have reported phosphorus doping in epitaxial Si. They achieved up to 2×10^{20} cm^{-3} using Si_2H_6 and PH_3 by gas source Si MBE. They observed a reduction in the Si growth rate under high P doping. Above 1×10^{20} cm^{-3}, a 3-D RHEED pattern was observed indicating roughened surface. They reported no significant change in selectivity with P doping.

4.6   SELECTIVE EPITAXIAL GROWTH (SEG) OF Si AND SiGe

Selective epitaxy is a growth process in which deposition occurs only on the bare Si surface in an oxide-patterned substrate. This is a unique feature of the chemical vapour deposition. It has attracted considerable interest due to its potential application in the self-aligned processing of ultra-high speed devices.

Optical and SEM examination of the oxide islands after the deposition is not likely to be adequate for detecting any submicron nucleation seeds formed on the oxide surface. Though TEM will reveal the submicron nucleation features, it is time-consuming and hence seldom used. Comparison of published data from different workers is hampered by the fact that criteria for assessing selectivity are not properly defined. Recent work on real-time LLS study of the oxide surface in an oxide-patterned substrate during LP-CVD suggests LLS may offer a simple yet very sensitive method of monitoring the poly-Si nucleation on the oxide surface [Leong et al, 1992].

Selective epitaxial growth (SEG) in Si has been widely demonstrated in a $SiH_2Cl_2/HCl/H_2$ based chemistry at high temperatures. With the recent advance in low temperature epitaxy, SEG has now achieved at growth temperatures much below 1000°C using chlorinated hydrides as well as chlorine-free hydrides such as disilane, silane and germane. Selectivity is affected by the oxide material. Kato et al (1990) have observed differences in the Si nucleation rate on the thermally grown oxide, and undoped and doped silicate glasses under the same deposition conditions. Doping during the deposition can also influence the selectivity (Zaslavsky et al, 1992).

4.6.1 Growth in Chlorine-free Ambience

Chlorine-free source gases are preferred in a stainless steel reactor system due to the highly corrosive nature of chlorine. There is also some concern over the effect of the Cl incorporation at the epi-substrate interface (Rahat et al, 1991). It is known that SEG with chlorine-free Si source chemistry is difficult (Regolini, 1989). This is due to the lack of an effective agent for etching off the Si nuclei formed on the oxide before they reach a critical size and coalescence upon which poly-Si deposition occurs.

Full SEG (i.e. unlimited epitaxial Si thickness in the window) using SiH_4/H_2 has only been observed at growth temperatures above 1100°C (Kobayashi et al, 1990). Etching of the SiO_2 mask was often observed when low deposition rates and high temperatures were used. At 960°C, the maximum Si SEG layer thickness reported so far is about 1μm deposited in a UHV-bakeable system under a SiH_4 pressure of 34 mTorr (Afshar-Hanaii et al, 1992). Lowering the deposition temperature apparently reduces the maximum obtained Si SEG thickness, for instance, ~400Å at 620°C in a UHV-CVD system (Leong et al, 1992), 600Å at 700°C in ultraclean CVD (Murota et al, 1989) and 1300Å at 800°C in ultra LP-CVD (Yew and Reif, 1989). This observation is consistent with the increase in the incubation period with improved cleanliness of the oxide surface (Murota et al, 1989). Baert et al (1992) used a mixture of SiH_4/SiF_4 to achieve selective epitaxy at 300-400°C in a rf plasma CVD reactor. The concentration of the Si etching F radicals dissociated from SiF_4 had to be carefully adjusted to balance with the Si depositing SiH_x precursors to give the selective deposition.

Hirayama et al (1988) have found disilane (Si_2H_6) to be more favourable for SEG at low temperatures than SiH_4 in gas source Si-MBE. Optimum conditions for SEG were low gas source flux and low substrate temperature. They suggested that lowering of the substrate temperature was more effective in reducing the dissociative adsorption of disilane on the SiO_2 surface than the desorption of the dissociated disilane fraction. The very low pressures in gas source Si-MBE should also lead to a lower H coverage of and more chemically active dangling bonds on the bare Si which promotes the dissociative adsorption of disilane on the bare Si and hence selectivity.

Hirayama et al (1990) and Aketagawa et al (1992) have also reported $Si_{1-x}Ge_x$ SEG by gas source Si-MBE and UHV-CVD respectively using Si_2H_6 and GeH_4. Selectivity could be maintained under different gas flow rates and substrate temperatures provided the Si_2H_6 flow did not exceed a critical impinging rate.

Dumin (1971) had reported SEG of Ge on Si by conventional CVD above 800°C using GeH_4. Selective deposition of polycrystalline and crystalline Ge on Si at between 330-530°C using GeH_4 was reported by Ishii et al (1985). Selectivity was only achieved for temperatures below 410°C due to some unknown contamination of previously deposited Ge in the reactor. Koyayashi et al (1990) later showed that by employing ultra-clean gases and chamber system, perfect selective Ge epitaxy was possible over the temperature range of 350 -500°C.

4.6.2 Growth in Chlorine-containing Ambience

Comprehensive studies of Si SEG using $SiCl_4$ and SiH_2Cl_2 with and without added HCl have been made since 1962 (Borland, 1987). As the growth temperature is reduced in order to minimise the thermal budget, SiH_2Cl_2 has become the preferred source gas due to its higher deposition rate. Though the addition of HCl is known to affect the facet formation and control selectivity, it is generally avoided due to the complexity and purity of the HCl source.

Regolini et al (1989) have reported full SEG using SiH_2Cl_2/H_2 without added HCl in the temperature range of 650-1100°C at reduced pressures (<2 torr). They have also observed selective growth by adding a small amount of HCl to SiH_4/H_2. They believed that the lower pressure regime used in their system enhanced HCl etching of Si thus promoting selective deposition.

Sedgwick et al (1989) reported selective Si epitaxy at 600-850°C in a ultra clean atmospheric pressure CVD system using SiH_2Cl_2/H_2. Considerable amount of effort was taken to purify the source gases to remove traces of oxygen and water. Their results also showed that atmospheric H_2 did not prevent epitaxial growth at 600°C as previously suggested.

Zhong et al (1990) observed significant enhancement in the selectivity when GeH_4 was added to SiH_2Cl_2/H_2. The transition from non-selective to selective deposition occurred at a flow rate ratio of GeH_4 to SiH_2Cl_2 between 0.1 and 0.2. They postulated that the improved selectivity might be due to the SiO_2 removal via a reaction with deposited Ge to form highly volatile GeO.

4.7   IN-SITU AND EX-SITU CHARACTERISATION

Understanding of the initial growth process is critical in achieving proper low temperature epitaxy. Diagnostic techniques capable of monitoring the growth process in-situ and/or in real-time are therefore highly desirable in both the research as well as production environment.

Due to the relatively high operating pressures ($>10^{-2}$torr) employed in a typical CVD process, in-situ high vacuum diagnostic techniques such as RHEED are not suitable. Most of the in-situ diagnostic techniques in CVD therefore involve some form of optical probe. The optical methods offer the unique advantage of siting the instrument outside the reaction chamber hence avoiding contamination and minimising system disturbance.

A number of optical techniques have been applied to study the low-temperature CVD Si and SiGe epitaxial growth process in-situ. Laser Light Scattering (LLS) (Robbins et al, 1988) and dual-wavelength ellipsometry (Pickering, 1991) are two good examples which have been proven to provide real-time monitoring of the pre- and post-epitaxy surface morphology and the layer structures. Spectroscopic Ellipsometry (SE) and photoluminescence (PL) have also been shown to be very powerful for analysing the SiGe/Si heterostructures revealing detailed information on the electronic and optical properties of layers. These techniques however do not provide an absolute measurement of the alloy composition which has to be determined from independent Rutherford Backscattering (RBS) and X-Ray Diffraction (XRD) measurements (Baribeau and Houghton, 1991).

The next section will described some of the above optical techniques which have been found to be extremely useful for characterising the CVD related growth of Si and SiGe epitaxial structures.

4.7.1 Laser Light Scattering (LLS)

A theoretical treatment of the principle behind Laser Light Scattering (LLS) analysis is given in reference (Robbins et al, 1988). A laser light beam incident on a substrate surface which is scattered in the non specular direction is detected by a photon counting system (consisting of a telephoto lens focused on the substrate surface and attached to a photo multiplier and pulse counter). The scattered light intensity reflects any changes in the surface topography during processing in real-time. The technique is sensitive to surface features with lateral dimension comparable to the laser wavelength used. Sub-nanometer changes in the vertical direction of the surface can also be detected. The same optical arrangement also facilitates the use of in-situ ellipsometry to provide additional and complementary information on the dielectric properties of the substrate surface.

Good correlation between substrate cleaning and the epitaxial quality of subsequent layer growth by Si-MBE with LLS monitoring has been reported by Pidduck et al (1989). The interpretation of LLS data during surface oxide desorption, thermal roughening of the substrate and subsequent epitaxial growth in VPE Si supported by simultaneous in-situ ellipsometry measurement has been reported by Pickering (1991).

The usefulness of LLS as an in-situ monitoring and diagnostic tool for Si epitaxy is best illustrated with the following examples. Figure 4.9 shows the LLS intensity plots with time during the oxide removal and the subsequent epitaxial

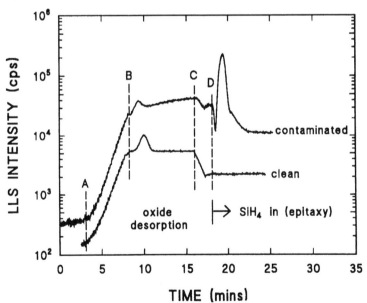

Figure 4.9 LLS monitoring of oxide removal and subsequent Si epitaxy on clean and contaminated substrates

deposition on two different substrates, one clean (I) and one contaminated (II). Between A-B, the substrate temperature was ramped from 620°C to 890°C in a H_2 ambient of 1.0 torr. The substrate was then maintained at 890°C to desorb the native oxide (B-C). The temperature was then lowered to 850°C before the H_2 ambient was reduced to 0.1 torr and SiH_4 admitted (D) into the reactor. The LLS intensity increase between A-B was due to the increase in the black body radiation as the substrate temperature was increased. The peak in the LLS plot seen just after B indicates the thermal desorption of the native oxide which proceeds in the form of minute pits at points of weakness. These pits enlarge with time to reveal the underlying bare Si. They eventually coalesce with neighbouring pits and smooth out leading to a fall in the LLS intensity. In the case of the clean substrate (I), the LLS intensity remained constant during the remaining part of the desorption cycle. With the contaminated substrate (II), however, the LLS intensity started to increase as the bare Si reacted with the surface contaminants to form nucleation sites (thermal roughening). When epitaxy was initiated (D), a very abrupt increase in the LLS intensity was observed with the contaminated substrate indicative of the 3D nucleation. The LLS intensity remained low and constant in the case of the clean substrate showing that a clean bare Si was achieved after the thermal desorption cycle.

4.7.2 Spectroscopic Ellipsometry

Spectroscopic ellipsometry (SE) is a rapid and non-destructive technique for analysing multilayered heterostructures. In SE, the amplitude ratio, $\tan(\psi)$, and phase difference $\cos(\Delta)$, in the reflected radiation from a circularly polarised monochromatic light incident on the sample surface as a function of wavelength are measured. The dielectric function, $<\varepsilon> = \varepsilon_1 + i\,\varepsilon_2$, is then calculated from the raw data using

$$<\varepsilon> = \sin^2\phi + [(1-\rho/(1+\rho)]^2 \sin^2\phi \tan^2\phi$$

where ϕ is the angle of incidence and $\rho = \tan\psi \exp(i\Delta)$.

To obtain the actual layer structure, a complicated calculation of the dielectric function expected from a modelled structure is carried out and compared with the measured data. Sophisticated iterative procedures for data fitting have been developed to ease the analysis of the structures.

Most of the SE studies on heterostructures have been on the III-V alloy systems such as AlGaAs/GaAs and InGaAs/AlInAs/InP. Though bulk cubic SiGe alloy and MBE grown Si:Ge superlattices have been studied by SE, Kamins (1991) first demonstrated the use of single-wavelength ellipsometry to determine the thickness and alloy composition in strained SiGe alloy layers. Recently, comprehensive SE measurements have been carried out by Ferrieu et al (1992), Pickering et al (1993) and Libezny et al (1993) on strained and relaxed $Si_{1-x}Ge_x$ layers grown by CVD with $0 < x < 0.25$. Good agreements of the alloy composition and layer thickness determined from the SE measurements and data obtained by RBS and XRD have been reported. These studies also indicated the requirement of taking the native oxide and surface roughness of the strained SiGe layers into account in order to model the structure accurately.

One of the recent exciting developments in SE is its application in real-time process control of the layer thickness and alloy composition during the deposition. Layer thicknesses and alloy compositions are two of the most fundamental parameters determining the device characteristics. Aspnes and co-workers (1990) demonstrated the first closed loop control system in which the near-surface composition x of $Al_xGa_{1-x}As$ was determined in real-time from the ellipsometric data ε_2 and used to control the source gas flows in organometallic MBE. A compositional control of better than 0.1% was claimed. Recently, they improved on the technique and grew a 200Å parabolic quantum well with x varying from 0.3 to 0 (Aspnes et al, 1992).

4.7.3 Modulation Spectroscopy (Photo-reflectance and Electro-reflectance)

A number of good reviews on the modulation spectroscopy of semiconductor microstructures have been published by Pollak and Glembocki (1988, 1990). In modulation spectroscopy, monochromatic light from a broad band source is incident on the sample and the modulated reflected light is measured as a function of the photon energy. The modulation can be achieved by applying either i) a pulsed voltage bias across a Schottky barrier or p-n junction in the sample (electroreflectance), or ii) a chopped illumination of the sample with above bandgap light to generate free electron-hole pairs in the depletion region at the surface/interface in the sample (photo reflectance). The second method relies on the build-in field at the sample surface or heterojunction, and therefore it does not require special sample contact preparation.

Any change in the electric field will change the dielectric constant, $<\varepsilon> = \varepsilon_1 + i.\varepsilon_2$ of the material which is detected as a change in the reflectance, R, given by

$$\Delta R/R = \alpha \Delta \varepsilon_1 + \beta \Delta \varepsilon_2$$

where $\Delta \varepsilon_1$ and $\Delta \varepsilon_2$ are the differential changes in $<\varepsilon>$, and α and β are the Seraphin coefficients.

Photoreflectance has been used to probe completed GaAs/GaAlAs heterojunction bipolar transistor (HBT) structures (Yan et al, 1992). The built-in fields in the GaAlAs emitter and GaAs collector regions were measured from the observed FKOs and correlated with the actual dc current gain of the device. The measurement also allowed the bandgap of the emitter (hence the Al composition) to be determined. Similar measurement has also been demonstrated in an InP/InGaAs HBT structure (Yin and Pollak, 1990).

Pearsall et al (1987) have studied the optical transitions by electroreflectance in the MBE grown Si-Ge superlattices consisting of 5 periods of 4 monolayers of Ge and 4 monolayers of Si on a Si (100) substrate. The measured transition energies were compared with the expected calculated transitions for a strained $Si_{0.55}Ge_{0.4}$ layer with the equivalent alloy composition and thickness. They demonstrated the Si-Ge superlattice has a basic energy level structure quite different from that of the alloy quantum well. They also observed optical transitions in the 1.0-2.0 μm range which they concluded were new transitions due be the loss of cubic symmetry in the 4x4 Si-Ge superlattice structure.

Attempts have been made to obtain unambiguous experimental proof of the quasi-direct transitions which have been theoretically predicted in the short period Si-Ge superlattice by a number of workers (Zachai, 1991). Menczinger et al (1991) have shown that photoreflectance analysis of these structures is complicated by the superposition of signals arising from the superlattice and the substrate. The appearance of interference oscillation due to the optical etalon within the superlattice in thick samples also hampers the observation of the weak signals expected from zone folded transitions in the spectral region below 1.5 eV (Churchill et al, 1991).

Few modulation spectroscopy studies have been made with CVD-grown strained SiGe/Si structures. This might be related to the fact that most effort on CVD growth has so far been concentrated on the thicker alloy layers which offer very promising prospects as the base layers in very high speed HBT. It is anticipated, however, as the effort on CVD growth is used to grow ever thinner alloy structures and Si-Ge superlattices, methods as PR and ER will be become more widely used.

Yin et al (1992) have done one of the few PR measurements on CVD grown strained SiGe alloy layers. They have examined the dependence of the E_0 and E_1 transition energies on the Ge composition in strained SiGe alloy layers at the near surface region (limited by the photon penetration depth to about ~100Å). The measured E1 transition energies which decrease rapidly with the Ge fraction, were essentially the same as in the unstrained material (Humlicek et al, 1989), in agreement with the deformation potential theory and data obtained by spectroscopic ellipsometry measurement on strained SiGe alloy layers (Pickering et al, 1993).

4.7.4 Photoluminescence

Photoluminescence (PL) is a non-destructive analytical technique which has been used to study optically active defects in Si since early 1960's (Davies, 1989). The semiconductor is excited at cryogenic temperatures by above bandgap photons to produce a high non-equilibrium concentration of electrons and holes. The electron-hole pairs (excitons) can decay directly to give free-exciton luminescence, or following capture by various impurities and defects, their radiative decay will give rise to impurity/defect specific luminescence features.

One difficulty with PL analysis of thin Si homoepitaxial layers is differentiating between luminescence from the grown layer and the underlying substrate. Though most of the excitation radiation is absorbed within a few μm from the top surface, the excitons and carriers can diffuse deep into the substrate. One convenient way of 'differentiating' the origin of the luminescence is to grow the layer on a substrate doped with the opposite type. Using this technique, it has been shown that the ratio of the principal TO bound exciton (BE) peaks associated with the dopant in the layer to that in the substrate is proportional to the integrated sheet carrier concentration in the layer provided the layer is thick enough such that direct excitonic generation is negligible in the substrate (Leong et al, 1990).

Inadvertent carbon incorporated in the epitaxial layer during growth which is optically inactive can be revealed by PL after high energy electron irradiation at room temperature (Robbins et al, 1985). The appearance of zero phonon line at 0.97eV (G-transition) and its local mode phonon sideband at 72meV to lower energy is indicated of C defect complex formed during the irradiation (Sauer and Weber, 1983).

Terashima et al (1990) were first to report near bandgap photoluminecence from thick strained epitaxial $Si_{0.96}Ge_{0.04}$ alloy grown by MBE at 650°C. Noel and co-workers (1990) have also studied PL from MBE SiGe alloys grown at 400°C covering the Ge fractions ranging $0.06 < x < 0.53$. The as-grown samples showed very weak luminescence. However, upon annealing at about 600°C, a strong broad band PL spectrum was observed. The peak energies of the broad bands were about 120meV below the calculated pseudomorphic alloy bandgap (Lang et al, 1985). The origin of this strong luminescence is still unclear. Recently, Noel et al (1992) suggested that this broad band luminescence seen only in MBE grown SiGe material might be related to small Ge rich interstitial platelets (<1.5nm) which formed exciton traps competing with the shallow dopants.

Sturm et al (1991) were first to report strong excitonic PL from strained $Si_{1-x}Ge_x$/Si multilayers grown by RT-CVD. The no-phonon (NP) line peak energies measured were quite close to the expected minimum bandgaps for the alloy with the given composition. Similar well-resolved excitonic luminescence was also been observed in strained $Si_{1-x}Ge_x$ alloy layers deposited by different CVD growth techniques such as RT-CVD (Dutarte et al, 1991), atmospheric pressure CVD (Rowell et al, 1992), UHV-VPE (Robbins et al, 1992), and gas source MBE (Hirayama and Tatsumi, 1990).

In general, well-resolved near bandgap luminescence has so far been consistently observed in CVD grown SiGe alloy layers. It is not clear if this is related to the higher growth temperature (>600°C) typically used in the CVD epitaxial process. A number of groups (Arbet-Engels et al, 1992, Spitzer et al, 1992, and Steiner et al, 1992) have recently reported band-edge PL from MBE grown SiGe/Si structures. Some of these studies apparently suggested that a high growth temperature (>600°C) might be a prerequisite to achieving well-resolved band edge PL in MBE grown SiGe alloy. Though Arbet-Engels et al (1992) have reported near bandgap luminescence from layers grown at 450°C, the PL intensities increased drastically after RTA at 500-700°C for 90s. Whereas Spitzer et al (1992) observed both the excitonic SiGe band edge luminescence and the broad band luminescence in layers grown below 450°C.

Recently, Wachter et al (1993) reported detailed PL studies of SiGe quantum wells grown by MBE at $350°C \leq T_g \leq 750°C$ to investigate the nature of the broad PL bands. They observed well-resolved band edge PL with either complete absent or weak broad band features in contrast to previous studies. They concluded that the defects or complexes responsible for the broad PL band is not inherent to the MBE growth technique.

Band edge electroluminescence (EL) up to 200K from strained $Si_{0.8}Ge_{0.2}$ alloy layers in a PIN diode was first reported by Robbins et al (1991). Correlations between the EL, PL and photoconductivity measurements from the same layer structure confirmed the EL originated from the near bandgap radiative recombination of the carriers in the alloy layer. Mi et al (1992) later showed that by increasing the Ge fraction in the alloy layers to above 0.35, much stronger carrier confinement effect was obtained in the alloy layer leading to the first demonstration of room temperature EL from the $Si_{0.65}Ge_{0.35}/Si$ quantum wells.

In sharp contrast to the sharp narrow EL and PL bands of the CVD grown SiGe layers, a strong broad band feature was seen in the EL and PL spectra of early Si-MBE grown strained $Si_{1-x}Ge_x$ alloy layers (Rowell et al, 1991). However, Fukatsu et al (1992) have recently reported band edge EL from strained SiGe/Si structures grown by Si-MBE at 620°C.

ACKNOWLEDGEMENT

The authors wish to acknowledge the invaluable contributions from our colleagues to the research work described in this chapter. This work has been carried out with the support of our respective national government bodies (DTI and MOD in U.K.) and the Commission of the European Communities (TIPBASE and MIDAS projects in the ESPRIT programme).

BIBLIOGRAPHY

Afshar-Hanaii N., Bonar J.M., Evans A.G.R., Parker G.J., Starbruck C.M.K. and Kemhadjian H.A. (1992), Microelectronic Eng., **18**, 237

Agnello P.D., Sedgwick T.O., Goorsky M.S. and Cotte J. (1992a), Appl. Phys. Lett., **60(4)**, 454

Agnello P.D., Sedgwick T.O., Bretz K.C. and Kuan T.S. (1992b), Appl. Phys. Lett., **61(11)**, 1298

Aketagawa K., Tatsumi T., Hiroi M., Niino T. and Sakai J. (1992), Jpn. J. Appl. Phys., **31**, 1432

Arbet-Engels V., Tijero J.M.G., Manissadijian A., Wang K. and Higgs V. (1992), Appl. Phys. Lett., **61**, 2586

Aspnes D.E., Quinn W.E. and Gregory S. (1990), Appl. Phys. Lett., **57**, 2707

Aspnes D.E., Quinn W.E., Tarmargo M.C., Pudensi M.A.A., Schwarz S.A., Brasil S.P., Nahory R.E. and Gregory S. (1992), Appl. Phys. Lett.,

Badoz P.A., Bensahel D., Guérin L., Perret P., Puissant C. and Regolini J.L. (1990), J. Electronic Mat., **19(10)**, 1123

Baert K., Deschepper P., Poortmans J., Nijs J. and Mertens R. (1992), Appl. Phys. Lett., **60**, 442

Baribeau J.M. and Houghton D.C. (1991), J. Vac. Sci. Technol., **B9**, 2054

Batsford K.O. and Thomas D.J.D. (1963), Electrical Communication, **38**, 354

Bloem J., and Gilling L.J. (1978), Current Topics in Materials Science, Vol I, E. Kaldis (ed), (North Holland Publishing Co., Amsterdam)

Borland J.O. (1987), Proc. 10th Int. Conf. CVD, PV87-8, Electrochem. Soc., p.307

Breaux L., Anthony B., Hsu T., Banerjee S. and Tasch T. (1989), Appl. Phys. Lett., **55**, 1885

Budde K (1992), Book of Abstracts of the 1st Int. Symp. on Ultra Clean Processing of Silicon Surfaces (UCPSS '92), Leuven, Belgium, September 1992, p.19

Buss R.J., Ho P., Breiland W.G. and Coltrin M.E. (1988), J. Appl. Phys., **63(8)**, 2808

Casel A., Kasper E., Kibbel H. and Sasse E. (1987), J. Vac. Sc. Technol., **B5(6)**, 1650

Caymax M.R., Poortmans J., Van Ammel A., Vandervorst W., Vanhellemont J. and Nijs J. (1992), Chemical Surface Preparation, Passivation and Cleaning for Semiconductor Growth and Processing MRS Proceedings, Vol 259, R J Nemanich, C R Helms, M Hirose and G W Rubloff (eds), (Materials Research Society Pittsburgh Pa), p.461

Caymax M., Poortmans J., Van Ammel A. and Nijs J.F. (1990), Proceedings of the Second International Conference on Electronic Materials, R.P.H. Chang, M. Geis, B. Meyerson, D.A.B. Miller and R. Ramesh (eds), (Mater. Res. Soc. Proc., Pittsburgh, PA), p.519-524

Churchill A.C., Klipstein P.C., Gibbings C.J., Gell M.A., Jones M.E. and Tuppen C.G. (1991), Semicond. Sci. Technol., **6**, 18

Coltrin M.E., Kee R.J. and Miller J. (1986), J. Electrochem. Soc., **133(6)**, 1206

Comfort J.H. and Reif R. (1987), Appl. Phys. Lett., **51**, 1536

Comfort J.H. and Reif R. (1989), J. Electrochem. Soc., **136(8)**, 2386

de Boer W. and Meyer D.J. (1991), Appl. Phys. Lett., **58(12)**, 1286

Davies G. (1989), Phys. Rep., **176**, 83

Donahue T.J., Burger W.R. and Reif R. (1984), Appl. Phys. Lett., **44**, 346

Donahue T.J. and Reif R. (1985), J. Appl. Phys., **57**, 2757

Donahue T.J. and Reif R. (1986), J. Electrochem. Soc., **133(8)**, 1691

Dumin D. J. (1971), J. Cryst. Growth, **8**, 33

Dutarte D., Bremond G., Souifi A. and Benyattou T. (1991), Phys. Rev. B., **44**, 11525

Fair R.B. (1981), "Impurity Doping Processes in Silicon", F.F.Y. Wang (ed), (North Holland)

Farrow R.F.C. (1974), J. Electrochem. Soc., **121**, 899

Ferrieu F., Beck F. and Dutartre D. (1992), Solid St. Comm., **82**, 427

Fukatsu S., Usami N., Chinzei T., Shiraki Y., Nishida A. and Nakagawa K. (1992), Jpn. Appl. Phys., **31**, L1015

Fukuda K., Murota J., Ono S., Matsuura T., Uetake H. and Ohmi T. (1991), Appl. Phys. Lett., **59**, 2853

Gates S.M. and Greenlief C.M. (1990), J. Vac. Sc. Technol., **A8(3)**, 2965

Gates S M and Kulkarni S K 1991 Appl. Phys. Lett. 38, 2963

Ghidini G. and Smith F.W. (1984), J. Electrochem. Soc., **131(12)**, 2934

Gibbons J.F., Gronet C.M. and Williams K.E. (1985), Appl. Phys. Lett., **47(7)**, 721

Glembocki O.J. (1990), SPIE, **1286**, 2

Gokoglu S.A. (1990), Proceedings of the XIth International Conference on Chemical Vapour Deposition CVD-XI, K.E. Spear and G.W. Cullen (eds), (The electrochemical Society) p.1

Gravesteijn D.J.,Van de Walle G.F.A., Pruijmboom A., and Van Gorkum A.A. (1991), Mat. Res. Soc. Symp. Proc., **220**, 3

Green M.L., Brasen D., Geva M., Reents W., Jr, Stevie F. and Temkin H. (1990), J. Electronic Mat., **19(10)**, 1015

Green M.L., Brasen D., Luftman H. and Kannan V.C. (1989), J. Appl. Phys., **65(6)**, 2558

Greenlief C. M., Gates S.M. and Holber P.A. (1989), Chem. Phys. Lett., **159(2,3)**, 202

Greve D.W. and Racanelli M. (1990), J. Vac. Sc. Technol., **B8(3)**, 511

Greve D.W. and Racanelli M. (1991), J. Electrochem. Soc., **138(6)**, 1744

Greve D W and Racanelli M (1992) J. Electronic Mat. 21, 593

Gruhle, A., Kibbel H., Konig U., Erben U. and Kasper E. (1992), IEEE Electron Device Lett., **13**, 206

Gupta R., Colvin V.C., Brand J.L. and George S.M. (1988), Deposition and Growth : Limits for the Microelectronics, Am. Inst. of Phys. Conf. Proc. nr 167, G.W. Rubloff (ed), (Am. Vac. Soc. NY), p.51

Hirayama H., Tatsumi T., Ogura A. and Aizaki N. (1987), App. Phys. Lett., **51**, 2213

Hirayama H., Tatsumi T. and Aizaki N. (1988), Appl. Phys. Lett., **52**, 1484

Hirayama H., Tatsumi T. and Aizaki N. (1988b), Appl. Phys. Lett., **52**, 2242

Hirayama H. and Tatsumi T. (1989), Appl. Phys. Lett., **55**, 131
Hirayama H. and Tatsumi T. (1990), Thin Solid Films, **184**, 125
Hirayama H., Hiroi M., Koyama K. and Tatsumi T. (1990), Appl. Phys. Lett., **56**, 1107
Hirose F., Suemitsu M. and Miyamoto N. (1991), J. Appl. Phys., **70(10)**, 5380
Hoyt J.L., Noble D.B., Ghani T, King C.A. and Gibbons J.F. (1990), Proceedings of the Second International Conference on Electronic Materials, R.P.H. Chang, M. Geis, B.S. Meyerson, D.A.B. Miller and R. Ramesh (eds), (Mater. Res. Soc. Proc., Pittsburgh, PA), p.551
Hull R. and Bean J.C. (1991), "Strained Layer Superlattices: Materials Science and Technology", Pearsall T P (Vol. ed) in : Semiconductors and Semimetals, Vol 33, R.K. Willardson and P.C. Beer (Eds), (Academic Press, Inc), p.9
Humlicek J., Garriga M., Alonso M.I. and Cardona M. (1989), J. Appl. Phys., **65**, 2827
Ishii H., Takahashi Y. and Murota J. (1985), Appl. Phys. Lett., **47**, 863
Ishizaka A., Nakagawa K. and Shiraki Y. (1982), Proc. of the 2nd Int. Symp. on MBE, Tokyo, August 1982, p.183
Ishizaka A. and Shiraki Y.(1986), J. Electrochem. Soc., **133(4)**, 666
Jang S.M. and Reif R. (1991), Appl. Phys. Lett., **59(29)**, 3162
Jang S.M. and Reif R. (1992), Appl. Phys. Lett., **60(6)**, 707
Jung K.H., Hsieh T.Y. and Kwong D.L. (1991), Appl. Phys. Lett., **58(21)**, 2348
Kamins T.I. (1991), Electronics Letters, **27**, 451
Kamins T.I. and Meyer D.J. (1991b), Appl. Phys. Lett., **59(2)**, 178
Kato M.,et al (1990), J. Cryst. Growth, **99**, 240
Kern W.A. and Puotinen P.A. (1970), RCA Rev., 31
King C.A., Gronet C.M., Gibbons J.F. and Wilson S.D. (1988), IEEE Electron Device Lett., **9(5)**, 229
King C.A., Hoyt J.L., Gronet C.M., Gibbons J.F., Scott M.P. and Turner J. (1989), IEEE Electron Device Lett., **10(2)**, 52
Kobayashi S., Ching M. L., Kohlhase A., Sato T., Murota J. and Mikoshiba N. (1990), J. Cryst. Growth, **99**, 259
Kubiak R. and Parry C. (1991), Mat. Res. Soc. Symp. Proc., **220**, 63
Lang D.V., People R., Bean J.C. and Sergent A.M. (1985), Appl. Phys. Lett., **47**, 1333
Leong W.Y., Robbins D.J., Young I.M. and Glasper J.L. (1988), UK IT 88 Conf. Publ., IED, Department of Trade and Industry, London , p.382
Leong W.Y., Canham L.T., Young I.M. and Robbins D.J. (1990), Thin Solid Films, **184**, 131
W. Y. Leong W.Y. (1992), unpublished work.
Lian S., Fowler B., Bullock D. and Banerjee S. (1991), Appl. Phys. Lett., **58**, 514
Libezny M., Poortmans J., Caymax M., Van Ammel A., Kubena J. and Vanhellemont J. (1993), Int. Conf. Spect. Ellipsometry, Paris (to be published in Thin Solid Films)
Liehr M., Greenlief C.M., Offenberg M. and Kasi S.R. (1990), J. Vac. Sci. Technol., **A8(3)**, 2960; Liehr M, Greenlief C M, Kasi S R and Offenberg M (1990) Appl. Phys. Lett. 56, 629

Liu W.K., Mokler S.M., Ohtani N., Zhang J. and Joyce B.A. (1992), Appl. Phys. Lett., **60**, 56

Matsuda A., Yoshida T., Yamasaki S. and Tanaka K. (1981), Jpn. J. Appl. Phys., **20**, L439

Matutinovic-Krstelj Z., Prinz E.J., Schwartz P.V. and Sturm J.C. (1991), IEEE Electron Device Lett., **12(4)**, 163

Menczigar U., Eberl K. and Abstreiter G. (1991), Mat. Res. Soc. Symp. Proc., **220**, 361

Meuris M., Verhaverbeke S., Mertens P., Heyns M.M., Hellemans L., Bruynseraede Y. and Philipossian A. (1992), Jpn. J. Appl. Phys., **31**, L1514

Meyerson B.S. and Olbricht W.L. (1984a), J. Electrochem. Soc., **131**, 2361

Meyerson B.S. and Yu M. (1984b), J. Electrochem. Soc., **131**, 2366

Meyerson B.S. (1986), Appl. Phys. Lett., **48**, 797

Meyerson B.S. and Jasinski J.M. (1987), J. Appl. Phys., **61(2)**, 785

Meyerson B.S., Uram K.J. and LeGoues F.K. (1988), Appl. Phys. Lett., **53(25)**, 2555

Meyerson B.S., Himpsel F.J. and Uram K.J. (1990a), Appl. Phys. Lett., **57**, 1034

Meyerson B.S., Himpsel F.J., Legoues F.K. and Wang P.J. (1990b), Proceedings of the Second International Conference on Electronic Materials, R.P.H. Chang, M. Geis, B.S. Meyerson, D.A.B. Miller and R. Ramesh (eds), (Mater. Res. Soc. Proc., Pittsburgh, PA), p.469

Meyerson B (1992) Proceedings of the I EEE 80, 1592

Mi Q., Xiao X., Sturm J.C., Lenchyshyn L.C. and Thewalt M.L.W. (1992), Appl. Phys. Lett., **60**, 3177

Moslehi M.M., Kuehne T., Yeakly R., Velo L., Najm H., Dostalik B., Yin D and Davis C.J. (1991), Mat. Res. Soc. Symp. Proc. 224, p. 143

Morita M., Ohmi T., Hasegawa E., Kawakami M. and Suma K. (1989), Appl. Phys. Lett., **55(6)**, 562

Murota J., Nakamura N., Kato M., Mikoshiba N. and Ohmi T. (1989), Appl. Phys. Lett., **54**, 1007

Nakanuma S. (1968), IEEE Trans. Electron Devices, **13(7)**, 578

Nishida S., Shimoto T., Yamada A., Karasawa S., Konagai M. and Takahashi K. (1986), Appl. Phys. Lett., **49**, 79

Noel J.P., Powell N.L., Houghton D.C. and Perovic D.D. (1990), Appl. Phys. Lett., **57**, 1037

Noel J.P., Rowell N.L., Houghton D.C. and Wang A. (1992), Appl. Phys. Lett., **61**, 690

Ohmi T (1992) Proceedings of the IES Conference, Nashville, USA, May 1992, 287

Patton G L, Harame D L, Stork J, Meyerson B S, Scilla G J and Ganin E (1989) IEEE Electron Device Lett. 10, 534

Patton G.L., Comfort J.H., Meyerson B.S., Crabbé E.F., Scilla G.J., De Frésart E., Stork J., Sum J., Harame D.L. and Burgharz J. (1991), IEEE Electron Device Lett., **11**, 171

Pearsall T.P., Bevk J., Feldman L.C., Bonar J.M. and Mannaerts J.P. (1987), Phys. Rev. Lett., **58**, 729

Pickering C. (1991), Thin Solid Films, **206**, 275

Pickering C., Carline R.T., Robbins D.J., Leong W.Y., Barnett S.J., Pitt A.D. and Cullis A.G. (1993), J. Appl. Phys., **73**, 239

Pidduck A.J., Robbins D.J., Cullis A.G., Gasson D.B., Pickering C. and Glasper J.L. (1989), J. Electrochem. Soc., **136**, 3083-3088

Pollock T.L., Sandhu H.S., Jodhan A. and Strausz O.P. (1973), J. Am. Chem. Soc., **95**, 1017

Pollak F.H. and Gelmbocki O.J. (1988), SPIE, **946**, 2

Poortmans J., Jain S.C., Caymax M., Van Ammel A., Nijs J., Mertens R.P. and Van Overstraeten R. (1992), Microelectronic Eng., **19**, 443

Poortmans J. (1993), "Low-temperature epitaxial growth of Si and strained $Si_{1-x}Ge_x$ layers and their application in bipolar transistors", PhD-Thesis (IMEC, Leuven)

Prinz E.J., Garone P.M., Schwartz P.V., Xiao X. and Sturm J.C. (1989), Intern. Electron. Devices Mtg Tech. Digest, p.639

Racanelli M. and Greve D. (1990), Appl. Phys. Lett., **56(25)**, 2524

Racanelli M. and Greve D. (1991), J. of Metals, **43**, 32

Racanelli M., Greve D., Hatalis M.K. and Van Yzendoorn L.J. (1991), J. Electrochem. Soc., **138**, 3783

Rahat I., Shappir J., Frase D., Wei J., Borland J.O. and Beinglass I. (1991), J. Electrochem. Soc., **138**, 2370

Regolini J.L., Bensahel D., Scheid E. and Mercier J. (1989), Appl. Phys. Lett., **54(7)**, 658

Regolini J.L., Bensahel D. and Mercier J. (1990), J. Electronic Mat., **19(10)**, 1075

Reif R. (1984), J. Vac. Sci. Technol., **A2**, 429

Rhodin T.N. and Adams D.L. (1976), "Treatise on Solid State Chemistry", Vol 6A, Surfaces I, Chapter 5, N. B. Hannay (ed), (Plenum Press, NY), p.376

Robbins D.J., Kubiak R.A.A. and Parker E.H.C. (1985), J. Vac. Sci. Technol., **B3**, 588

Robbins D.J. and Young I.M. (1987), Appl. Phys. Lett., **50(22)**, 1575

Robbins D.J., Pidduck A. J., Pickering C., Young I.M. and Glasper J.L. (1988), SPIE, **1012**, 25

Robbins D.J., Glasper J.L., Pidduck A.J. and Cullis A.G. (1990), Proceedings of the Second International Conference on Electronic Materials, R.P.H. Chang, M. Geis, B.S. Meyerson, D.A.B. Miller and R. Ramesh (eds), (Mater. Res. Soc., Pittsburgh, PA 1990), p.477-482

Robbins D. J., Calcott P. and Leong W.Y. (1991), Appl. Phys. Lett., **59**, 1350

Robbins D.J., Glasper J.L., Cullis A.G. and Leong W.Y. (1991), J. Appl. Phys., **69(6)**, 3729

Robbins D.J., Canham L.T., Barnett S.J., Pitt A.D. and Calcott P. (1992), J. Appl. Phys., **71**, 1407

Roksnoer P.J., Maes J.W.F.M., Vink A.T., Vriezema C.J. and Zalm P.C. (1991a), Appl. Phys. Lett., **58**, 711

Roksnoer P.J., Maes J.W.F.M., Vink A.T., Vriezema C.J. and Zalm P.C. (1991b), Appl. Phys. Lett., **59**, 3297

Rowell N.L., Noel J.P., Houghton D.C. and Buchanan M. (1991), Appl. Phys. Lett., **58**, 957

Rowell N.L., Noel J.P., Wang A., Namavar E., Perry C.H. and Soref R.A. (1992), J. Appl. Phys., **71**, 6201

Rudder R.A., Fountain G.G. and Markkunas R.J. (1986), J. Appl. Phys., **60**, 3519

Sauer R. and Weber J. (1983), Physica, **116B+C**, 195

Schulze G. and Henzler M. (1983), Surf. Sci., **124**, 336

Sedgwick T.O., Berkenblitz M. and Kuan T.J. (1989), Appl. Phys. Lett., **54**, 2689

Sedgwick T.O., Agnello P.D., Nguyen Ngoc D., Kuan T.J. and Scilla G. (1992), Appl. Phys. Lett., **58(17)**, 1896

Sinniah K., Sherman M.G., Lewis L.B., Weinberg W.H., Yates J.T. and Janda K.C. (1989), Phys. Rev. Lett., **62(5)**, 567

Smith F.W. and Ghidini G. (1982), J. Electrochem. Soc., **129(6)**, 1301

Spitzer J., Thonke K., Sauer R., Kibbel H., Herzog H.J. and Kasper E. (1992), Appl Phys. Lett., **60**, 1729

Steiner T.D., Hengehold R.L., Yeo Y.K., Godbey D.J., Thompson E. and Pomrenke G.S. (1992), J. Vac. Sci. Technol., **B10**, 924

Sturm J.C., Schwartz P.V. and Garone P.M. (1990), Appl. Phys. Lett., **56(10)**, 961

Sturm J.C., Garone P.M. and Schwartz P.V. (1991a), Appl. Phys. Lett., **59(1)**, 542

Sturm J.C., Schwartz P.V., Prinz E.J. and Manoharan H. (1991b), J. Vac. Sci. Technol., **B9(4)**, 2011

Sturm J.C., Manoharan H., Lenchyshyn L.C., Thewalt M.L.W., Rowell N.L., Noel J.P. and Houghton D.C. (1991c), Phys. Rev. Lett., **66**, 1362

Suzuki S. and Itoh T. (1983), J. Appl. Phys., **54**, 6385

Terashima K., Tajima M. and Tatsumi T. (1990), Appl. Phys. Lett., **57**, 1925

Townsend W.G. and Uddin M.E. (1973), Solid State Electron., **16**, 39

van Opdorp C. and Leys M.R. (1987), J. Cryst. Growth, **84**, 271

Venkataraman V. and Sturm J.C. (1991), MRS-Proceedings, **220**, 391

Verhaverbeke S., Meuris M., Mertens P.W., Schmidt H.F. and Heyns M.M. (1992), Book of Abstracts of the 1st Int. Symp. on Ultra Clean Processing of Silicon Surfaces (UCPSS '92), Leuven, Belgium, September 1992, p.6

Verhaverbeke S., Meuris M., Mertens P., Heyns M.M., Philipossian A., Gräf D. and Schnegg A. (1991), IEDM Tech. Digest, p.71

Wachter M., Schaffler F., Herzog H.J., Thonke K. and Sauer R. (1993), Appl. Phys. Lett., **63**, 376

Wang P.J., Fang F.F., Meyerson B.S., Mocera J. and Parker B. (1989), Appl. Phys. Lett., **54**, 2701

Xie Y-H, Wu Y.Y. and Wang K.L. (1986), Appl. Phys. Lett., **48**, 287

Yamada A., Jia Y., Konagai M. and Takahashi K. (1989), Jpn. J. Appl. Phys., **28**, L2284

Yan D., Pollak F.H., Boccio V.T., Lin C.L., Kirchne P.D., Woodall J. M., Gee R. C. and Asbeck P. (1992), Appl. Phys. Lett., **61**, 2066

Yew T.R. and Reif R. (1989), J. Appl. Phys., **65(6)**, 2500

Yin X. and Pollak F.H. (1990), Appl. Phys. Lett., **56**, 1278
Yin Y., Pollak F.H., Auvray P., Dutartre D., Pantel R. and Chroboczek J.A. (1992), paper presented at E-MRS, Strasbourg 1992
Zachai R. (1991), Mat. Res. Soc. Symp. Proc., **220**, 311
Zaslavsky A., Grutzmacher D.A., Lee Y.H., Ziegler W. and Sedgwick T.O. (1992), Appl. Phys. Lett., **61**, 2872
Zhong Y., Ozturk M.C., Grider D.T., Wortman J.J. and Littlejohn M.A. (1990), Appl. Phys. Lett., **57**, 2092

5

Materials Properties of (strained) SiGe layers

J. Poortmans, S.C. Jain, J. Nijs and R. Van Overstraeten

5.1   INTRODUCTION

Although $Si_{1-x}Ge_x$ alloys have been the subject of study since the late fifties (Braunstein et al., 1958) it is only since the last ten years that these layers have gained a considerable technological relevance. This renewed and enhanced interest came together with the advent of new growth methods like Molecular Beam epitaxy (MBE), Very Low Pressure Chemical Vapor Deposition combined with Ultra-High Vacuum techniques (UHV-VLPCVD) and Rapid Thermal Chemical Vapour Deposition (RT-CVD). The main common feature of these techniques is their capability of growing epitaxial layers at low temperatures (300-700°C). This feature is not only important for the steepness of dopant profiles, but also allows to combine *lattice-mismatched materials in a coherent way*.

More specifically, when alloying Si and Ge (these two materials are ideally miscible in each other over the complete concentration range), the freestanding alloy has a diamond-type crystal lattice. The lattice constant of this alloy follows a nearly linear dependence between the values for Si and Ge, which differ about 4.17% at room temperature. Dismukes et al. (1964) have shown that the deviation of linearity (Vegard's law) remains small.

$$f_m = - \left(\frac{a_{Si} - a_{Si_{1-x}Ge_x}}{a_{Si}} \right) = (4.17 \times 10^{-2} \, x) \, C_m, \qquad (5.1)$$

a_{Si} and $a_{Si_{1-x}Ge_x}$ are the lattice constant of the unstrained lattices and f_m is called the lattice mismatch parameter. The mixing factor C_m is between 0.9 and 1 over the whole range. When such a SiGe alloy layer is epitaxially grown on a substrate with a different lattice constant, for instance a Si-substrate, the growing layer will in first instance try to adapt its in-plane lattice constant as to form a *coherent interface* with the underlying substrate. The term *pseudomorphic growth* is often used for this type of growth. This is schematically illustrated in fig.5.1a for a simplified cubic lattice.It can be easily understood that under these conditions a large elastic strain will be built in the *commensurate layer* . In the case of a Si substrate the alloy layer will be under compressive strain since the lattice constant of the free-standing alloy is larger.

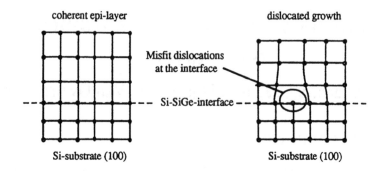

Fig. 5.1 Schematic illustration of (a) pseudomorphic growth ; (b) after relaxation of the coherent interface

The layer can however not continue growing in this perfectly commensurate way since there exists a trade-off between the *elastic strain energy* accumulated in the layer and the energy, connected with *misfit dislocations which relieve* partially the strain, which is shown in fig. 5.1b. This will lead to the so-called concept of *critical thickness* which is the thickness above which the layer will partially or completely relax to its freestanding lattice constant by the creation of misfit dislocations (plastic deformation) at the interface with the original substrate. An extensive body of experimental and theoretical work exists on this issue. The whole matter is reviewed thoroughly by S.C. Jain et al (1991a). The improvement of the theories concerning critical thickness, equilibrium strain relaxation in layers exceeding the critical thickness and strain relaxation rate form the subject of section 5.2.

The new epitaxial growth techniques mentioned in the beginning of this chapter allow the growth of these strained layers on Si-substrates to a thickness which is well *above* the theoretically calculated critical thickness of the Matthews-Blakeslee theory (J.C. Bean, 1985). It is this property, together with *strain-induced changes in the electronic properties* of the alloy, which has triggered the practical use of such layers in for instance the base of a bipolar transistor (Double Heterojunction Bipolar Transistor). It has to be kept in mind however that such layers when thicker than the critical thickness are in a *non-equilibrium* situation, which places strong restrictions on the thermal budget these *metastable* layers can support during or after their deposition.

The capability of growing such pseudomorphic strained layers (metastable or not) would however be quite idle if it were not accompanied by drastic changes in the electronic structure of the alloy. The exploitation of these strain-induced effects, which might look unwanted at first sight, was first worked out in a paper by Gordon Osbourn (1982). The discussion of these changes in band offset's and indirect bandgap forms the subject of chapter 5.3 and 5.4. In chapter 5.5 we will revise the influence of the strain on the transport properties of holes and electrons. This will provide a sound base for the understanding of the applications of monocrystalline $Si_{1-x}Ge_x$ layers. Since the Heterojunction Bipolar Transistor is currently the most important device application we will mainly emphasize these properties which are most relevant for this application.

5.2   STRUCTURE AND STABILITY

Equilibrium theories are needed to calculate the structure of stable epilayers. If the layers are metastable, strain relaxation by the introduction of dislocations occurs with aging or on heating the layers. The layers tend to approach the final equilibrium configuration. Equilibrium theories are also required to calculate the rate of strain relaxation in metastable layers.

5.2.1   Critical layer thickness

If a layer of a semiconductor is grown on another semiconductor, and if the lattice constants of the semiconductors are only slightly different from each other, initially the atomic planes of the layer are in perfect registry with those of the substrate, the growth is coherent and the layer is pseudomorphic. If the lattice constant of the layer is bigger than that of the substrate, the lattice mismatch is accommodated by tetragonal compression of the layer. As the thickness of the layer exceeds a certain thickness known as the critical layer thickness h_c, the energy of the system is lowered by the introduction of dislocations and relaxation of strain. For growth on a (100)-substrate and when the Ge fraction is not large, only 60° dislocations are observed experimentally. At large Ge fractions however, strain is relaxed by the introduction of 90° dislocations. The principle of energy minimization predicts that as the thickness increases, the number of dislocations also increases; for each thickness there is a definite number of dislocations which must be present in the layer for its energy to be minimum. Lowest energy is obtained when there are two perpendicular arrays of dislocations and the distribution of dislocations in each array is periodic with spacing p between neighbouring dislocations. The number of dislocations per unit area is then 2/p.

The total energy of the system consists of the energy due to homogeneous strain and the dislocation energy (Ball and Van der Merwe 1983, Jain et al. 1992a). The equilibrium configuration is obtained by minimization of the total energy with respect to $1/p$. For determining the critical thickness h_c, p is then allowed to go to infinity. The expression for h_c for an epilayer with uniform composition and with free surface at the top comes out to be (Ball and Van der Merwe 1983, Jain 1990, Jain and Hayes 1991a),

$$h_c = \frac{b^2(1-\nu\cos^2\alpha)}{8\pi f_m(1+\nu)b_1} \ln\left(\frac{h_c}{q}+1\right), \qquad (5.2)$$

where b is the Burgers vector, $b_1 = b \sin\alpha \sin\beta$ is the active component of b, q is the radius of the core of the dislocations, $\alpha = \tan^{-1}(1/\sqrt{2})$ and $\beta = 60°$ for $60°$ dislocations and $\alpha = \beta = 90°$ for $90°$ dislocations (Jain 1993) for growth on a (100) crystalline substrate. ν is the Poisson coefficient of the strained layer.

The above expression can be derived using the force balance given by Matthews and Blakeslee (1974). It must be pointed out that the method of energy minimization and force balance are equivalent. If a Newtonian body is in equilibrium, we say that its energy is minimum. We can also say that all forces acting on the body cancel out and the net force is zero. The principles of energy balance and force balance can not constitute two different theories. Nevertheless several authors keep treating them as two different theories. The expressions for dislocation energy used by different authors in deriving this equation are slightly different and the final expression for h_c are also different. Numerically the differences are significant only when f_m is large and h_c is small, comparable to q. In this case, the dislocation core energy becomes important and the linear theory of elasticity breaks down.

Since the separation between dislocations is much larger than h when h is not much bigger than h_c, interactions between the dislocations are ignored in the theories developed by Van der Merwe and Matthews and Blakeslee. Jain et al. (1992a) have shown that dislocation-dislocation interactions are important even when the spacing between the dislocations is two orders of magnitude larger than the thickness of the layer. They found that for 60 degree-dislocations, the actual critical layer thickness is smaller than the critical thickness obtained using the old theories but the difference between the two critical thicknesses is small.

Eqn. (5.2) for h_c is exactly valid for $90°$ dislocations. It is also perfectly valid for $60°$ dislocations if interactions between dislocations are ignored or if the distribution of the dislocations is irregular even if the interactions are included in the theory. If the dislocations are periodic and interacting, a two-dimensional array of dislocations is introduced in the layer at a thickness which is somewhat lower than the value of h_c obtained using above equation. Rigorously speaking, the critical thickness for the onset of dislocations is now smaller but the difference is only a few angstroms and is negligible. The above equation for h_c

for strained semiconductors can therefore be used in all cases.

The calculated and experimental values of h_c are shown in Fig. 5.2. Curve 1 is for 90° dislocations, curve 2 for 60° dislocations and curve 3 represents a close fit to the experimental data of critical layer thickness of Ge_xSi_{1-x} layers grown at 550°C by MBE by Bean et al. (1984a). The discrepancy between theory and experimental results is very large. Similar discrepancy is found in the case of InGaAs/GaAs layers (Fritz et al. 1987). The discrepancy arises because of two reasons. Firstly, if the layers are not grown at sufficiently high temperatures and at sufficiently low growth rates, the layers are not in equilibrium. Because of an energy barrier, the introduction of dislocations is a thermally activated process. At low temperatures and fast growth rates, a sufficient number of dislocations required by thermodynamics do not set in the layers and the layers are metastable (Jain et al 1990). The experimental critical thickness appears to increase. Secondly, many techniques used to measure strain relaxation are not sensitive and they do not detect a small amount of strain relaxation (Jain et al 1990). If the temperature of growth is sufficiently high and if the onset of dislocations is measured by a sensitive technique such as TEM, good agreement is found between the observed values of h_c and the values calculated using Eqn. (5.2). For the long term reliability of a device, the thickness of a strained layer must be smaller than h_c.

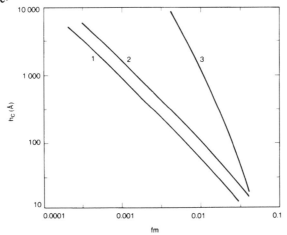

Fig. 5.2 Critical layer thickness h_c plotted as a function of the misfit parameter f_m. Curve 1 is for 90° dislocations, curve 2 for 60° dislocations and curve 3 represents the experimental data of Bean et al. (1984). Jain et al. (1993) (reprinted from Solid-state Electronics)

There are two other cases of practical interest, one is the case of an epilayer in which the Ge-profile is graded (Patton et al. 1990) and the other when the strained layer is covered by a Si capping layer on top of the strained layer. In

Double Heterostructure Bipolar Transistors (Patton et al 1990), optical detectors (Karunasiri et al. 1991) and in many other devices, the strained layer is sandwiched between the substrate and the capping layer. The critical thickness has been calculated in both cases. In the first case of graded layers, the critical thickness can be obtained by using average values of f_m in Eqn. (5.2) (Jain et al 1991b). Since the dislocation energy remains unchanged, the other quantities in this equation are unaffected. If there is a thick capping layer at the top, the missing half-plane constituting the dislocation terminates at the upper interface constituting a dislocation dipole. The energy of the dislocation increases on this account. Detailed calculations for this case have been made by Jain et al (1992b) and their results are plotted in Fig. 5.3. At low Ge fractions, the critical thickness of the capped layer is about twice as large as that of an uncapped layer, it becomes 4 times as large at higher Ge fractions.

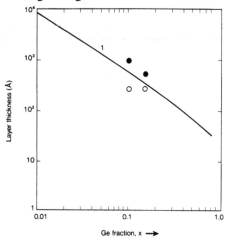

Fig. 5.3   Critical layer thickness of a capped layer plotted as a function of Ge-fraction x. Solid and open circles are the experimental data of Houghton et al. (1989).

During the growth process, the surface of the layer is always free. However in many cases the layers are metastable after growth. If a capping layer is now placed on top of the layer, its stability will be determined by the critical thickness of the capped layer. Even if its thickness is more than the critical thickness given by Eqn. (5.2), it will not degrade in the long run if the thickness is smaller than that of a capped layer as given in Fig. 5.3.

A pseudomorphic superlattice is defined by two different critical thicknesses. Each individual layer has to satisfy the condition that $h<h_c$, as discussed above. Additionally overall thickness of the superlattice should not be greater than the critical thickness of an alloy layer of the same composition as the average composition of the superlattice taking all layers into account. If a superlattice is

symmetrically strained (i.e. individual members of a period are strained equally but one is under compression and the other, in tension), the second condition is relaxed (see the review by Jain et al. 1990). Symmetrically strained superlattices with sharp interfaces have been grown and studied (Kasper *et al.* 1988).

5.2.2 Equilibrium strain relaxation in thicker layers

If the equilibrium theory (principle of energy minimization or method of force balance) is used and the interactions between dislocations are neglected, the following relation between critical points (h_e,p_e) giving the stable configuration can be easily derived (Matthews and Blakeslee, 1974),

$$h_e = \frac{b^2(1-\nu\cos^2\alpha)}{8\pi(f_m- b_1/p_e)(1+\nu)b_1} \ln\left(\frac{h_c}{q} +1\right), \qquad (5.3)$$

This equation has been extensively used to interpret experimental results. Jain et al. (1992a) have shown that in this equation, unlike Eqn. (5.2), neglecting the interactions between dislocations causes large errors in Eqn. (5.3). They calculated the total energy E_T of the epitaxial layers for $x=0.1$ and for several values of h including dislocation interactions. The results of these calculations are shown in Fig. 5.4 for $60°$ dislocations.

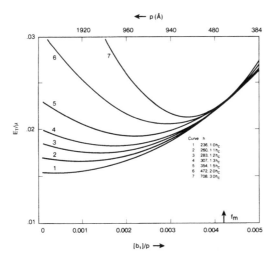

Fig. 5.4 Values of total energy E_T/μ of $Si_{0.9}Ge_{0.1}$ strained layers vs b_1/p are plotted for seven different values of h (From Jain et al. (1993) reprinted from Solid-State Electronics)

The minima in these curves give the values of (h_e, p_e). It is seen that for $h=h_c$, the minimum in the E_T vs b_1/p plots is at $b_1/p=\infty$ as expected. As the thickness increases, the minimum shifts to values of equilibrium concentration $1/p_e$ of the dislocations. A comparison of these results with those obtained neglecting interactions shows that for a given h, the value of b_1/p_e is larger when interactions are included. As h goes to ∞, b/p_e approaches f_m asymptotically. Note that at every minimum (h_e, p_e), the homogeneous strain is not completely eliminated, the complete relaxation of strain occurs only at very large values of h.

Until recently the actual mechanism of strain relaxation in thick layers was obscure. For any given thickness, the observed number of dislocations was always less than that predicted by Eqn. (5.3). The discrepancy becomes worse if interactions are included in the theory and results of Fig. 5.4 are used. This discrepancy is not limited to GeSi strained layers, it is observed in almost all solids including metal-alloys mismatched systems (Ball and Van der Merwe, 1983) and III-V semiconductor mismatched systems (Fritz 1987). A partial answer to this long standing problem was provided by the recent work of Jain et al (1993). The authors noted that distribution of dislocations in the arrays in thick strained layers is never periodic, it is highly irregular. The energy of these highly irregularly distributed dislocations in a unit area is considerably more than the energy of the same concentration of dislocations distributed regularly. The calculated number of dislocations and strain relaxation is considerably reduced for the real system, bringing the result closer to the observed values.

A part of the lack of strain relaxation must be attributed to the fact that the activation energy for the process to occur is rather large and the process is very sluggish at most growth temperatures (Jain 1990). It is therefore rare to find experimental results of strain relaxation which will agree with the predictions of equilibrium theory. The results of the equilibrium theory, nevertheless are important. If the thickness of the strain layer is below the calculated value of h_c, we know that the device will have a long term stability. In thicker layers, the difference between measured and calculated strain relaxation (see section 5.2.3) determines the driving force which allows the calculation of strain relaxation.

5.2.3 Strain relaxation in metastable layers

If the crystals are subjected to sufficiently large external stress, they are deformed by the generation of dislocations. The theory of plastic flow in semiconductors was developed by Alexander et al. (1968). Dodson et al modified the theory to make it applicable to strained layers. The rate of increase of dislocation density N by multiplication is given by (Dodson 1988),

$$\frac{dN}{dt} = NVK\tau_{eff}(t) \tag{5.4}$$

where τ_{eff} is the excess or effective stress due to misfit strain and is given by

$$\tau_{eff}(t) = 2\mu\frac{(1+v)}{(1-v)}[P_e - P(t)], \tag{5.5}$$

where $P(t)=b_1/p(t)$ ($P_e = b_1/p_e$) and V is the glide velocity,

$$V = B\tau_{eff}(t)\exp\left(\frac{-U}{kT}\right), \tag{5.6}$$

where U is the energy for the the thermally activated glide of dislocations. The effective stress τ_{eff} differs from the actual stress in that it allows for the fact that in thermodynamic equilibrium there is a residual strain f_m-b_1/p_e present in the epilayer. A part of the total stress, corresponding to this strain is required by the thermodynamics and does not take part in the driving force for plastic relaxation to take place. The excess stress is the actual driving force for the plastic relaxation to occur. If the dislocations interactions are not included in the theory, the equilibrium strain f_m-b_1/p_e can be obtained From Eqn. (5.3).

Dodson and Tsao solved the above equation numerically. For fixed thickness h, an analytical solution of these equations can be written in the following form (Jain et al 1992a), relating time and strain relaxation.

$$t = \frac{1}{C\mu^2(P_e-P_0)^2}\left\{\ln\left(\frac{P_0+P(t)}{P_e-P(t)}\right) + \left(\frac{P_0+P_e}{P_e-P(t)}\right) - \ln\left(\frac{P_0}{P_e}\right) - \left(\frac{P_e+P_0}{P_e}\right)\right\}, \tag{5.7}$$

with P_0 the relaxation at t=0.

$$C = 4BKe^{-U/kT}\left(\frac{1+v}{1-v}\right) \tag{5.8}$$

The inclusion of interactions changes the dislocation energy and P_e, and hence the effective stress, plastic flow and strain relaxation are also affected by the interactions. An analytical solution for plastic flow and strain relaxation can not be obtained if the dislocation interactions are included in the theory.

The plastic flow theory has been fitted with the experiments of Bean et al.

(1984) both ignoring interactions (Dodson 1987) and including interactions (Jain et al 1992a). Values of the constants C and P_e obtained in both cases are shown in Table 5.1. Including interactions changes considerably the value of the parameter C. No independent reliable values of this parameter are available to which these values could be compared.

	$C\mu^2(s^{-1})$	P_e
Dodson (1987)	46	3×10^{-5}
Present theory	14	9×10^{-4}

Table 5.1   Comparison of the values of parameters C and P_e for h=50 nm obtained by fitting the present plastic flow theory with interactions and the theory of Dodson et al. (1987) without interactions to the experimental data

5.3   THE Si-STRAINED SiGe BAND ALIGNMENTS

5.3.1   Experimental results

It is most instructive to begin this paragraph by briefly describing how insights have grown in this matter. The topic deals with the *relative position of the valence and conduction band edges* at the interface between two different semiconductors in contact. We will show how the seemingly contradictory experimental results obtained on structures with strained $Si_{1-x}Ge_x$-layers around the mid eighties allowed the building of a consistent theory for the band offsets at the Si-SiGe interface and how these are influenced by the actual strain conditions. These experiments were oriented towards *two-dimensional carrier gases confined at these interfaces*. It is quite obvious that the formation of such carrier gases is strongly related to the question of the band offset's.

Around 1984 People et al. (People, 1984) reported on their experiments on two-dimensional hole gases at the interface between unstrained Si and a pseudomorphic $Si_{0.8}Ge_{0.2}$-layer. The first indication that holes were confined at the interface was the different temperature dependence of the mobility between layers which were homogeneously doped (10^{18} cm^{-3}) and modulation doped structures where only the Si cladding layers were intentionally doped. In the second case the mobilities were consistently higher and no carrier freeze-out could be observed at low temperatures. The mobility was about 3300 cm^2/Vs at 4.2K. This gave strong support to the view that holes were transferred from the Si to the SiGe and that the holes were confined at the SiGe-side of the hetero-interface. By performing magnetoresistance measurements on these devices they prooved unambiguously the presence of a two-dimensional carrier gas. They found indeed the Shubnikov-deHaas oscillations expected in such a two-dimensional carrier system. However, when doping the Si cladding layers with phosphorus, no indication for the formation of a two-dimensional electron gas was found. From these experiments they concluded that the energy of the valence band edge in the strained layer was significantly higher than in Si, while the conduction band offset was small.

Around the same time Jorke and Herzog (Jorke, 1985) reported on enhanced electron mobilities in Si$_{.55}$Ge$_{.45}$/Si structures, which were grown on a relaxed Si$_{.75}$Ge$_{.25}$ buffer layer. The most surprising was however the dependence of the electron mobility on the position of the doping spike ; the mobility was maximal when the doping spike was positioned at the centers of the Si$_{.55}$Ge$_{.45}$-layers, but no enhancement was seen when the spike was positioned in the Si-layer. In addition, Abstreiter et al. (1985) derived from cyclotron resonance measurements an effective mass of about 0.2m$_0$ for the electrons in the two-dimensional electron gas. This is the same value found for twofold degenerate electrons in the inversion layer of a MOSFET. Both findings can only be explained consistently if one assumes a band line-up where the conduction band in Si is significantly lower than in the strained Si$_{1-x}$Ge$_x$ which will cause electron transfer from the SiGe into the Si.

These apparently contradictory results can only be understood when one considers the particular strain situation in the two cases. In the first case (People and Bean, 1984) it was an *asymetrically strained* structure. Since the Si$_{.8}$Ge$_{.2}$-layer is grown directly on the Si-substrate, it will be under a *biaxial compressive* strain, while the much thicker Si-substrate and the Si cladding layers remain practically unstrained since for pseudomorphic growth the in-plane lattice constant remains the same for all layers. For the second case the situation is quite different since there the active layers are not grown directly on the Si-substrate, but on the relaxed Si$_{.75}$Ge$_{.25}$ buffer layer, which forms a virtual substrate with a lattice constant larger than for Si. Therefore the Si$_{.55}$Ge$_{.45}$-layer will again be under compressive strain, but the Si-layer will now be under *biaxial tensile strain*. Such biaxial strain can always be decomposed in a *hydrostatic and a uniaxial component*, which is perpendicular to the growth plane. The hydrostatic component does not change the symmetry properties of the crystal. *The uniaxial component however does lift the essential degeneracies*, since the lattice does not remain perfectly cubic, but undergoes a tetragonal deformation. More specifically, in presence of such a uniaxial strain the sixfold degenerate conduction band splits into a fourfold and a twofold degenerate component. For the valence band on the other hand there is a splitting between heavy and light hole band, while the split-off band goes down in energy.

In the asymmetric case these phenomena only take place in the compressively strained $Si_{0.8}Ge_{0.2}$-layer. In this case four conduction band valleys are lowered in energy and we end up with a situation where the conduction band offset remains small and most of the difference in bandgap is taken up by the valence band offset, which is often indicated as a band-lineup of type I (see fig. 5.5a). In the symmetrically strained case also the Si-layers are strained. Due to the biaxial tensile strain in the Si layer there are now two conduction band valleys which are shifted downwards, while the other four move up in energy. Since the shift for the twofold degenerate valleys is *twice as large* as for the fourfold degenerate level, it can be easily understood that in such circumstances the lowest conduction band level is situated in the Si layers and the band-lineup is called type II (see fig. 5.5b). In this way the observed phenomena were explained in a consistent way.

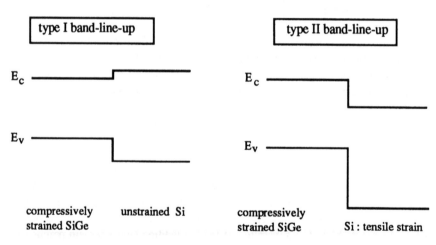

Fig. 5.5 Schematic definition of the different band line-ups and the associated strain configurations

5.3.2 Theoretical calculations and experimental validation

The picture developed until now has the advantage of giving a quite elegant explanation for the experimental data. At the same time several research groups tried to theoretically calculate the band alignment. This represents clearly a challenging theoretical problem. We will shortly revise them as to show the reader that the results of these calculations confirm the trends observed in the experiments.

The most prominent work in this direction has been performed by Van De Walle and Martin (1985, 1986). They did self-consistent calculations based on local density functional and ab-initio pseudopotentials to find the band line-ups. They performed these calculations for the case of three extreme conditions : growth of compressively strained Ge on a Si substrate, growth of tensilely strained Si on a Ge substrate and growth on a $Si_.6Ge_.4$-substrate where both the Si and Ge layers will be strained. The result of their calculations is shown in fig. 5.6.

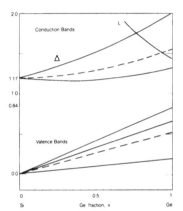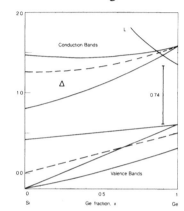

Fig. 5.6 The energetic evolution of the valence and conduction band edges in the case of a) the growth of strained $Si_{1-x}Ge_x$-alloys on Si (001) or on b) Ge (001). The dashed lines refer to the weighted averages of the valence bands or the (100) conduction band minima (Van de Walle and Martin, 1985 reprinted from Phys. Rev.B)

Although the estimated accuracy of the calculations was not better than about 100 meV, we can deduce from this figure the same tendencies as remarked in the experiments. *The valence band offset is clearly much larger than the conduction band offset* which is in agreement with the findings of People and Bean (1984) for the asymmetrical strain case. Whether band alignment will be type I or II in this case cannot be derived from these calculations due to the limited accuracy. They also extended their calculations to the symmetrical strain case where they found unambiguously a type II conduction band offset of about 130 meV, which explains the two-dimensional electron gas observations by Jorke and Herzog (1985).

The matter has been revised in a recent paper by L. Colombo et al. (1991) who find that the valence band offset is mainly determined by the strain far from the interface and is independent of the details of the interface (abruptness, interface strain). The last finding is important since it allows to compare the experimental results of different research groups without having to worry too much about interface-related features as micro-roughness (as long as these do not disturb the strain configuration). Their calculations also indicate a somewhat lower valence band offset for the growth of strained Ge on a Si--substrate (0.75 eV instead of 0.84 eV). This is confirmed in a paper by J.M. Bass and C.C. Matthai (1991) where the same value was found. The same authors also report that, in order to find the exact values for the valence band offset at a $Si/Si_{1-x}Ge_x$-interface, the linear interpolation of the value found for the Si/Ge interface (People, 1986) produces a result which is somewhat larger than their theoretical result.

These conclusions have since been confirmed by a number of independent analysis techniques and devices. In 1988 there was a paper by Ni et al. (Ni, 1988) who studied the band offset's by X-ray photo-emission. Their study included different strain configurations. Their method is based on the assumption that the Fermi-level will nearly coincide with the valence respectively conduction band edge in highly p-type respectively n-type doped strained $Si_{1-x}Ge_x$-layers covered by one monolayer of In or Sb to avoid problems by band bending at the surface (Fermi-level pinning by the surface states). Bandgap narrowing effects in these highly doped layers and uncertainties about complete activation of the dopant species limit the accuracy of this method to about 100 meV (60 meV was the estimated accuracy in their paper). Nevertheless their measurements nearly confirm the calculations for the valence band offset by Van de Walle and Martin (1986) at a Ge-content of 25% for the asymmetrical strain configuration. They found however a type II alignment with $\Delta E_c = 20$ meV. Their values of valence and conduction band offset for the symmetrical strain case in which a $Si_{.75}Ge_{.25}$ relaxed buffer layer was used as virtual substrate are lower by about 40 meV than the theoretical predictions. Yu et al. (1990) obtained by X-ray photospectroscopy a valence band offset of about 0.83 eV at the Si/Ge interface where the Ge was pseudomorphically grown on the Si. This result is very close to the expected theoretical value of 0.84 eV. A very recent result, obtained by admittance spectroscopy of the thermionic emission carrier transport over the barrier formed created by the valence band offset (Nauka, 1991), provided further direct confirmation of the valence band offset being much larger than the conduction band offset in the asymmetrical strain case. For x=0.35 they found a value of about 270 meV. They also remarked that the valence band offset is strongly reduced when the layer is partially relaxed. Northop et al. (1992) also got strong evidence for a type I line-up from the hydrostatic pressure dependence of the photoluminescence of strained SiGe-layers.

Another indirect proof of the type I alignment in this case comes from frequency (Slotboom, 1991) and temperature measurements (Prinz, 1991, Pruymboom, 1991) on Heterojunction Bipolar Transistors. In both cases a type I band alignment will lead to strong degradation of the device performance when the Si/SiGe conduction band offset is located in the neutral base. This was indeed experimentally verified. This subject will be dealt with in detail in the part about the Heterojunction Bipolar Transistor (chapter 6).

When comparing all these experimental and theoretical data it can be concluded that the type II band line-up is well established for the symmetrical case. For the asymmetrical case there is no doubt that the valence band offset largely exceeds the conduction band offset. There is also strong indication that the conduction band in the strained $Si_{1-x}Ge_x$-layer is somewhat lower than in the unstrained Si for x<0.3 and hence the band line-up is of type I.

5.3.3 Heavy-doping effects on the band offset's

The band offsets as discussed in the previous paragraph are valid as long as the strained $Si_{1-x}Ge_x$-layers are relatively lowly doped. There were however experimental indications (Tsaur, 1991) that the situation changes quite drastically when combining a strained p^+- $Si_{1-x}Ge_x$-layer with a p-type Si-substrate as in a long-wavelength infra-red detector (LWIR) based on *internal photo-emission over a barrier*. Under such conditions the band offset is modified by the shifts of the valence and conduction band edges caused by the heavy-doping effects.

To calculate these band shifts Jain et al. (1992c) used the following simple closed form expressions for the heavy-doping induced band edge shifts (Jain, 1991c):

Exchange shift : $\qquad \dfrac{\Delta E_{x(majority)}}{R} = 1.83 \dfrac{L}{N_b^{1/3}} \dfrac{1}{r_s}$,   (5.9a)

Correlation shift : $\qquad \dfrac{\Delta E_{cor(minority)}}{R} = \dfrac{0.95}{r_s^{3/4}}$,   (5.9b)

Impurity shifts : $\qquad \dfrac{\Delta E_{i(majority)}}{R} = \dfrac{1.57}{N_b r_s^{3/2}}$,   (5.9c)

$\qquad\qquad\qquad\qquad \dfrac{\Delta E_{i(minority)}}{R} = \dfrac{R_{(minority)}}{R} \dfrac{1.57}{N_b r_s^{3/2}}$,   (5.9d)

Majority (minority) stands for the majority (minority) band. R is the Rydberg energy for the majority carrier band $=13.6m_d/\varepsilon^2(eV)$, $r_s=r_a/a$, $a=0.53\varepsilon/m_d$ (Å), ε is the dielectric constant $=11.4+4x$, N is the doping concentration, $r_a^3=3/(4\pi N)$ and m_d= the effective density of states mass of holes for the majority band and of electrons for the minority band. As suggested by Jain and Roulston, $N_b=2$, L=0.75 and the electron effective mass = 0.33 m_o (m_o is the free electron mass) for the Ge_xSi_{1-x} strained layers. N_b is a constant taking into account the valence band degeneracy and L is used to describe the coupling between the light and heavy hole band.

The values of the valence band offsets for the strained layers doped with 1×10^{20} cm^{-3} boron atoms and for Ge fraction x in the range 0 to 0.5 calculated using the above equations are shown in fig. 5.7. The values for these highly doped layers are very different from those for the intrinsic layers given by People (1985), also shown in the figure.

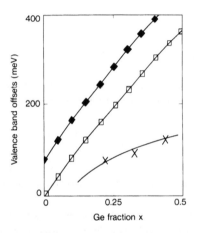

Fig. 5.7. The values of the valence offsets for strained layers doped with 10^{20} cm^{-3} boron-doping (after Jain et al., 1992c). The squares represent the valence band offset between lowly-doped Si and strained $Si_{1-x}Ge_x$ (taken from People, 1985), the diamonds represent the offset in case of a p^+-$Si_{1-x}Ge_x$/p-Si junction. The lowest curve is the effective barrier that the holes have to cross at such a junction (see also fig. 5.8). Also shown are experimental results (crosses) on LWIR-detectors (Tsaur, 91). (From Jain, 1992c, reprinted from Microelectronic Engineering)

ΔE_{v2}, as defined in fig. 5.8, is equal to the barrier height for the holes to go from the strained layer to Si and determines the cut-off wave-lengths of the LWIR detectors.

These barrier heights have been determined experimentally by Tsaur et al (Tsaur, 1991) by measuring the quantum efficiency of LWIR optical detectors as a function of temperature for three values of Ge fraction x. Their experimental results are also shown in fig. 5.7. Agreement of the experimental results with our theory is good. It can be seen from fig. 5.7 that the values of the band offsets if heavy doping effects are ignored are more than two times larger than the experimental values of Tsaur et al. (1991).

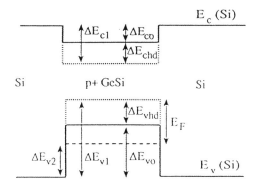

Fig. 5.8 Definition of the different offset's at a p^+-$Si_{1-x}Ge_x$/p-Si junction. ΔE_{vo} is the band offset between lowly-doped Si and strained $Si_{1-x}Ge_x$ (see for instance Van De Walle, 1986). ΔE_{v1} is the offset corrected for high-doping effects. ΔE_{v1} - E_f is the actual barrier that the holes have to cross when going from the p^+-$Si_{1-x}Ge_x$ to the lowly-doped p-Si (From Jain, 1992c reprinted from Microelectronic Engineering).

5.4   THE INDIRECT BANDGAP OF STRAINED $Si_{1-x}Ge_x$

Until now the discussion was focused on the question of the band line-up's. For certain applications it is also important to know the indirect bandgap as for Infra-red detectors, Heterojunction Bipolar Transistors. In principle one can derive the bandgap by adding the band offset's, as they were found by calculations and experiments (see previous section), but there is more direct experimental and theoretical evidence available, which will be discussed in this section.

Already in 1985, there was a paper by People (1985) in which an attempt was made to calculate the indirect bandgap of strained $Si_{1-x}Ge_x$ layer. He used the phenomenological deformation potential theory of Kleiner and Rothe to describe the motion of the band edges under the influence of strain. The indirect bandgap can then be seen to consist of four different contributions. First of all, there is the *bandgap of the unstrained alloy*. These values were already measured by Braunstein et al. (Braunstein, 1958). The second contribution stems from the *hydrostatic component* in the biaxial strain. This hydrostatic component changes the volume of the unit cell, but the symmetry properties are conserved in that the cubic lattice remains cubic. In his calculation People assumed a linear dependence of the hydrostatic deformation potential between the values for Si and Ge. This assumption can be considered to be quite crude, since the conduction band remains Si-like up to Ge-contents of about 85 %. The third and fourth contribution are due to the splittings in both conduction band and valence band states by the *uniaxial component of the strain*. This can be understood quite easily by remembering that this uniaxial strain leads to a tetragonal deformation of the lattice (see for instance fig. 5.1.a). Since the symmetry properties of the lattice are changed it is not difficult to see that the degeneracies at both conduction and valence band will be lifted. In his calculation of the splitting and shifts of the six different conduction band valleys People used the value for the uniaxial deformation potential of Si up to Ge-concentrations of 85%. For the valence uniaxial deformation potentials on the other hand he used the linear interpolation between the Si and Ge values. In the valence band the heavy and light hole band are splitted, while the split-off band sinks away deeper in the valence band. Taking these contributions together, the indirect bandgap can be seen to vary as a function of the Ge-concentration as in fig. 5.9.

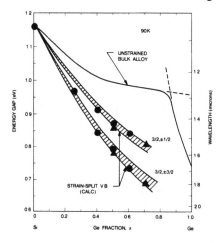

Fig. 5.9 Theoretical (R. People, 1985) and experimental results (D.V. Lang, 1985) showing the strong reduction of the indirect bandgap in the strained $Si_{1-x}Ge_x$-alloys (reprinted from Applied Physics Letters)

It has to be remembered that these calculations only dealt with the growth of strained $Si_{1-x}Ge_x$ layers on an unstrained Si-substrate. Under different strain configurations the situation will be different due to the different values of hydrostatic and uniaxial shifts.

The most striking feature of fig. 5.9 is the much faster decreasing of the indirect bandgap of the strained alloy when compared to the indirect bandgap of the unstrained alloy, as determined by Braunstein et al. (1958). At 295 K the resulting indirect bandgap can be written as a function of the Ge-concentration, where x is the relative Ge-content.

$$E_g(x) = 1.1 - 1.02x + 0.52x^2 \quad [eV] \qquad (5.10)$$

Although the assumptions made in People's calculation seem to be quite simplifying, these results were confirmed in the same year. D.V. Lang et al. (1985) performed photocurrent spectroscopy measurements at 90 K to derive the absorption edge in the strained $Si_{1-x}Ge_x$ layers. Their experimental results confirmed not only the theoretical calculation of People, but they also got experimental evidence about the splitting in the valence band. Since then, numerous measurements have confirmed this strain-induced narrowing of the indirect bandgap. For instance Dutartre et al. (1991) derived the *fundamental* bandgap of strained layers at 6 and 90 K by adding the free exciton dissociation energy to the observed no-phonon exciton peaks. For the exciton dissociation energies they assumed a linear interpolation between the values for Si and Ge. The results of their photoluminescence experiments for the indirect bandgap can be presented under the form of a parabolic equation valid for 0<x<0.22 at T=6 K.

$$E_g(x) = 1.17 - 1.01x + 0.835x^2 \quad [eV] \qquad (5.11)$$

In their experiments the strained layers were capped with a Si layer to prevent signal loss by excessive surface recombination. Another interesting feature of their photoluminescence spectra is the presence of a strong no-phonon recombination peak which under normal circumstances is very weak. Due to the alloying, which weakens the normal selection rules, this peak is strongly enhanced. This is also reported by other research groups (Sturm, 1991, Robbins, 1991).

5.5 TRANSPORT PROPERTIES AND EFFECTIVE DENSITY OF STATES IN STRAINED $Si_{1-x}Ge_x$ LAYERS

In this chapter the transport properties will be revised. We will focus our attention to the usual *three-dimensional carrier systems*. For the sake of completeness we have to mention that the last two years, impressive results have been published on the mobilities in two-dimensional carrier gases confined at the band offset's as described in section 5.3.

Thanks to improvements in the growth of relaxed buffer layers by grading the Ge-concentration from 0% to the wanted concentration (30% in the mentioned case) Schaffler et al. (1992) were able to reduce the threading dislocation density in the upper active layers to very low values. In such remotely-doped structures they reported an electron mobility of about 175000 cm^2/Vs at T=1.5 K which is about twenty times higher than in comparable Si-MOSFET's. Even at room temperature the mobility is still twice as high.

Also for two-dimensional hole gases in optimized MODFET-structures high mobilities are reported. An ultra-high field-effect hole mobility of 9000 cm^2/Vs at T=77K in a 2-D hole gas at a Ge-$Si_{0.5}Ge_{0.5}$-interface was reported by Murakami (1991). The valence band discontinuity at the interface was optimized by changing the composition of the relaxed buffer layer on which the structure was grown.

In the next section we will deal with the transport properties and the three-dimensional effective density of states at the same time. It will become clear how both properties are *intimately related through the strain-induced changes in the band curvature.*

5.5.1 Changes in the valence band and their influence on hole mobility

The strain not only changes the relative energetic position of the valence and conduction band edges and introduces splittings in normally degenerated levels, but also modifies the elctronic structure of the valence band. The calculations of the band curvature and structure are excellently reviewed in an article by F. Pollak (1990) where he gives a concise treatment on the effects of homogeneous strain. One can deduce from this that the warping of the heavy and light hole band spheres is strongly reduced. In fact, the heavy and light hole band not only split but they also become *ellipsoids of revolution* around the uniaxial strain axis. Hence the band curvature becomes direction-dependent. It must be remembered that the usual meaning of heavy and light hole band is lost in the presence of strain. Therefore as a kind of definition, the higher energy band is called the heavy hole band and the lower energy state the light hole band (Pikus, 1960).

In the case of (001) biaxial strain which is the usual strain configuration one can define an *in-plane mass* which is valid in the growth plane and an *out-of plane mass* perpendicular to this plane. In his article, F. Pollak also indicates that in diamond-type semiconductors the in-plane mass will be lower than the out-of plane component. Since these effective masses are direction dependent it can be easily understood that this will be reflected in the anisotropy of the hole mobility.

First experimental evidence of the influence of strain on the heavy hole band curvature was provided in 1984 by People et al. (People et al., 1984) in their paper on the properties of the two-dimensional hole gas at the interface of a $Si/Si_{0.8}Ge_{0.2}$ p-type modulation doped structure (see also section 5.3). From the temperature dependence of the magnetoresistance results they were able to deduce an effective mass of about $0.3m_0$ which was lower than expected from a linear dependence between the heavy hole masses of Si and Ge.

In a recent paper Manku and Nathan (1991a) performed detailed calculations on the valence band of strained $Si_{1-x}Ge_x$ layers as a function of the energy, taking into account spin-orbit splitting and the effect of the biaxial strain. From this they were also able to deduce the effective density of states mass as a function of energy (Manku, 1991b). They restricted their calculations to energy levels between E_V and E_V-0.065 eV since below this level the calculations become prohibitively complicated. One can see clearly in fig. 5.10a and b that the valence band is strongly non-parabolic, both in the heavy-hole and light-hole band ; in the parabolic case $m^*_{DOS}(E)$ would be independent of energy. This is true for both Si and strained $Si_{1-x}Ge_x$ layers. The energy dependence of the valence band density of states for Si in their paper corresponds well to earlier calculations (Madarasz, 1982).

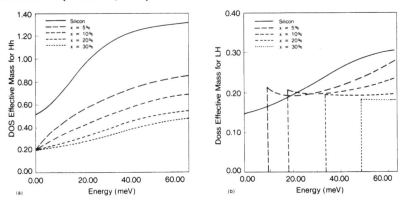

Fig. 5.10 The effective valence band density of states mass (normalized to the free electron mass) in Si and strained $Si_{1-x}Ge_x$-alloys for the a) heavy-hole band and b) the light-hole band (after Manku, 1991b, reprinted from Applied Physics Letters)

The most striking feature however is the strong reduction of the density of states effective mass due to the strain. At a Ge-content of 30% the density of states effective mass is lowered by a factor of three. This reduction will not only influence the position of the Fermi-level as a function of the acceptor concentration, a topic which will be further discussed in the chapter about heterojunction bipolar transistor (see chapter 6), but also the scattering events in iso-energetic scattering transitions.

From this detailed image of the valence band structure Manku and Nathan (1991c) tried to calculate the *lattice mobility of holes* (for an intrinsic material) in these strained layers. They considered elastic acoustic phonon scattering, non-polar optical phonon scattering and alloy scattering. Manku et al. used a value of the interaction potential which was chosen to fit the trends observed for Hall mobilities in unstrained SiGe alloys (A. Levitas, 1955). The final result of their calculations is shown in fig. 5.11. They predict an enhancement of the in-plane hole mobility when compared with the hole mobility in Si. The in-plane hole mobility is also larger than the out-of plane mobility for all Ge-concentrations. The in-plane hole mobility at 20% of Ge is a factor 1.3 higher than in Si. Both components are also substantially higher than the lattice hole mobility in unstrained alloys.

In their paper they also compare these results with a full Monte-Carlo simulation done by Hinkley et al. (Hinckley et al., 1990). The latter found an even stronger enhancement of the mobility by a factor of 1.4, but this is probably related to the lower value of the interaction potential that was used (0.2 eV instead of 0.27 eV). In a recent paper Chun et al. (1992) calculated the directional density-of-states effective masses of the valence bands in strained $Si_{1-x}Ge_x$-layers on a Si or relaxed $Si_{1-y}Ge_y$ substrate. From this they calculated the mobility as function of dopant concentration, temperature and strain configuration. At low doping concentrations they found a clear enhancement of both the transversal and longitudinal hole mobility. The enhancement was of the same order as in the paper of Hinckley (1990). The lower density of states in the strained layers is however reflected in a less effective screening of the ionized impurities. Therefore the enhancement of the hole mobility becomes less pronounced at high doping levels. At high doping levels they found that the hole mobility (both the transversal and longitudinal component) in strained and relaxed layers becomes comparable due to the dominance of ionized impurity scattering. They also mention that the enhancement of the mobility in the strained alloys will be stronger at low temperatures, because most carriers are then located near the top of the valence band, whose degeneracy has been lifted. This reduces strongly the interband scattering.

There is not much experimental information available on the hole mobility in strained $Si_{1-x}Ge_x$-layers at low doping levels, but Grivickas (Grivickas et al., 1991) tried to examine the carrier dynamics by a transient grating method. They found an ambipolar diffusion coefficient which was nearly 50% higher compared with the value found for Si in the case of a $Si_{0.5}Ge_{0.5}$ strained alloy. The most probable explanation is an enhancement of the in-plane hole mobility, because the in-plane electron mobility is expected to be lower (see for instance Hinkley, 1989). Further evidence of this higher in-plane hole mobility is also provided by base sheet resistance measurements on heterojunction bipolar transistors (Poortmans, 1992) and measured drift mobilities (McGregor, 1992). The latter found an enhancement of the drift mobility in a strained $Si_{.8}Ge_{.2}$ by a factor

1.55 at a B-doping level of about 2×10^{19} cm^{-3}.

These theoretical and experimental results for the in-plane hole mobility in strained $Si_{1-x}Ge_x$-alloys, which is most relevant for their use in a Heterojunction Bipolar Transistor, are shown in fig. 5.11.

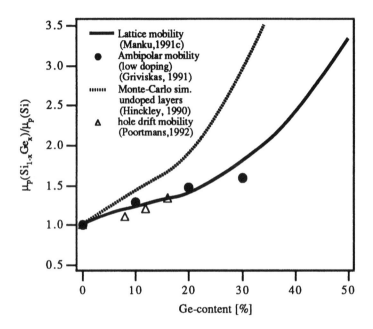

Fig. 5.11 Plot of theoretical predictions and simulations together with experimental results for the ratio of the in-plane hole drift mobility in strained SiGe-layers to the isotropic hole mobility in Si

5.5.2 Changes in the conduction band and its influence on the electron mobility

The six conduction band minima in Si are located out off the center in the Brillouin zone. They are located along the <100> axes in k-space and their minimum is at $0.85(2\pi/a_o)$, where a_o is the lattice constant of the cubic unit cell. The constant energy surfaces are ellipsoids, characterized by a longitudinal effective mass m_l $(=0.98m_o)$ and a transverse effective mass $m_t (=0.19m_o)$.

It was already indicated by Braunstein in 1958 that the conduction band in unstrained SiGe alloys remains Si-like in unstrained alloys up to Ge-concentrations of 85%. In the case of asymmetric strains configurations (growth of a strained layer on an unstrained Si-substrate) this is probably true uptill 100% of Ge (Van de Walle, 1986). We already indicated that under the influence of strain, an energetic splitting will be induced between these equivalent ellipsoids. But in contrast with the quite drastic changes of the band curvature in the valence band the *shape of the constant energy ellipsoids remains unchanged* to first order in strain (People, 1985, Hinckley, 1989, Manku, 1992).

In a situation where the splitting between the conduction band minima exceeds largely the thermal energy, which can be written as

$$3kT < 0.6x \quad [eV] \tag{5.12}$$

the conduction band effective density of states in the strained layer is proportional to the density of states in Si.

$$N_c(x,T) = \frac{2}{3} N_c(0,T) \tag{5.13}$$

In the more general case (at low Ge-content or high temperature) where the above-mentioned condition is not satisfied the relation can be written as :

$$N_c(x,T) = \frac{N_c(0,T)}{3} \left(2 + \exp(\frac{-0.6x}{kT})\right) \tag{5.14}$$

Already in 1987 Smith (Smith, 1987) reported on the opportunities for Heterojunction Bipolar Transistors created by a possible improvement of the out-of plane electron transport. He predicted a strong enhancement of the mobility in the perpendicular direction. This can be readily understood when visualizing the movement of the electrons in k-space (see fig. 5.12).

Perpendicular transport is equivalent to motion of the carriers along the <001> axis in k-space. The energetic splittings in the conduction band lead of course to a redistribution of the electrons over the different valleys, with the lower valleys being highly populated. One can see clearly in fig. 5.12 that most of the electrons when they are in the lower valleys will move along the short axis of the ellipsoids in the case of perpendicular transport. Hence the inverse effective conductivity mass which has to be used to describe this motion is not longer equal to $1/3 \ (2/m_l + 1/m_t)$ but is rather given by $1/m_t$. Following the same line of thinking it is not difficult to see that the in-plane electron transport will be deteriorated compared with the value in the unstrained alloy. One can also remark that, just as in the case of the hole mobility the electron mobility becomes anisotropic.

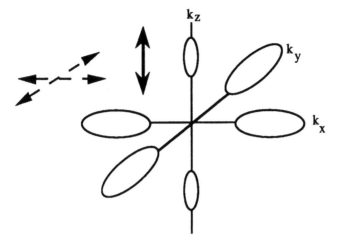

Fig. 5.12 Schematic drawing of the movement through k-space of the electrons in the conduction band ellipsoids. The dashed arrow shows the in-plane direction. The full arrow indicates the transversal movement. The smaller ellipsoids are the two conduction band valleys which have been shifted upwards.

These qualitative findings were confirmed in a Monte-Carlo simulation done by Hinckley et al (1989) and Pecjinovic et al. (1988). They retraced indeed the deterioration of the in-plane transport, while the perpendicular mobility was enhanced. The same authors also remarked that the splitting reduces the *intervalley scattering*.

In recent calculations by Manku and Nathan (1992), an analytical approach was chosen to calculate the drift mobility of electrons in strained and unstrained SiGe alloys as a function of the dopant concentration. They only found an improvement of the out-of-plane mobility for doping levels exceeding 10^{17} cm^{-3}. For lower doping levels both in-plane and out-of plane components are lower than the mobility in Si. They attribute this to the alloy scattering which will dominate at the lower doping levels where the effect of ionized impurity scattering is relatively small. But their results however strongly depend on the value of the interaction potential, which remains an important factor of uncertainty. Their approach also does not include explicitly the reduced intervalley scattering. Nevertheless their results agree very well with the Monte-Carlo simulations of minority and majority carrier electron mobilities done by L.E. Kay et al. (1991). They found that at low doping levels, even in the presence of strain, the mobility is decreased due to the predominance of alloy scattering. It is only at a dopant concentration of 10^{19} cm^{-3} that they found an improvement for the room temperature minority carrier mobility. In addition the improvement is rather small compared to the value for Si (1.2 times higher at a Ge-content of 10%). At 77 K the improvement can already be expected at a doping level of 10^{17} cm^{-3} but remains limited to a factor 1.2. At very high doping concentrations of 10^{20} cm^{-3} the ionized impurity scattering completely dominates the mobility which becomes then nearly independent of the Ge-concentration. They also remark that the enhancement of the mobility is preserved at high values of the electrical field (100 kV/cm). This conclusion is particularly relevant for graded base transistors where large electrical fields are found in the base. They also mention that the ratio of the minority to majority carrier mobility remains nearly the same as in Si. For Si the reader is referred to two recent papers where for the first time a first-principle-based model was developed for the doping and temperature dependence of minority and majority carrier mobility in Si (Klaassen, 1992 a and b). Finally L.E. Kay et al. (1991) mention the existence of a certain Ge-content at higher doping levels for which the mobility enhancement is maximum. This maximum is temperature-dependent. This maximum is related to the competition between alloy scattering and decreasing upper valley population. The maximum occurs at the Ge-content where the splitting in the conduction band becomes so high that the upper valleys are empty. Since the equilibrium energy of an electron is a linear function of temperature, it is clear that at higher temperatures the Ge-content to obtain the maximal mobility has to be larger.

There is not much experimental evidence about the minority and majority carrier mobility of electrons in strained $Si_{1-x}Ge_x$-layers. A result which diverges from the views developed in the text is the enhancement of the in-plane electron majority carrier mobility in $Si/Si_{0.85}Ge_{0.15}$ multilayer structures, reported by Manasevit (Manasevit, 1982) for doping levels between 5×10^{15} cm^{-3} and 10^{17} cm^{-3}, but this could be due to the fact that the total thickness of his structure was very large (5 μm), putting the intermediate Si layers under tensile strain. This explanation is supported by his finding that in thin $Si_{0.85}Ge_{0.15}$ layers no enhancement was seen.

Patton et al. (1988) concluded from their measurements on minority electron transport through the base of a heterojunction bipolar transistor that the minority carrier electron diffusivity was decreased by about 20 to 30% at a Ge-content of 12%. King et al. (1989) reported a mobility which was about 86% of the value in Si in a transistor with a $Si_{0.78}Ge_{0.22}$-base and a doping level of 2.5×10^{18} cm^{-3}. Pruymboom et al. (1991) concluded that the diffusion constant in a transistor with $Si_{0.8}Ge_{0.2}$-base was nearly the same as in Si. Poortmans et al. (1992) however found an enhancement of about 1.35 when the Ge-content in the base is 16% and the doping level around 10^{18} cm^{-3}. The discrepancy between the different experimental results arises mainly from the differents assumptions concerning the effective density of states in the valence band. The main results for the out-of plane electron mobility (which is the most relevant for the electron transport through the base of a Heterojunction Bipolar Transistor) are summarized in fig. 5.13. We will come back on this in chapter 6.

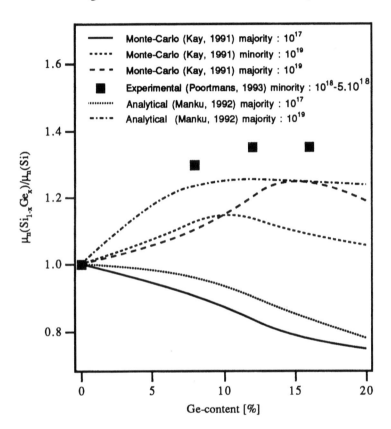

Fig. 5.13 Plot of the ratio of the transversal electron (minority and majority) carrier mobility in strained SiGe-layers to the electron mobility in unstrained Si. Results are from Monte-Carlo simulations, analytical models and experimentally extracted values

5.6 REFERENCES

Abstreiter G, Brugger, H, Wolff, T, Jorke, H and Herzog, H.J. 1985, *Phys. Rev. Lett.*, **54** 2441

Ball C.A.B and Van der Merwe J.H *The growth of dislocation-free layers, in Dislocations in Solids*, edited by F.R.N. Nabarro (North Holland 1983), Chap. 27 122

Bass J M and Matthai C C 1991 *Semicond. Sci. Technol.* **6** 109

Bean J C, Feldman L C, Fiory, A T, Nakahara S and Robinson I K 1984 *J. Vac. Sci. Technol. A2* 436

Bean J C, 1985 in *Proc. of the 1st International Symposium on Silicon Molecular Beam Epitaxy*, edited by J.C. Bean (Electrochemical Soc., Pennington, NJ, 1985) 337

Braunstein R, Moore A R and Herman F 1985 *Phys. Rev.*, **109** 695

Chun S K and Wang K L , 1992 *IEEE Trans. El. Dev.* **39**(9), 2153

Colombo L, Resta R and Baroni S, 1991 *Physical Review B (Condensed Matter)* **44** 5572

Dismukes J P, Ekstrom L and Paff R J, 1964 *J. Phys. Chem.* **68**, 3021

Dodson B W and Tsao J Y, 1987 *Appl. Phys. Lett.* **51**, 1325. For correction see Dodson B W and Tsao J Y, 1988 *Appl. Phys. Lett.* **52** 852

Dutartre D, Bremond G, Souifi A and Benyattou T, 1991 *Physical Review B (Condensed Matter)* **44** 11525

Fritz I J, 1987 *Appl. Phys. Lett.* **51** 1080

Grivickas V, Netiksis V, Noreika D, Petrauskas M, Willander M, Ni W.-X, Hasan M.-A, Hansson G V and Sundgren J.-E, 1991 *Journal of Appl. Phys.* **70** 1471

Hinckley J M, Sankaran V and Singh J, 1989 *Appl. Phys. Lett.* **55** (19) 2008

Hinckley J M and Singh J, 1990 *Phys. Rev. B* **41** (5) 2912

Jain S C, Bullough R and Willis J, 1990 *Advances in Phys.* **39** 127

Jain S C, and Hayes W, 1991a *Semiconductor Science and Technology* **6** 547

Jain S C, Balk P, Goorsky M S and Iyer, S.S., 1991b *Microelectronic Engineering* **15** 131

Jain S C and Roulston D J 1991c *Solid-State Electronics* **34** 453

Jain S C, Gosling T J, Willis J R, Totterdell D H J and Bullough R, 1992a *Phil. Mag. A* **65** 1151

Jain S C, Gosling T J, Willis J R, Bullough R and Balk P, 1992b *Solid-State Electronics* **35** 1073

Jain S C, Poortmans J, Nijs J, Mertens R and Van Overstraeten R, 1992c *Microelectronic Engineering* **19** 439-442

Jain U, Jain S C, Nijs J, Willis J R, Bullough R, Mertens R P and Van Overstraeten R, 1993 *Solid-st. Electron.* **36** 331

Jorke H and Herzog H J, 1985 Proc. *1st Int. Symp. Silicon Molecular Beam Epitaxy* (Ed. J.C. Bean, Pennington, N.J., Electrochemical Soc., 1985) 352

Karunasiri R P G, Park J S and Wang K L, 1991 *IEEE Trans. El.Dev.* **38** 2708

Kasper E, Kibbel H, Jorke H, Brugger H, Friess E and Abstreiter G, 1988, *Phys. Rev. B* **38** 3599

Kay L E and Tang T.-W, 1991 *Journal of Applied Physics* **70** 1483

King C A, Hoyt J L and Gibbons F J, 1989 *IEEE Trans. El. Dev.* **36** 2093

Klaassen D B M., 1992a *Solid-St. Electron.* **35** 961 and also Klaassen D B M, 1992b *Solid-St. Electron.* **35** 953

Kleiner W H and Roth L M, 1959 *Phys. Rev. Lett.* **2** 334

Lang D V, People R, Bean J C and Sergent A M, 1985 *Appl. Phys. Lett.* **47** 1333

Levitas A, 1955 *Phys. Rev.* **99** 1810

Manku T and Nathan A, 1991a *Phys. Rev. B (Condensed Matter)* **43** 12634

Manku T and Nathan A, 1991b *Journal of Applied Physics* **69**, 8414

Manku T and Nathan A, 1991c *IEEE El. Dev. Lett.* **12** 704

Manku T and Nathan A, 1992 *IEEE Trans. El. Dev.* **39** 2082

Madarasz F L, Lang J E and Hemeger M, 1981 *J. Appl. Phys.* **52**, 4646

Matthews J W and Blakeslee A E, 1974 *J. Cryst. Growth* **27** 118

McGregor J M, Manku T and Nathan A, 1992 *Proceedings of the Electronics Materials Conference* Boston

Murakami E, Nakagawa K, Nishida A and Miyao M, 1991 *IEEE El. Dev. Lett.* **12** 71

Nauka K, Kamins T I, Turner J E, King C A Hoyt J L and Gibbons J F, 1992 *Appl. Phys. Lett.* **60** 195

Ni W.-X, Knall J, and Hansson G V, 1988 *Proc. 2nd. Int. Symp. Silicon MBE* (ed. J. C. Bean and L. J. Schowalter), Proc. Vol. 88-8, Electrochemical Society Pennington, New Jersey 68

Northrop, GA, Morar J F, Wolford, D J and Bradley J A, 1992, *Appl. Phys. Lett.* **61** 192

Osbourn G C, 1982 *J. Appl. Phys.* **53** 15860

Patton, G L, Iyer S S, Delage S L, Tiwari S and Stork J M C, 1988 *IEEE Elec. Dev. Lett.* **9** 165

Patton G L, Comfort J H, Meyerson B S, Crabbé E F, Scilla G J, de Frésart E, Stork J M C, Sun J Y C, Harame D L and Burghartz J N, 1990 *IEEE El. Dev. Lett.* **11** 171

People R, Bean J C and Lang D V 1985a *Proc. of the 1st International Symposium on Silicon Molecular Beam Epitaxy*, edited by J.C. Bean (Electrochemical Soc., Pennington, NJ, 1985a) 360

People R, Bean J C, Lang D V, Sergent A M, Stormer H L, Wecht K W, Lynch R T, and Baldwin K, 1984 *Appl. Phys. Lett.* **45** 1231

People R, 1985 *Phys. Rev. B* **32** 1405

People R and Bean J C, 1986 *Appl. Phys. Lett.* **48** 538

Pejcinovic B, Tang T and Navon D H, 1988 *Proc. IEEE 1988 Bipolar Circuits & Technology Meeting*, (September 12-13, 1988, Minneapolis, Minnesota) 46

Pikus G E and Bir G L, 1960 *Soviet Phys. - Solid State* **1** 1502

Pollak F H , 1990 *Semiconductors and Semimetals* **32** (edited by T. P. Pearsall) 17

Poortmans J, Caymax M, Van Ammel A, Libezny M and Nijs J, 1992b *MRS Symp. Proceedings Vol 281* (Ed. Charles W. Tu) 409

Poortmans J, 1993, *Ph. D. thesis* Katholieke Universiteit Leuven

Prinz E J, Garone P M, Schwarz P V, Xiao X and Sturm J C, 1991 *IEEE El. Dev. Lett.* **12** 42

Pruymboom A, Slotboom J W, Gravesteijn D J, Fredriksz C W, Van Gorkum

A A, Van De Heuvel J M, Van Rooij-Mulder J M L, Streutker G and Van de Walle G F A., 1991 *IEEE El. Dev. Lett.* **12** 357

Robbins D J, Canham L T, Barnett S J, Pitt A D and Calcott P, 1992, 1992 *J. Appl. Phys.* **71** 1407

Schäffler F, Többen D, Herzog H-J, Abstreiter G and Holländer B, 1992 *Semicond. Sci Technol.* **7** 260

Slotboom J W, Streutker G, Pruijmboom A and Gravesteijn D J, 1991 *IEEE El. Dev. Lett.* **12** 486

Smith C and Welbourn A D, 1987 *Proceedings of the Bipolar Circuits and Technology Meeting* 57

Sturm J C, Manoharan H, Lenchyshyn L C, Thewalt M L W, Rowell N L, Noel J.-P and Houghton D C, 1991 *Phys. Rev. Lett.* **66** 1362

Tsaur B.- Y, Chen C K and Marino S A, 1991 *IEEE El. Dev. Lett.* **12** 293

Van de Walle C G and Martin R M, 1985 *J. Vac. Sci. Technol. B3* 1256

Van de Walle C G and Martin R M, 1986 *Phys. Rev. B* **34** 5621

Yu E T, Croke E T and McGill T C, 1990 *Appl. Phys. lett.* 56 569

6

SiGe heterojunction Bipolar applications

J. Poortmans, S.C. Jain and J. Nijs

6.1   INTRODUCTION

In the previous chapter the electronic structure of strained $Si_{1-x}Ge_x$-layers was described with special emphasis on those properties which are related to their use as a narrow-bandgap material in the base of a Heterojunction Bipolar Transistor (HBT). We will always use the term HBT, although the device may contain two heterojunctions (at the base-emitter and base-collector junction). It is indeed this device which has attracted the *main body of applied research* efforts in the field of strained $Si_{1-x}Ge_x$-layers between 1988 and 1993. This has led to impressive results for both internal device speed (Patton, 1990a and Gruhle, 1993) where a cut-off frequency of 75 GHz, respectively 91 GHz was reported at room temperature as well as for f_{max}, where a value of 59 GHz (Harame, 1993) was reached. Also for SiGe pnp-transistors one has made considerable progress in achieving a peak f_T of 55 GHz (Harame, 1991). Before discussing these results in a more comprehensive way it is important however to shortly point out why this application has been the object of such intense interest.

First of all, it has to be noted that the physics and the technology behind the HBT is *closest to the nowadays device structures,* which one can still describe by quite classical views. Although the MODFET has been the experimental test vehicle by which such important heterojunction related quantities as band offset's have, at least qualitatively, been determined, it is clear that this device had the drawback of being too far from practical application in a field which at that time was completely dominated by GaAs-based materials. Moreover, it must be clear that, although the speed of CMOS-circuits is rapidly increasing, *silicon bipolar devices are still expected to remain dominant in high speed devices* and applications where a high drive current, low-voltage-swing operation capability or excellent threshold voltage control is needed (Yano, 1991). III-V-based material systems possess clearly a higher intrinsic speed, but they suffer the disadvantage of being completely *incompatible* with the existing, extraordinarily mature Si-technology. Attempts are being made to circumvent this drawback by combining the two materials (GaAs on Si - see for instance J. De Boeck, 1991), but it remains largely uncertain whether these techniques will meet general acceptance, certainly if one considers some specific problems of III-V-based HBT's such as the lack of adequate passivation. This compatibility requirement is clearly much better satisfied in the case of SiGe-alloys.

Therefore the preferred technology for LSI and VLSI high speed circuits will remain based on Si-based bipolar transistors. To increase the frequency performance of the Si-homojunction bipolar transistor one deals however with a *difficult trade-off.* In order to reduce the emitter-collector transit time, which is the dominant component of for instance an ECL-gate delay, except at very high current densities (see fig. 6.1), a further reduction of the base width is necessary.

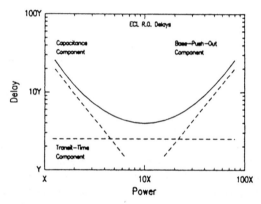

Fig. 6.1 Schematic representation of the power delay curve for an ECL-circuit. Epitaxial base technologies provide possibilities to improve each of these components. Reduction of the base transit time dominated regime is obtained by vertical scaling and the use of graded SiGe-base HBT's. (Comfort et al., reprinted from IEDM Techn. Dig. 1990, ©1990 IEEE)

This reduction of base width should be realized without increasing the base sheet resistance. This can only be achieved by a higher doping level in the base, which at the same time also reduces the risk for punchtrough. But the higher doping level in the base will lead to low current gain and therefore increased charge storage in the emitter.

A possible way to get around this compromise is the use of *a high-bandgap material in the emitter.* Several possibilities have been proposed : a-Si:H (Ghannam, 1984, Symons, 1987), n+-microcrystalline Si (Sasaki, 1987), β-SiC (Sugii, 1988, Sugii, 1992)) and Oxygen-Doped Si epitaxial films (Takahashi, 1987). All these approaches suffer however from high bulk and contact resistivities, ruining the transconductance of the device. Moreover, it seems difficult to realize ideal or at least reproducible base currents with these materials (Symons, 1987 and Bonnaud, 1992).

A much more promising approach is the use of a *narrow- bandgap material in the base, like for instance strained Si$_{1-x}$Ge$_x$-alloys* (see section 5.4). This allows higher doping levels in the base without decreasing the injection efficiency. At the same time low-temperature operation of Si-based bipolar transistors becomes feasible, because of the bandgap difference between base and emitter (Crabbé, 1990 and Cressler, 1991).

One can even further extend the benefits of the bandgap engineering in the base by *grading the Ge-concentration in the base* (Kroemer, 1985, Sturm, 1991 or Patton, 1990). The position-dependent bandgap in the base will give rise to additional drift fields, which will aid the minority electrons (in case of an npn-transistor) to travel even faster through the base.

In view of these possibilities it is not surprising that the use of pseudomorphic Si$_{1-x}$Ge$_x$-layers in the base of bipolar transistors has been oriented mainly towards fast digital applications. But other properties of Si$_{1-x}$Ge$_x$ HBT's could also be beneficially used in analog-like applications. The first has to do with the much higher current gain-Early Voltage product possible in HBT's (Prinz, 1991a). The use of high doping levels in the base is also expected to reduce the base resistance, which reflects positively on the noise figure of the bipolar transistor.

In the following sections both the theories describing DC- and AC-behaviour of the heterojunction bipolar transistor will be reviewed. At the same time we will also refer to the theories concerning the homojunction bipolar transistor and their modifications to take into account the phenomena in the double-heterojunction case. It is in the authors' opinion that such an approach will allow the reader to get a better idea about the performance improvement, made possible by using Si-based heterojunctions in the bipolar device. In these sections the authors will also emphasize these effects which deserve *special attention* when designing bipolar transistors with Si$_{1-x}$Ge$_x$-heterojunctions. In the following sections we will mostly deal with the case of the npn-transistor. In section 6.2 our discussion will be focused on the DC-behaviour of the HBT. In section 6.3 the AC-behaviour will be discussed. The case of the pnp-transistor will be also dealt with as far as it deviates from the theory of the npn-transistor. In the last section we will discuss some specific technological items when trying to incorporate epitaxial strained layers into advanced bipolar processing sequences.

We will not discuss in detail the low-temperature operation (liquid nitrogen-temperature) of these heterojunction transistors, but section 6.2 will provide enough material to understand which is the physical background for their superior characteristics at low temperature. In homojunction transistors the current gain at low temperatures is reduced due to the bandgap narrowing in the heavily doped emitter. Nevertheless, attempts have been made to transfer the heterojunction concept to the homojunction case. For instance, Yano et al.

(1991) have reported on their sidewall base-contact structure (SICOS) with a heavily doped base (10^{19} cm^{-3}) and moderately doped emitter (10^{18} cm^{-3}). They observed a current gain of 100 at 77K, which was 5 times higher than at room temperature. This is due to the additional heavy-doping induced bandgap narrowing in the base. This approach is often called the *pseudo-HBT* concept, an item not further discussed in this chapter.

6.2 DC-BEHAVIOUR OF THE DOUBLE HETEROJUNCTION BIPOLAR TRANSITOR

6.2.1 Factors determining the enhancement of the injection efficiency in an HBT with strained $Si_{1-x}Ge_x$-base

In this section we will focus on the collector current, which is equal to the electron current, injected over the base-emitter junction in case of negligible neutral base recombination and negligible reverse leakage of the base-collector junction.

As a starting point it is interesting to shortly review the equations governing the collector current in a bipolar transistor and to discuss in detail how the use of a heterojunction affects this behaviour. In modern bipolar transistors the base widths can be kept very small (50-150 nm), whereas the doping level in homojunction transistors will seldom exceed 5×10^{18} cm^{-3}. Under these conditions *the recombination in the base can be neglected*. The collector current, which isan injection current can then be written under the following general form (Huang, 1991, Kroemer,1985):

$$J_n = \frac{q \, (p.n \, (y{=}0) - p.n \, (y{=}W_b))}{\displaystyle\int_0^{W_b} \frac{N_a \, dy}{D_{nb}}} \qquad (6.2)$$

In this formula D_{nb} represents the minority carrier diffusion constant in the base. In principal, the base width must be large compared to the carrier mean free path (otherwise the notion of diffusion loses its physical meaning). N_a represents the base doping level. $y{=}0$ and $y{=}W_b$ are the base edges of the base-emitter and base-collector depletion layers. W_b is the neutral base width. The p.n-product is in the case of homojunctions equal to n_{ib}^2, which is the intrinsic carrier concentration in the base region.

The dependence of the minority carrier concentration at the base edge (y=0) of the base-emitter junction as a function of the forward voltage, in case of Boltzmann statistics and *in the absence of additional barriers in the conduction band at the base-emitter junction* is given by:

$$n(y=0) = \frac{n_{ib}^2}{N_a} \exp(\frac{qV_{be}}{kT}) \tag{6.2}$$

These injected carriers will move further through the base by diffusion and drift (dopant gradients induce additional electrical fields). In case of a constant intrinsic carrier concentration in the base (no Ge-grading in the base or heavy-doping induced bandgap narrowing gradients), eq. 6.2 reduces to the well-known expression (J. L. Moll and I. M. Ross, 1956):

$$J_n = J_c = \frac{q\, D_{nb}\, n_{ib}^2\, \exp(q\, V_{be}\,/\,kT)}{W_b} \tag{6.3}$$
$$\int_0^{W_b} N_a(x)\, dx$$

Based on this formula one can expect a strong enhancement of the collector current density in HBT's with strained $Si_{1-x}Ge_x$-layers. Indeed the intrinsic carrier concentration n_i^2 for any semiconductor can be written as :

$$n_{ib}^2(x,T) = N_c(x,T)\, N_v(x,T)\, \exp(\frac{-E_{gb}(x,T)}{kT}) \tag{6.4}$$

where $N_c(x,T)$ and $N_v(x,T)$ represent the effective density of states in conduction, respectively valence band of the strained $Si_{1-x}Ge_x$-layer at a temperature T. $E_{gb}(x,T)$ is the indirect bandgap of the semiconductor in the base. We have shown in chapter 5.4 how the bandgap depends on the Ge-content and the strain configuration. As rule of thumb one could use a bandgap shrinkage of about 8 meV/%Ge for 0<x<0.15. From this one could expect an enhancement of the collector current by nearly a factor 30 at room temperature for a base layer containing 10% of Ge.

Such enhancement will however not be obtained, because, as shown in chapter 5.5, the effective density of states is reduced in both conduction and valence band. The ratio of the $N_c.N_v$-product in the strained layer to the $N_c.N_v$- product in Si is shown in fig. 6.2 as a function of Ge-content at room temperature and 77K. The data for N_v are based on a paper by Manku (Manku, 1991). They are to be compared with a calculation of the $N_c.N_v$-product made by Prinz (Prinz, 1989).

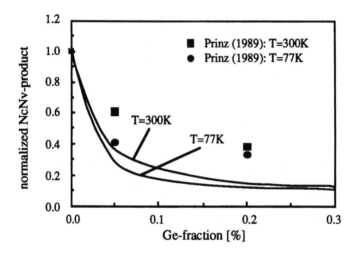

Fig. 6.2 Ratio of the $N_c.N_v$-product in a strained $Si_{1-x}Ge_x$-alloys to the $N_c.N_v$-product in Si at T=77 and 300K. The squares are from Prinz (1989) for layers containing 5 or 20% of Ge, while the full line is based on the values for the valence band density of mass given by Manku (1991b)

At this point however we should critically revise the condition to derive expression 6.2, where we assumed *the absence of any additional barrier besides the p-n junction built-in voltage* for electron transport. When looking to the band-line up between Si and strained $Si_{1-x}Ge_x$ one can remark that in case of an abrupt transition, a small ($\Delta E_c << \Delta E_g$) spike in the conduction band is formed because of the type I band line-up at the heterojunction (see section 5.3 on band offset's). This may lead to a lower excess minority carrier density at the base edge of the base-emitter depletion layer than expected on the basis of expression 6.2, depending on the position of the Si-SiGe-transition relatively to the metallurgical junction. To calculate this excess carrier density the details of the carrier transport over this spike should be taken into account. Since $\Delta E_c < kT/q$ for Ge-contents between 0 and 30% the influence of the spike will however be minor at room temperature.

When the p-n transition and heterojunction are exactly coinciding and $\Delta E_c << V_{jp}$, V_{jp} being the electrostatic potential difference (the band bending) over the p-type region, the influence of the spike will be very small and the excess carrier concentration is still given by the classical Boltzmann equation (6.2). The determing transport factor is then the diffusion through the base. This condition will be met in case of high emitter concentrations, at low forward base-emitter voltages, low Ge-content and/or non-abrupt $Si-Si_{1-x}Ge_x$-transitions. Under these conditions one can write down the following expression for the ratio of the collector current in a heterojunction transistor to the collector current in a homojunction transistor with equal integrated base doping.

$$\frac{I_c(HBT)}{I_c(Homo)} = \frac{N_{cSiGe}\, N_{vSiGe}\, D_{nSiGe}}{N_{cSi}\, N_{vSi}\, D_{nSi}}\, \exp(\frac{\Delta E_g}{kT}) \qquad (6.5)$$

At this moment it is instructive however to critically review the use of Boltzmann statistics for the majority carriers in the derivation of the before-mentioned formulas. Indeed, when the density of states in the valence band is strongly reduced (see section 5.5), the question arises whether this approach remains valid. In case of equal free hole density at equal temperature, the term E_f-E_v will be lower in p-type strained $Si_{1-x}Ge_x$ layers than in Si. This is illustrated in fig. 6.3 for 4 different Ge-concentrations at 300K (Poortmans, 1993a).

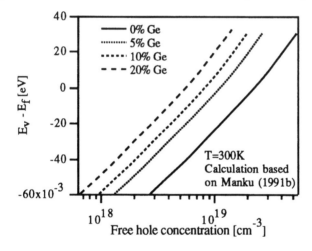

Fig. 6.3 Theoretical calculation of the position of the Fermi-level relative to the valence band edge at T=300K versus doping concentration (p-type) for 4 different Ge-concentrations. The calculation is based on the energy dependence of the valence band density of states mass as reported by Manku (1991b). (After J. Poortmans, 1993a)

Under such circumstances the use of *Fermi-Dirac statistics* is mandatory to correctly describe the situation. To account for this expression (6.5) should be rewritten.

$$\frac{I_c(HBT)}{I_c(Homo)} \approx \frac{N_{cSiGe}\, D_{nSiGe}\, \exp(\frac{\Delta E_{eff}-\Delta E_f}{kT})}{N_{cSi}\, D_{nSi}} \qquad (6.6)$$

ΔE_f is the short-hand notation for the following expression :

$$\Delta E_f = (E_v-E_f)_{Si} - (E_v-E_f)_{SiGe} \qquad (6.7)$$

where the two terms are equal to the distance between the Fermi-level and the upper valence band edge. Due to the reduction in effective density of states in the strained layer ΔE_f is a negative quantity. In case of $E_v-E_f > 3kT$ the reader can easily verify that expression 6.6 reduces to 6.5 again. One should also notice that at these high dopant concentrations N_v is not only a function of the Ge-content and the temperature, but also becomes a function of the *dopant concentration itself*, because of the non-parabolicity of the valence band (Manku, 1991). The quantity ΔE_f used in expression 6.7 is also dependent on temperature. This further complicates the interpretation of the temperature dependence of the collector-current ratio in expression 6.5 , in that it leads to an underestimation of the actual ΔE_{eff} by 10 to 15 meV, for doping levels exceeding 10^{19} cm^{-3} and between 77 and 300 K (Poortmans, 1993a).

Even though equation 6.6 might look general since it takes into account the presence of additional barriers for electron flow from emitter to base, the differences in effective density of states and the additional necessity of using Fermi-Dirac statistics at high base doping levels, we still overlooked possible differences in heavy-doping induced bandgap narrowing. Most researchers in this field assume it to be the same as in Si. A priori, this is a far-reaching assumption considering the drastic strain-induced changes in electronic structure and therefore it certainly requires further investigation. Based on existing analytical models describing the different rigid shifts of the band edges due to heavy doping effects (Jain, 1991), Poortmans (1992a) performed theoretical calculations on the bandgap narrowing in highly p-type doped strained Si$_{1-x}$Ge$_x$-layers and found that it was 5-10 meV larger than in Si for doping levels between 10^{18} and 10^{19} cm^{-3} and Ge-contents between 0 and 20%.

In the same paper the authors also present an expression to calculate the *apparent bandgap narrowing*, which describes the influence of the bandgap narrowing on the pn-product of a highly-doped semiconductor and which is the parameter of interest for the collector current. This expression can be written with ΔE_{gapp} the apparent bandgap narowing and ΔE_{real} is the real bandgap narrowing which is defined as the sum of the shifts of valence and conduction band edges. This expression is a more general version for the apparent bandgap narrowing expression derived in the paper by S.C. Jain and D.J. Roulston (1991).

$$\Delta E_{gapp} = kT \ln\left(\frac{N_a}{N_v(x,T)}\right) + (E_f - E_v)(x,T,N_a) + \Delta E_{real}(x,N_a) \qquad (6.8)$$

In the paper of Poortmans (1992) it was shown theoretically and experimentally that between 10^{18} and 10^{19} cm^{-3} *the difference in apparent heavy-doping induced bandgap narrowing* between Si and strained Si$_{1-x}$Ge$_x$-layers remains small for $0 < x < 0.2$. The theoretical curves for $x=0$, 0.1 and 0.2 are shown in fig. 6.4. However, based on extensions of their calculations it was expected that the apparent heavy-doping induced bandgap narrowing will be *smaller* in strained layers than in Si for doping levels exceeding 10^{19} cm^{-3} and $x > 0.1$. This can be explained physically by the Fermi-level sinking deeper under the valence band in the case of strained layers due to the strong reduction of the effective density of states in their valence band. This so-called *Fermi-Dirac correction* will reduce the minority carrier concentration enhancement which would result from the heavy doping induced bandgap narrowing.

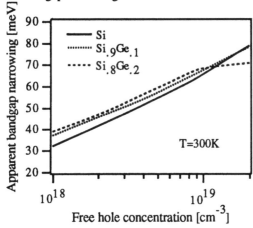

Fig. 6.4 The apparent heavy-doping induced bandgap narrowing as a function of doping and Ge-concentration T=300K. (After Poortmans 1993b reprinted with permission from Solid-State Electronics)

Although direct experimental evidence for this reduced apparent bandgap narrowing is missing at this moment, there are some reports supporting this view. For instance Shafi (Shafi, 1991a) reported on experiments with HBT's with a base doping level of 5×10^{19} cm^{-3}. He deduced an effective bandgap shrinkage between the reference homojunction transistor and the transistor with Si$_{.85}$Ge$_{.15}$-base of 73 meV. For 15% Ge about 120 meV was expected since no indication of relaxation in their strained layers is found. This result can partially be explained when remembering that the Fermi-level is about 35 meV deeper under the valence band edge for the Si$_{.85}$Ge$_{.15}$-base compared to Si in the case of such high doping level. This effect will suppress the minority carrier concentration in the highly doped base. Further study is required to elucidate the whole matter of heavy-doping induced bandgap narrowing in these strained layers.

6.2.2   Parasitic barriers and graded-base HBT's

In the previous section it was shown how the conduction band offset gives rise to a possible barrier for the electron flux from the emitter to the base. In this respect, the reader should keep in mind that this barrier in HBT's with strained $Si_{1-x}Ge_x$ base is not only present at the base-emitter junction but also at the base-collector junction, in contrast with AlGaAs-GaAs HBT's where the heterojunction is only present at the base-emitter junction. Therefore it is of utmost importance to optimize layer structure, doping profile and thermal budget to reduce the risk for and impact of these barriers on the device performance as much as possible. We will discuss these items now into some detail to provide some insight in the *specific* problems arising when designing HBT's with strained $Si_{1-x}Ge_x$-base.

Grossly speaking, one can distinguish among two possible solutions to keep this conduction band spike deep in the depletion layer. In the first solution *undoped or lowly doped spacers* are placed in the base-emitter (Tang, 1989, Lu, 1990 and Patton, 1990a) and base-collector depletion layers. The second solution consists of growing *gradual transitions* from Si to strained $Si_{1-x}Ge_x$. Both approaches can of course be combined.

In the beginning of this chapter 5 we mentioned how the advent of new epitaxial low-temperature growth methods allows principally the growth of very sharp dopant profiles. Nevertheless there are several factors which affect the abruptness of these profiles adversely. First of all, already during deposition segregation of dopant atoms will cause a broadening of the profiles. This effect is especially severe for n-type dopants like Sb in MBE (Jorke, 1988) or P in CVD-systems (Leong, 1988) when PH_3 is used as precursor.

In addition to this phenomenon occurring *during* base layer growth, most processes to complete the complete transistor structure need at least one step at high temperature for the activation of base- and emitter contact implants. This step is mostly done in rapid thermal annealing facilities, but the thermal budget connected with this step is certainly not negligible. To get an idea about the allowable outdiffusion Prinz (Prinz, 1991b) simulated the case of a heterojunction device with x=0.2 and base doping 10^{20} cm^{-3} and an emitter doping level 10^{17} cm^{-3}. They found that, when the B in the base diffuses about 2.5 nm towards the emitter, a barrier of about 80 meV was created at the base-emitter junction, thereby ruining completely the enhancement of the collector current by the low bandgap base-material. This can be seen in fig. 6.5, where the ratio of the collector current in the HBT to the collector current in the homojunction reference transistor with same base Gummel number is shown as a function of the inverse absolute temperature. In case of structures with a spacer the activation energy (which equals the effective bandgap shrinkage of the strained $Si_{1-x}Ge_x$-base - see above) is clearly larger than when no spacer is used. The barrier can be nearly as high as ΔE_g because the Fermi-level position is fixed close to the valence band, neglecting differences in the valence band density of states.

Although the simulated case might look quite extreme it illustrates well the sensitivity of the HBT-behaviour to these details. In the same paper the authors experimentally verified this predicted reduction of the collector current. When adding 10 nm wide undoped SiGe-spacers at *both ends* of the doped base, they were able to completely restore the expected collector current enhancement, when no parasitic barrier is present. The introduction of these spacers can also play a role in reducing the base-emitter tunneling current, when both areas are highly doped (Matutinovic-Krstelj, 1991). Finally, also the emitter-base and base-collector capacitances are reduced, which affects favorably the frequency performance (Patton, 1990b). Of course, the width of the spacer needs to be tailored to the specific thermal budget of the complete process.

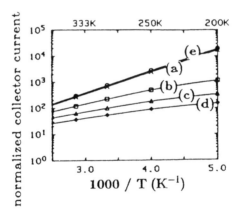

Fig. 6.5 Simulated normalized colector current versus inverse temperature for varying outdiffusion. Curve a ; no outdiffusion, Curve b ; 2.5 nm outdiffusion, Curve c ; 5 nm outdiffusion, Curve d ; 7.5 nm outdiffusion, Curve e ; 2.5 nm outdiffusion with 10 nm spacers (After Prinz, 1991 reprinted from Electron Device Letters, ©IEEE 1991)

Until now we focused on the detrimental influence of eventual barriers at the *base-emitter junction*, but the same holds true for the *base-collector junction*. One (Cotrell, 1990, Pruymboom, 1991) has reported extensively on this issue. Due to channeling effects during the phosphorous-implant of the emitter, the latter found a strong reduction of the collector current due to a parasitic barrier at the base edge of the base-collector junction. This was confirmed unambiguously by the dependence of this reduction of collector current on the inverse base-collector voltage. At higher reverse voltage, they found that the detrimental effect could be partially or completely eliminated. This is easily understood by remembering that at larger inverse base-collector voltage the base-collector depletion layer will extend further into the base, burying the conduction band spike deeper into the depletion layer. Slotboom et al. (1991) developed an analytical model which describes the influence of such barriers at the base-collector junction. It can be seen in fig. 6.6 that the normal triangular minority carrier distribution in the neutral base with negligible recombination changes into a box-like profile in the presence of a barrier in the neutral base. Assuming that the height of the barrier ΔE is much larger than kT and recombination is still negligible, he derived the following equation for the electron flux through the base at low and intermediate injection conditions :

$$J_n = \frac{q\,D_n\,n_{ib}^2(x,T)}{N_b\,W_b} \left(1 - \frac{\Delta W}{W_b} + \frac{\Delta W}{W_b}\exp(\frac{\Delta E}{kT})\right) \qquad (6.9)$$

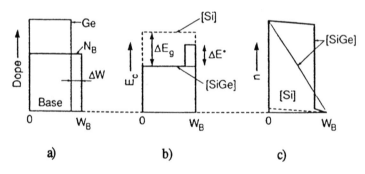

Fig. 6.6   Schematic model representing the effect of a parasitic barrier at the base-collector junction. In a) the Ge- and B-profile is shown, whereas in b) the corresponding energy diagram is shown. In c) it is shown that due to this barrier the minority electron density has a boxlike distribution instead of being triangular (After Slotboom, 1991 reprinted from Electronic Device Letters, ©IEEE 1991)

It is also clear that in these circumstances the charge storage in the neutral base will considerably increase, thereby seriously increasing the base transit time. We will treat this in detail in the chapter on the AC-properties of the SiGe HBT.

At this point it should also be remembered that even if the parasitic barrier does not affect the characteristics at low and intermediate collector current, it can still be important under large injection conditions. In case of inappropriate collector doping level the base pushout shifts the base edge of the base-collector depletion layer towards the collector, which makes the parasitic barrier show up again.

Until now we restricted our discussion to npn-HBT's, but the problem of the parasitic barriers is much more severe in the case of pnp strained $Si_{1-x}Ge_x$-base HBT's (Harame, 1990) since the difference in bandgap is mainly taken up by the valence band offset in the asymmetrical strain configuration (see section 5.3). This makes the minority holes pile-up in the base (Patton, 1990b). As stated before the heterojunction must be buried much deeper than in the base-collector junction to avoid this excess hole storage. This may however lead to problems with impact ionization in the lower bandgap part of the base-collector depletion layer, affecting adversely the output conductance and the breakdown voltage of the transistor. This makes the optimization of pnp SiGe HBT's a much more complicated matter than for their npn-counterpart. This is well illustrated by the attempts of Harame et al. (1990a, 1990b) to use a strained n-type $Si_{1-x}Ge_x$-layer in the optimization of pnp-transistors. By combining a graded-base and a retrograde Ge-profile in the base-collector junction they obtained a cut-off frequency of 30 GHz. By careful optimization however they succeeded in increasing this to 55 GHz (Harame, 1991).

There exists a second solution to avoid spikes in the conduction band (Kroemer, 1982). If one grows a graded transition from Si to strained $Si_{1-x}Ge_x$ alloys, it is quite readily understood that the spike in the conduction band will not occur since the properties of the conduction band are a *continuous* function of the alloy concentration. Hence, the problem of parasitic barriers can be avoided by grading the Ge-concentration over the base-emitter and base-collector depletion layers of the transistor (Harame, 1991). This is not the only benefit which makes this solution so attractive. In modern bipolar transistors the collector doping can be quite high (5×10^{16} to $5 \times 10^{17} cm^{-3}$) to reduce base push-out at very high collector current densities. But this leads to high electrical fields in the base-collector depletion layer. Crabbé (Crabbé, 1990) proved by means of of a hydrodynamical model for the carrier flow and Monte-Carlo simulations that impact ionization in this very high electrical field is not determined by the maximal electrical field, but is mainly determined by the carrier energy and is therefore a non-local phenomenon. Since the treshold for impact ionization depends on the bandgap it is quite obvious that this will be particularly severe when the lowest-bandgap part extends deep in the depletion region. Hence, to reduce the effect of the impact ionisation on the output conductance of the bipolar transistor, the low bandgap material should not extend far into the base-collector depletion layer. But this can come into conflict with the requirement to avoid parasitic barriers. Therefore the best solution is to use a graded transition from the Si-collector to the strained $Si_{1-x}Ge_x$-base. Patton (1990b) used this approach and their calculations showed that there was no statistical difference found between the multiplication factor (M-1) of the Si and SiGe-base npn devices in case of such a base-collector design.

The use of graded transitions is not limited to the depletion layers, but is also attractive in the neutral base. By grading the Ge-content in the neutral base downwards from the collector to the emitter, a drift field is created by the redistribution of holes in the base, creating a small excess hole concentration at the collector side. This will create an electrical field in the base. Hence, the minority carriers will not only travel through the base by diffusion, but they will also acquire an additional drift velocity on their travel towards the collector. The electrical field will be nearly equal to the bandgap difference at the emitter and the collector side of the base, when the density of states in the valence band is constant throughout the base. Iyer (1989) developed developed an expression for the collector current in such a graded base. J.C. Sturm (1991a) calculated an "effective base Gummel number" G_{Beff} to describe the electron flux in such graded-base situation by a classical homojunction terminology. He derived the following analytical formula for G_{Beff} as function of temperature in the case of linear grading and assuming a constant $N_c.N_v$-product in the base :

$$G_{Beff}(T) = \frac{W_b\,N_b\,kT}{\Delta E_{Ge}-\Delta E_{Gc}}\,\exp(\frac{-\Delta E_{Gc}}{kT}\,(1-\exp(\frac{-(\Delta E_{Ge}-\Delta E_{Gc})}{kT}))) \qquad (6.10)$$

where ΔE_{Ge} and ΔE_{Gc} are the bandgap reductions compared to Si at respectively the emitter and collector side of the neutral base. The effect of grading in the base on the base transit time will be discussed in the chapter on the AC-behaviour of strained $Si_{1-x}Ge_x$-base HBT's. Nevertheless, one can expect that it will lead to some overestimation of this enhancement since the assumption of constant $N_c.N_v$-product in the base is not really valid. We already described this in chapter 6.2.1. This was indeed the case for their experimental results which turned out to be consistently lower than their predictions.

6.2.3 Neutral base recombination

In our analysis until now we always assumed that recombination in the base can be neglected. The end of this chapter on DC-characteristics is the appropriate moment to critically revise this assumption. There is also practical relevance in doing so in that a substantial increase in base recombination can compromise the enhancement of the current gain (Shafi, 1991a). Although the base widths of HBT's will practically never exceed 100 nm because of the restrictions imposed by the critical thickness, one should not forget that the doping level in the base often largely exceeds 10^{18} cm^{-3}. At such doping levels Auger recombination becomes dominant in the recombination processes (see for instance the excellent review on bipolar transistors by P.A.H. Hart, 1981).

Not much experimental work on the recombination issue has been published, whereas the interpretation of the data is further complicated by the presence of

eventual zero-dimensional (point defects) or one-dimensional (dislocations) defects in the used device structures and the way these are influenced by the various thermal treatments. Nevertheless it is possible to deduce some general trends. King mentioned in his pioneering paper of 1989 that, based on the temperature dependence of the ratio of the base current in the HBT to the collector current of the reference homojunction transistor, an important portion of the base current was caused by neutral base recombination. Their layers were deposited in a *Limited Reaction Processing reactor (LRP)*. Afterwards it was pointed out that this important neutral base recombination was caused by the high oxygen content of about 10^{20} cm^{-3} in their layers. Sturm reported in 1991 (Sturm, 1991a) that, by using a load-lock on their *RT-CVD system*, they were able to reduce the oxygen content in their epitaxial layer to below the SIMS detection level ($<10^{18}$ cm^{-3}). It was then found that the base current of homo- and heterojunction transistors processed in these layers were the same, indicating that the recombination lifetimes in Si and strained $Si_{1-x}Ge_x$-layers are comparable.

For strained $Si_{1-x}Ge_x$-layers grown in UHV-CVD environment (Meyerson, 1988), Patton (1989) found identical base currents in Si BJT's and SiGe HBT's. This agrees with the results obtained on layers grown in RT-CVD systems. This result is quite logical since the basics of the layer growth is the same for both deposition methods : heterogeneous decomposition of the precursor gases at the heated silicon surface and the important role of hydrogen in the reaction kinetics and even more important, its influence on the surface state.

For *MBE-grown layers* the situation seems to be quite different. In a paper by Johansson (Johannson, 1989) very short lifetimes in p-type $Si_{0.85}Ge_{0.15}$ (doped 8×10^{17} cm^{-3}) of about 0.15 ns were deduced from measurements of the total admittance of Si/SiGe heterostructure p-n junction diodes. Because of these very short lifetimes it was necessary to use frequencies in the micro-wave range and appropriate de-embedding techniques for the lifetime extraction. The authors attribute this very short lifetime to dislocations in their MBE-grown layers. In 1988, Willander (1988) measured a hole lifetime of about 0.8 µs in n-type SiGe doped about 3×10^{17} cm^{-3}, which is again much shorter than in Si with a comparable doping level. Shafi (Shafi, 1991a) reported on his study of the dependence of the base current in stable and metastable devices on the rapid thermal treatment conditions. In the *stable devices* (below the critical thickness) they were able to reduce the emitter-base recombination to such an extent that neutral base recombination was dominant. From basewidth modulation measurements he found that the neutral base recombination was much stronger in the device with the $Si_{.85}Ge_{.15}$-base than in the homojunction reference transistor. He was only capable of explaining these results by assuming that near the base-collector depletion layer the recombination lifetime was extremely small i.e. 3×10^{-13}s (Shafi, 1991b). Even taking into account that the base doping level in their base was 5×10^{19} cm^{-3} this is an extremely low value. The authors claim this could not be explained by possible enhanced recombination at a parasitic barrier at the base-collector junction, because the inverse transistor characteristics showed that the base-collector junction was as good as the base-emitter junction.

The enhanced recombination in the MBE-grown SiGe base layers could be caused by the presence of a large number of point defects in MBE-grown layers. Luminescence data, which show a broad PL-peak around 0.8-0.9 eV in MBE-grown layers and caused by interstitial-type features(Noel, 1992) indeed support this hypothesis. The density of these features they found was in the order of 10^9 cm^{-2}.

As a conclusion it can be stated that reported results for the recombination lifetime in MBE-grown layers show a reduction of minority carrier lifetime in strained $Si_{1-x}Ge_x$-layers for both types of carriers.

6.2.4 Early voltage-current gain product of double-barrier HBT's

We devote here some attention to the output conductance of the HBT since it is one of the elements which make this device attractive for analog applications. In a homojunction one cannot optimize both Early voltage (V_A) and current gain (β) at the same time, since high V_A requires a high base doping level (compared to the collector doping level), whereas high current gain necessitates a low doping level in the base. In a HBT one can shift this compromise to much higher levels, since the high current gain can be conserved at high base doping level thanks to the increased injection efficiency. In case of a homogeneous Ge-profile in the base it is straightforward to show that the $\beta \times V_A$-product can be much higher in HBT's and that the improvement is proportional to $\exp(\Delta E_g/kT)$ with ΔE_g the bandgap reduction between a homojunction and a heterojunction base.

At this moment it is instructive to derive what will happen in case of a graded-base transistor where the bandgap is non-uniform over the base. This has been described in a paper by Prinz (Prinz, 1991a), where, starting from the commonly known expressions for the current gain and Early voltage, the authors derive the following expression for the $\beta \times V_A$-product.

$$\beta V_A = \frac{q}{J_{Bo} C_{BC}} n_i^2(W_b) D_n(W_b) \qquad (6.11)$$

J_{Bo} is the base saturation current density, which is independent from the actual dopant and Ge-profile in the base. C_{BC} is the base-collector capacitance which will be determined by the collector doping level since in most typical cases it will be lower than the base doping level. From formula (6.11) one can deduce that, when grading the Ge-profile in the base, the Ge-content should be maximal at the collector side of the base. Physically, this can be understood easily by considering two different cases of grading : in the first one the Ge-content is maximal at the emitter side and graded down towards the collector, in the second case the Ge-profile is reversed. The second Ge-profile is normally used since the drift field in the base will have the right direction for reducing the base transit time. In that case the barrier height for electrons will be maximal at the base-emitter junction (see fig. 6.7a), making the collector current insensitive to the base-collector voltage. It can be seen in fig. 6.7b that in case of a reverse Ge-profile the heighest barrier is at the base-collector region, thereby seriously deteriorating the output conductance of the device. The experiments of Prinz (Prinz, 1991a) confirmed these views and they reported on $\beta x V_A$-products as high as 100000V in devices with a cut-off frequency, expected to be around 30 GHz.

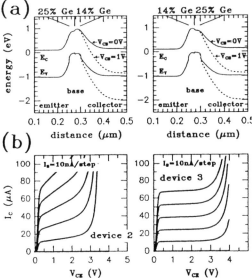

Fig. 6.7 Calculated band diagrams (a) and experimental characteristics (b) showing the effect of the grading and the position of the highest bandgap region in the base on the output conductance of a graded SiGe-base transistor (After Prinz, 1991b reprinted from Electronic Device Letters, ©IEEE 1991)

In the derivation of (6.11) the classical expression for the Early voltage was used. But for modern bipolar transistors the electrical field in the base-collector depletion region makes the carriers travel at saturation velocity. Under such circumstances the dependence of the collector current on the base-collector voltage will be reduced. E_B describes the influence of eventual drift fields in the base, whereas v_s is the electron saturation velocity in the base-collector depletion layer.Following D.J. Roulston (Roulston, 1990), the enhancement of the Early voltage F_E is given by :

$$F_E = \sqrt{1 + \frac{D_n\, E_B}{v_s\, W_b}} \qquad (6.12)$$

This enhancement factor accounts for the experimental values of $\beta x V_A$-product which were even higher than calculated by equation 6.11.

6.3   FREQUENCY AND CIRCUIT PERFORMANCE OF HBT'S WITH STRAINED $Si_{1-x}Ge_x$-BASE

In this section we will first focus on the intrinsic transistor speed and more specifically how it is affected by the heterojunctions at the base-emitter and base-collector junctions. The intrinsic speed of the device is best reflected in the emitter-collector transit time τ_{EC}. Although this quantity is often cited as a figure-of-merit, it must be realized that it is not sufficient to completely characterize the transistor operating as a switching element in a digital circuit, like in an ECL-gate. ECL-gates are indeed expected to be the mainstream technology for the application of bipolar transistors in VLSI (Goto, 1990). This section will be completed with an analysis of the relation between the gate delay of this kind of circuit and the intrinsic parameters of the bipolar device.

6.3.1 The different contributions to the emitter-collector transit time

In most textbooks about bipolar transistors it is explained how the different transit times, making out the total transit time, are related to the different charges present in the device (see for instance the excellent review by H.C. De Graaff, Elsevier, 1986). Therefore the question about the influence of the heterojunction on these transit time reduces then to acquiring insight in the influence of the heterojunction on these different charges, present in the device.

The effect of the heterojunction is most pronounced on the *emitter charge storage time i.e. the diffusion charge present in the neutral emitter due to the holes injected from the base*. This transit time τ_e can be written with W_e the emitter depth (or the hole diffusion length if the latter is shorter than the emitter depth) and D_{pe} the hole minority carrier diffusion length:

$$\tau_e = \frac{W_e^2}{2D_{pe}} \frac{G_b}{G_e} \exp(\frac{-\Delta E_g}{kT})$$

(6.13)

Taking the rule of thumb that $\Delta E_g(x)$ is given by 8 meV/%Ge the reader can easily check that τ_e becomes negligible for the commonly used Ge-concentrations. In the case of a transistor with a moderately doped emitter (1.8×10^{17} cm^{-3}) and about 30% of Ge in the base τ_e would be lower than 0.1 ps which can be neglected in comparison with the other time constants involved (Kamins, 1989). In fig. 6.8 one can also see the simulation results for monocrystalline emitters as reported by Patton (1990b)

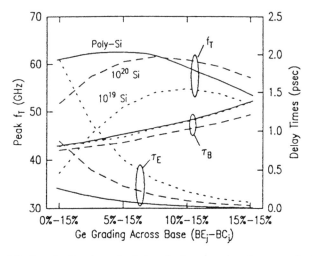

Fig. 6.8 Results for the calculation for base- and emitter transit times in a bipolar transistor with epitaxial emitter or poly-emitter. It is clearly shown that in case of a poly-emitter the emitter transit time τ_e is already small compared to the base transit time (After Patton, 1990b, reprinted from IEDM Techn. Dig., ©IEEE1990)

The relevance of this reduction of the emitter charge storage time should be looked at in connection with the emitter formation technology. In case of a monocrystalline epitaxial emitter (Kamins, 1989, Gruhle, 1992), the charge storage in the emitter is appreciable and therefore the base-emitter heterojunction can play a substantial role in improving the cut-off frequency of these transistors by the reduction of τ_e. However, polysilicon emitters form the mainstream in nowadays Si bipolar technology. In these transistors the diffusion charge stored in the emitter will be already much smaller, especially when the monocrystalline part of the emitter is restricted to a very shallow zone, outdiffused from the implanted polysilicon. This zone can be as small as 20-40 nm. In a theoretical paper by Suzuki (1991) it was shown how τ_e depends on polysilicon thickness, emitter junction depth, bandgap narrowing in base and emitter for homojunctions and the presence of a thin interfacial oxide at the poly-monocrystalline interface. He showed that at the usual polysilicon thicknesses used (about 200 nm) and emitter junction depths below 20 nm, τ_e is lower than 1 ps. This is confirmed in simulations, reported by Patton (1990b). One can see that τ_e forms only a small part of the total transit time (about 20%). It is clear from fig. 6.8 that in case of poly-emitters with very small monocrystalline junction depths the frequency performance gain by using a strained narrow bandgap layer in the base will remain marginal.

The second component in the total transit time is formed by the *base transit time*. When there is no Ge-grading in the base and the base doping profile is flat the base transit time is still given by :

$$\tau_b = \frac{W_B^2}{2D_{nB}} \qquad (6.14)$$

The effect of the heterojunction on this transit time is only indirect. Since the heterojunction allows to use high doping levels in the base, the base width can be made smaller than in conventional homojunction transistors, where the minimal base width is limited by punchthrough problems. Hence, W_b can be made very small - base widths as small as 15 nm are reported at base doping levels between 10^{19} and 10^{20} cm^{-3} (Kamins, 1989)-. They calculated a base transit time of about 1 ps for their transistor with a base doping level of 7×10^{18} cm^{-3} and a Ge-content of 31%. He neglected however the influence of the retrograde field due to the lower doping in the emitter.
There is however some experimental evidence that the minority carrier diffusion constant increases as a function of Ge-content (Poortmans, 1992b), which contributes then to a further reduction of the base transit time (see section 5.5).

In the paragraph about the emitter charge storage time it was shown that in case of a polysilicon emitter technology, the use of a Si-Si$_{1-x}$Ge$_x$-heterojunction at the base-emitter transition does not provide a real strong frequency performance gain. Therefore a more efficient approach in such case is to grow a graded Ge-profile in the base where the Ge-content is graded down from the collector to the emitter. The additional electrical field in the base speeds up the minority electrons travelling through the base. Kroemer (1985) and Patton (1990b) gave the following expression for the ratio of the base transit times between a homojunction and graded heterojunction bipolar transistor in case of linear grading and $E_{ge}-E_{gc}>kT$ ($E_{ge(c)}$ is the indirect bandgap at the emitter, respectively collector side of the base) :

$$\frac{\tau_b \text{ (SiGe)}}{\tau_b(\text{Si})} = \frac{2kT}{\Delta E_{gec}}(1-\frac{kT}{\Delta E_{gec}}) \qquad (6.15)$$

ΔE_{gec} is the indirect bandgap difference between the emitter and collector side of the base. The base transit time has a linear dependence on the bandgap grading. Therefore, at room temperature the grading should be at least over 10% in the base to have an appreciable effect. In this way the base transit time can be reduced by nearly 40% (see Patton, 1990b).

McGregor (McGregor, 1991) showed that linear grading of the bandgap provides minimal base delay *in case of uniform base doping*. Since the relation between bandgap and Ge-content is nearly linear, this means that the Ge-profile in the base must be nearly triangular. Researchers of the same group also showed that the emitter delay time is minimal in case of no base bandgap-grading. (D.J. Roulston, 1992).

Uniform base doping is however a condition which is normally not met in advanced transistors. We showed in the section about parasitic barriers that the B-doping is kept low near the junctions (lowly doped spacers avoid parasitic barriers and reduce at the same time tunneling and the junction capacitances). At the base-emitter junction this creates a retarding field for the electrons, while at the collector side the electrical field is directed towards the collector. Starting from Kroemer's expression (Kroemer, 1985) for the base transit time in a nonuniform bandgap base, Gao and Morkoc (Gao, 1991a) derived an analytical expression for the base transit time in the presence of these boron-grading induced electrical fields. By breaking up the base doping profile in three parts (exponential boron-tails at emitter and collector and in between a flat B-profile) they found the following expressions for the transit times in these three zones :

$$t_{b1} = \frac{\exp((b_e-a)x_1) + (a-b_e)x_1 - 1}{D_n(a-b_e)^2} \qquad (6.16a)$$

$$t_{b2} = \frac{\exp(-a(x_2 - x_1)) + a(x_2-x_1) - 1}{D_n a^2} \qquad (6.16b)$$

$$t_{b3} = \frac{\exp(-(a+b_c)(W_b-x_2)) + (a+b_c)(W_b-x_2) - 1}{D_n(a+b_c)^2} \qquad (6.16c)$$

In these expressions b_e and b_c characterize the exponential decay length of the B-profile at the emitter, respectively collector side of the base. a is the slope of the linear Ge-profile. x_1, x_2 and W_b are the depth coordinates of the edges of the respective zones. When grading down the B-profile from 10^{19} cm^{-3} to below 10^{17} cm^{-3} near the emitter, the retarding field contribution becomes dominant in the total transit time (>80% of the total delay). By careful combination of Boron- and Ge-grading they predicted base transit times as low 0.4 ps, which in combination with electron velocity overshoot in the base-collector depletion layer would allow for cut-off frequencies of over 100 GHz, at the expense of a lower maximum oscillation frequency f_{max} however.

In the discussion of the base transit time we implicitly assumed the absence of any energetic barriers to the electron flux through the base. In the section about the DC-behaviour we discussed the necessary conditions to keep these barriers away from the neutral base region. It is readily understood that when such barriers are present the diffusion charge in the base will start to increase by nearly a factor of two (the minority carrier profile changes from triangular to rectangular). Based on the basic equations for electron transport at low and intermediate injection, Slotboom (1991) derived an analytical formula (6.17) to calculate the dramatic increase of the base transit time in the presence of a barrier at the base-collector junction. In this formula (6.17) the different terms have the same meaning as in formula (6.11).

$$\tau_b = \frac{W_B^2}{2D_{nb}} \left(1 + 2\left(\frac{\Delta W}{W_b}\right)\left(1 - \frac{\Delta W}{W_b}\right)\left(\exp\left(\frac{\Delta E}{kT}\right) - 1\right)\right) \qquad (6.17)$$

When $\Delta E > kT$, τ_b will start dominating the other transit times, which will result in a strong degradation of the cut-off frequency.

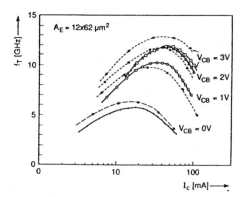

Fig. 6.9 Plot showing the effect of the inverse base-collector voltage on the cut-off frequency of a $Si_{0.8}Ge_{0.2}$-base transistor in the presence of a parasitic barrier close to the base-collector junction. At higher inverse voltages the detrimental effect of the barrier is reduced (After Slotboom, 1991, reprinted from Electron Device Letters, ©IEEE 1991)

The barrier height ΔE being a function of the base-collector voltage, this will also be the case for f_T. This dependence was experimentally verified on the transistors, described in the paper of Slotboom. Their experimental results are shown in fig.6.9. By increasing the inverse base-collector voltage from 0 to 3V, the f_T in their transistors increased from 5 to 13 GHz.

Even when the heterojunction is buried deep enough in the base-collector depletion layer it was observed that the roll-of of the current gain and cut-off frequency at high collector current densities was faster in strained base $Si_{1-x}Ge_x$ DHBT's than in comparable single HBT's (for instance AlGaAs/GaAs HBT's). For homojunction bipolar transistors the roll-off is explained by a shift of the base-collector junction into the base-collector space charge region (P.A.H Hart Bipolar Transistors and Integrated Circuits p.121-122, 1981) making the neutral base wider, thereby increasing the base transit time. In the case of HBT's however, the injection of holes into the collector is suppressed by the large discontinuity in the valence band leading to a build-up of localized holes at this interface. Together with the non-negligible charge density of injected mobile electrons this leads to *an electrical field which* is opposite to that caused by the base-collector bias. Gao et al. (Gao, 1991b) developed an analytical model which takes also into account the lateral flow caused by the built-up of excess carriers under the active device area. This lateral flow can be described by a current dependence of the effective area through which the collector current flows. Taking both 2D and barrier effects into account the cut-off frequency can then be written as :

$$f_T = [f_{peak}^{-1} + \frac{2\pi (I_c/I_k-1)^2 S_E^2}{4nD_{nb}} + \pi \frac{W_b}{v_s} \exp(\frac{\Delta E}{kT})]^{-1} \qquad (6.18)$$

where S_E is the emitter stripe width and I_k is the Kirk-effect limited current. f_{peak}^{-1} is the peak cut-off frequency without the high injection effects. The authors found indeed by comparing the experimental data with their analytical model that this effects starts to play before the onset of the actual Kirk effect, thereby explaining the observed inset of roll-off at lower collector current densities. It must be noted at this point that this reduced injection of holes into the collector can be used advantageously in circuits where the transistors are to be operated *in saturation*. Ugajin and Amemiya (1991) studied the carrier storage under such conditions and found it to be two orders of magnitude smaller than in homojunction transistors with identical dopant profile, making the recovery time of the base-collector diode much smaller. Strained $Si_{1-x}Ge_x$-base HBT's are therefore also a good candidate for *LSI-circuits other than ECL-circuits*.

The *collector transit time* as such is not really influenced by the use of a double heterojunction structure. To the authors' knowledge there is no experimental or theoretical work indicating that the saturation velocity in strained $Si_{1-x}Ge_x$-layers would be different from the values in Si. This is also not to be expected since the saturation velocity is the result of scattering events at energy levels high in the band. Nevertheless it is possible to go beyond this saturation velocity limit in the nowadays very fast bipolar transistors. For these transitors with high doping levels in the collector and very narrow base widths with very large electrical fields in the base-collector depletion layer one can find a deviation from the simple, theoretical picture as developed by Middlebrook (1963). Since it takes some time before the scattering events will bring the carriers to their saturated drift velocity, the electrons will exhibit *significant velocity overshoot* (Crabbé et al., 1990) in the base-collector depletion layer. The effects of such phenomena can be verified by simulating the inelastic scattering processes in Monte-Carlo simulations. Based on such simulations, Crabbé et al. reported a reduction of the base-collector transit time by nearly 20%.

Besides these three different transit times the overall transit time is of course also determined by the *RC-constants caused by the junction capacitances*. This is not a specifically heterojunction related topic and therefore it is left out from this discussion. For more specific information the reader is referred to dedicated articles (for instance H.C. De Graaff, 1986).

6.3.2 The strained $Si_{1-x}Ge_x$-base HBT as a switching element

"Although the transit time provides a lot of information concerning the intrinsic speed of the transistor it lacks a clear relation to its performance as a switching element in a digital circuit" (Stork, 1988). Since ECL-gates form the mainstream bipolar VLSI-technology for very fast circuits (Goto, 1990), it is logical that strained $Si_{1-x}Ge_x$-HBT's have mainly been tried out in these circuits. Before presenting these experimental ECL- gate delays into more detail in the next section (6.4) about $Si_{1-x}Ge_x$ related bipolar technology issues it is useful to see what the relation is between the intrinsic device speed of a bipolar transistor and its performance in an ECL-gate.

The relation between gate delay and the internal transistor parameters was first developed by using computer-based sensitivity analyses (Tang, 1979). Such approach is however quite cumbersome.

Several analytical models linking gate delays and internal transistor speed have been reported since 1988. By splitting the ECL gate delay in two components, the first of which is related to the response of the collector current to the input voltage (T_b) while the second is related to the load (T_c), Stork (1988) derived an analytical formula to express the time needed for the output signal to rise to halfway its final value. This analytic expression depends only on T_c and T_b. Optimal performance can be expected when these two constants are close to each other. In case of $T_b = T_c$ and the load resistance being equal to the resistance of the emitter follower, the half-rise time $T_{0.5}$ can be written as :

$$T_{0.5} = 1.7\sqrt{\tau_{EC}(1.+ 2\alpha R_b\frac{I_s}{V_s}) \frac{V_s}{I_s} (3C_{BC}+C_{CS}) + \tau_{EC} (2 + \alpha R_b\frac{I_s}{V_s})} \quad (6.19)$$

In this expression R_b is the base resistance. I_s and V_s are the switching current, respectively the switching voltage and α, which is about 0.7, is a constant describing the high-frequency effect on the base resistance. Stork predicted, based on this expression, in 1988 gate delays below 20 ps. Such results have indeed been achieved since then (Harame, 1993) and the results will be discussed in the last section.

M. Ghannam (1990) presented a model valid at low and intermediate collector currents, where he splits the time response of the ECL circuit in 3 components. The first is the the transit time delay which is a function of both the response time of the collector current T_b and collector circuit response time T_c which are basically the same as in Stork's paper. In his model the delay due to the junction capacitances is also included, while the third term represents the delay of the emitter follower. This analytical model for the gate delay was then applied to strained $Si_{1-x}Ge_x$ HBT's (Ghannam, 1991). In that paper it was shown in a case study how the use of graded-SiGe base HBT's can shift the relation between ECL gate delay, cut-off frequency and intrinsic base sheet resistance. The results of his model calculations are shown in fig. 6.10.

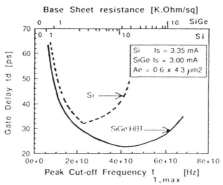

Fig. 6.10 Plot showing the effect of using a graded SiGe-base HBT on the relation between ECL-gate delay, cut-off frequency and intrinsic base sheet resistance (After Ghannam, 1991, reprinted from Techn. Dig. of the BTCM Meeting, ©IEEE1991)

Although the intrinsic speed of HBT's (τ_{EC}) is clearly higher, there will not be such a strong difference in the gate delays of circuits with comparable geometrical layout. Several factors attribute to this behaviour. *Area reduction and reduction of parasitics* contribute more to the reduction of the gate delays than increasing further the intrinsic speed. From this arises the need for self-aligned (and preferably selective processes - see Meister, 1992 and 1993). Moreover, the tendency to increase transit times, leads to very small base widths (high R_b) and high collector doping levels (high base-collector capacitances) which adversely affects the gate delay as shown by equation 6.19.

The conclusion of this short analysis of the relation between cut-off frequency and ECL-gate delay must therefore be that optimization of the internal device should be done in a moderate way and must be accompanied by reduction of extrinsic elements as to get full profit of the enhancement of the intrinsic device speed. Hence, not only technological improvements are the driving force behind shifts of the frequency performance of the bipolar transistor. In addition improved circuit designs can also provide significant speed improvement (Goto, 1990). An example of such approach is the *active pull-down for the emitter-follower circuit* (Toh, 1989). This has made possible the realization of ECL-gates with a delay as short as 13.2 ps (Toh, 1991).

6.4   INCORPORATION OF STRAINED $Si_{1-x}Ge_x$-LAYERS IN ADVANCED BIPOLAR TECHNOLOGIES

6.4.1   Introduction

The previous sections dealt mainly with the physics of SiGe-based HBT's. The incorporation of these strained layers in state-of-the-art bipolar technologies allows a quantitative comparison between the different figures-of-merit, characterizing the internal bipolar transistor or his function as a switch in a circuit (the intrinsic base sheet resistance, the cut-off frequency, the maximal oscillation frequency and the NTL- or ECL-gate delay). In this way it will become clear to which extent the use of SiGe-based heterojunctions can contribute to the further development of Si-based bipolar technologies.

The Si-based bipolar technology has been characterized by an enormous progress in the eighties. An excellent overview of these new processing sequences is given by Goto (1990). Improved implantation techniques, eventually accompanied by pre-amorphization and diffusion from polysilicon sources to form base and emitter profiles must be mentioned in this frame (Ehinger, 1988). The use of low-temperature epitaxial techniques has added additional freedom to tailor the dopant (and alloy) profiles to increase performance. However, the use of metastable (when thicker than the critical thickness -see section 5.2) strained layers like pseudomorphic $Si_{1-x}Ge_x$-alloys puts additional constraints on the thermal budgets used in these processes. We will show how the different research groups tried to fullfil this requirement.

6.4.2 Perimeter and surface passivation

Surface passivation and more specifically perimeter passivation (remember that modern transistors are characterized by a large perimeter-to-area-ratio of the base-emitter junction) is not a difficult problem in classical Si-technology, since the thermal oxide of Si has excellent passivation properties. Thermal oxidation however represents a large thermal budget and leads to oxidation enhanced diffusion, which might compromise the steep dopant profiles, pursued in these devices. It must be mentioned however that oxidation-enhanced diffusion is partially suppressed in SiGe strained layers because of the interaction between the Si-interstitials created during the oxidation and the compressive strain (Legoues, 1989a)).

A much more severe problem is that thermal oxidation of SiGe-alloys leads to segregation of Ge atoms under the oxide (Legoues, 1989a and b), leading to unacceptably high interface state densities. In the base-emitter depletion layer this will lead to a high non-ideal recombination perimeter component in the base current. Principally, the problems related to the thermal oxidation can be eliminated by covering the strained $Si_{1-x}Ge_x$-base with a Si capping layer. This does not only increase the stability (see section 5.2) of the layer (Noble, 1989), but also the interface state density will remain low as long as the upper Si is not completely consumed in the oxidation process. This solution has gained even stronger relevance since oxidation with low thermal budget has become possible by the introduction of the HIPOX-technique, which is basically an oxidation at high pressure and low temperature. Thicknesses of more than 10 nm can be grown at 700°C with a pressure of 10 atmosphere. This technique has been used in several advanced processes (see for instance Burghartz, 1991).

Some groups have reported on HBT's without any passivation. Gruhle (1993) used a very agressive processing sequence to demonstrate the frequency performance of MBE-grown devices. The emitter was selectively etched (Narozny, 1990) with the Pt/Au emitter-contact acting as a mask. The emitter and base contacts were air-bridged to reduce parasitic capacitances. Although the performance (f_T and f_{max}) of their device is the best for MBE-devices (see table 6.1) and still could be enhanced by using a graded base, the non-ideality of the base current will hinder the use of such transistor schemes for practical applications.

6.4.3   Emitter formation in strained $Si_{1-x}Ge_x$ HBT's

It was mentioned to which extent the low-temperature epitaxial techniques have triggered the optimization of the base-layer dopant and alloy profile. Extending this technique towards the emitter might look then the most straightforward approach, since emitter and base layers can then be grown together and low emitter resistance can be expected.

Epitaxial emitter HBT's are mainly made by MBE (Slotboom, 1991 or Gruhle, 1992) or RT-CVD (Kamins, 1989). This is not purely coincidental. Most low-temperature epitaxial techniques suffer from problems connected to n-type doping. Segregation effects of Sb (Jorke, 1988), smear-out of P-profiles (Leong, 1988), background contamination by PH_3 (Poortmans, 1993) and reduction of the growth rate upon addition of PH_3 (Poortmans, 1993) seriously complicate the growth of n-type epitaxial layers. Atmospheric-pressure CVD at low temperature has been shown to give high-quality epitaxial layers with high active As doping levels (Agnello, 1992). In MBE-systems the n-type doping problems can be solved by growing the layer at temperatures below 400°C to exploit the kinetical limitation of Sb-segregation (Jorke, 1988) or to use doping by secondary implantation (Jorke, 1985). In RT-CVD one can alleviate the problem by growing each separate layer at its optimal growth temperature and in between the depositions heating cycles can be inserted to reduce problems with PH_3- or AsH_3-contamination.

	Emitter width [nm]	Emitter doping $[cm^{-3}]$	Ge-content in the base [%]	f_T [GHz]	f_{max} [GHz]
Kamins (1989)	/	1.8×10^{17}	31	29	35
Slotboom (1991)	250	7×10^{17}	20	13	/
Gruhle * (1993)	300	2×10^{18}	21	91	65

* the base doping was 3×10^{19} cm^{-3} of Boron which together with a base thickness of 40 nm yielded a base sheet resistance of about 2KΩ/square

Table 6.1 : summary of results obtained on HBT's with epitaxial emitters

The main problem however with fully epitaxial emitters is their complete incompatibility with conventional processing sequences. Therefore a lot of effort has been devoted to incorporating epitaxial layers in processes with poly-emitters. The fastest device as far as internal speed is concerned was based on a non-self-aligned As-implanted polysilicon -emitter process (Patton, 1990a), who reported a cut-off frequency of 75 GHz at room temperature. A cross-section and SIMS-profile of this transistor is shown in fig. 6.11. In their approach the thermal budget, which consisted of a combination of furnace and rapid thermal annealing steps, was however still quite high due to the low diffusivity of As in Si. Since Boron diffuses much faster than As, the lightly doped spacers at the junctions loose part of their effect and the neutral base width becomes larger. Therefore it is a logical choice to go to phosphorus-doped poly-emitters (Crabbé, 1992), since the diffusivity of P is much closer to B. The thermal budget, associated with the annealing of the P-emitter implant, can then be substantially reduced to a 15 min furnace anneal at 750°C, followed by a RTA-step at 860°C for 10s to form the single crystal emitter and increase the dopant activation. This has to be compared to the As-emitters, which require a furnace anneal at 850°C and RTA at 970°C (no times were given). An additional advantage was the lower polysilicon sheet resistance (35% lower). The better preservation of the spacers in case of the P-doped polysilicon emitters reduced the leakage current of the base-emitter junction by a factor of 60 at a reverse bias of -1V. The low annealing temperatures however were unable to fully activate the B in the extrinsic base region, which limited f_{max} to 26 GHz (see table 6.2).

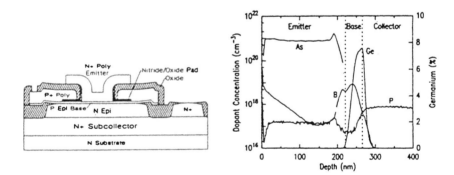

Fig. 6.11 Cross-section and SIMS-profile of the 75 GHz bipolar transistor. With a non-self-aligned technology a cut-off frequency of 75 GHz at room temperature was obtained (After Patton, 1990a, reprinted from Electron Device Letters, ©IEEE1990)

When going to very high doping levels in the base in the range between 10^{19} and 10^{20} cm^{-3} the optimization of the thermal cycle associated with the polysilicon emitter formation process becomes nearly impossible. The thermal budget must be low enough to prevent the formation of an n$^+$-p$^+$-junction, but at the same time it must be high enough to provide a low emitter resistance and prevent junction non-idealities and non-reproducible Si-poly-Si interfaces. A possible solution for this difficult trade-off was proposed by Comfort et al. (1991). In their process the single monocrystalline emitter is formed by epitaxially growing a thin n-type Si-cap of about 30 nm on top of the heavily doped strained $Si_{1-x}Ge_x$-base. The emitter is then completed by depositing an in-situ doped n^+-polysilicon emitter on this single crystalline cap, needed for the reduction of the emitter charge storage and for a low emitter and contact resistance. An important advantage also stems from the fact that the emitter cap is grown immediately after the base deposition, which makes the process less sensitive to interface- and diffusion-related phenomena during the outdiffusion from the polysilicon. To enhance the performance, the Ge-content in the base was graded down from 9% at the collector to 3% at the emitter side of the base. One can remark that the Ge-content at the emitter side of the base was not reduced to 0% in order to keep a certain barrier for injection of holes to reduce the charge storage in the single crystalline emitter portion. Nevertheless, the authors report only a partial success since they observed a large base leakage current which was due to tunneling. ECL-gates made by this method and with a silicided poly-layer to reduce the extrinsic base resistance, showed a gate delay as low as 24 ps for the SiGe HBT's which was about 14% lower than for Si-pseudo HBT's (because of their high base doping level) which underwent nominally identical processing.

	Emitter dopant species	base-doping [cm^{-3}]	Ge-grading [%]	f_t [GHz]	f_{max} [GHz]	ECL delay [ps]	R_{bi} [kΩ/]
Patton (1990)	As (poly)	5×10^{18}	0-8	75	/	/	17
Crabbé (1992)	P (poly)	5×10^{18}	0-9	73	26	34	16
Comfort (1991)	5×10^{17} epi-emitter	3×10^{19}	3-9	43	40	24	8
Haramé (1992)*	As (poly)	3×10^{18}	0-7.5	50	59	18.9	12
Burg-harz (1991)	As (poly)	6×10^{18}	18	63	/	24.3	8.3

* The HBT's were incorporated in an ECL BiCMOS Technology. This represents the highest integration and performance level in a SiGe-base technology until half 1993

Table 6.2 : Summary of reported results as a function of emitter technology and base doping

An interesting new technique is the so-called Selective Epitaxy Emitter Window (SEEW) technology, developed by Burgharz et al. (1991). The processing sequence is shown schematically in fig. 6.12. It is based on epitaxial overgrowth over a thin nitride/oxide pad to form the extrinsic base region, which defines at the same time the emitter window. This guarantees the link-up between the intrinsic and extrinsic base. The latter is a very difficult problem in submicron transistors where the base is formed by implantation and the extrinsic base is outdiffused from p^+-polysilicon (see for instance Chuang, 1987). Remark that the sidewalls of the emitter window are formed with the HIPOX-technique. In addition, it makes possible deep submicron transistors with emitter widths down to 0.2 μm without the necessity to go for e-beam lithography. They combined this with epitaxial Si or strained $Si_{1-x}Ge_x$-layers. With these transistors minimum ECL gate delays of 24.3 ps were measured in unloaded ring oscillators (Burgharz, 1990). The differences between Si and SiGe-base transistors must be called marginal since the difference in gate delay was only about 1 ps, although the cut-off frequency of the graded-base HBT's were significantly (25%) higher.

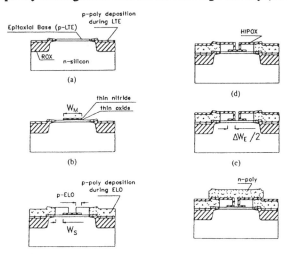

Fig. 6.12 Processing sequence for the SEEW-technology (After Burghartz, 1991, reprinted from IEEE Transaction on Electron Devices, ©IEEE 1991)

6.4.4 HBT's and selective deposition schemes of $Si_{1-x}Ge_x$-alloys

In section 6.3.2 the relation between internal and extrinsic transistor elements and the ECL-gate delay was dealt with. It became clear then how extrinsic elements like base resistance and the extrinsic base-collector capacitance will obscure the advances on the level of internal speed. Kamins (1989) mentioned that in their HBT-design 70% of the base-collector capacitance resides in the extrinsic part. Therefore reduction of these extrinsic elements is of outmost importance if one goes for a further improvement. Several solutions have been proposed like for instance a *selectively implanted collector* (SiC) where the collector doping is increased only in the active region of the transistor by implantation in the emitter window. Careful optimization is needed however to avoid problems at the edges of the active area when operating the device at high collector current densities (Taft, 1991).

Although the selective epitaxial growth of the base layer of bipolar transistors is attractive, the first attempts by Burghartz et al. (1988) were not really successfull due to remaining polysilicon nucleation around the perimeter opening. A real breaktrough for SiGe HBT's based on selective growth of the base epitaxial layer was reported in 1992 by Sato (Sato, 1992). By careful optimization of the base thickness and grading (0-15%) in their selective base growth process, they obtained a cut-off frequency of 51 GHz and an maximum oscillation frequency of 50 GHz. The very low resistance in the base-link region was obtained by outdiffusion from a BSG-film which contains up to 10 mole % of Boron (Bianco, 1991).

6.5   CONCLUSIONS

In this review chapter it was shown which opportunities are offered by the use of strained $Si_{1-x}Ge_x$-layers in the base of bipolar transistor. In the first section dealing with the collector current it was explained which peculiarities can arise due to heavy-doping effects in these layers and the double heterojunction at both base-emitter and base-collector diode. In the part about the speed performance it was made clear how the *internal speed* of the bipolar transistor is clearly improved by the bandgap engineering in the base which allows the creation of additional electrical fields by grading the Ge-concentration over the base. This improvement of the internal speed however is not proportionally translated in a performance gain at the circuit level. Therefore it is necessary that the incorporation of these strained layers in the base must be accompanied by process improvements which reduce the parasitic elements in the switching circuit. Additional features of the epitaxial growth of SiGe-alloys, like the great ease to grow them selectively, might therefore be a key element to bring these double heterojunction bipolar transistors to a stage of being an attractive extension of the Si-based bipolar technology. The extent to which this will turn out to be true, will however also depend on *reliability aspects*, a topic not covered by this review, since no systematic reports about this issue are available.

6.6 REFERENCES

Agnello D, Sedgewick T O, Goorsky M S and Cotte J, 1992 *Appl. Phys. Lett.* **60** 454

Bianco M, Ehinger K, Hantke B, Klose H and Philipsborn v H, 1991 *Proceedings of the 21st ESSDERC-conference, Microelectronic Engineering* **15** 525

Bonnaud O, Sahnoune M, Solhi A and Lhermite H, 1992 *Solid-State Electronics* **35** 483

Burghartz J N, Ginsberg B J, Mader S R, Chen T.-C and Harame D L, 1988 *IEEE El. Dev. Lett.* **9** 259

Burghartz J N, Mader S R, Ginsberg B J, Meyerson B S, Stork J M C, Stanis C L, Sun J Y.-C and Polcari M R, 1991 *IEEE Trans. El. Dev.* **38** 378

Burghartz J N et al. , 1990 *IEDM Techn. Dig.* 297

Chuang C T, Tang D D, Li G P and Hackbarth E, 1987 *IEEE Trans. El. Dev.* **34** 1519

Crabbé E F, Stork J M C, Baccarani G, Fischetti M V and Laux, S E, 1990 *IEDM Techn. Dig.* 463

Comfort J H, Crabbé E F, Cressler J D, Lee W, Sun J Y.-C, Malinowski J, D'Agostino M, Burghartz J N, Stork J M C and Meyerson B S, 1991 *IEDM Techn. Dig.* 857-860

Crabbé E F, Comfort J H, Lee W, Cressler J D, Meyerson B S, Megdanis A C, Sun J Y.-C, Stork J M C, 1992, *IEEE El. Dev. Lett.* **13** 259-261

Cressler J D, Comfort J H, Crabbé E F, Patton G L, Lee W, Sun J Y.-C, Stork J M C and Meyerson B S, 1991, *IEEE El. Dev. Lett.* **12** 166-168

De Boeck, J. , 1991, Ph.D. Thesis Katholieke Universiteit Leuven

De Graaff H C, 1986 *Process and Device Modeling* (Ed. W. L. Engl.), Elseviers Science Publishers B. V. (North-Holland)

Ehinger K, Kabza H, Weng J, Miura-Mattausch, Maier I, Schaber H and Bieger J, 1988 *Journal de Physique* Colloque C4 supplément au n°9 Tome 49 109

Gao G.-B and Morkoc H, 1991a *El. Lett.* **37** 1408

Gao G.-B, Fan Z.-F and Morkoc H, 1991 *Appl. Phys. Lett.* **58** 2951

Ghannam M, Nijs J, Mertens R and Dekeersmaecker R, 1984 *IEDM Techn. Dig.* 746

Ghannam M Y, Mertens R P, Van Overstraeten R J, 1990 *IEEE Trans. El. Dev.* **37** 191

Ghannam M Y, 1991 *Tech. Dig. of the Bipolar Circuits and Technology Meeting* Minneapolis Minnesota 166

Goto H., 1990 *from the book 'High-Speed Digital IC-technologies'* (Ed. Marc Rocchi), Edited by Artech House 57-85

Gruhle A, Kibbel H, König U, Erben U and Kasper E, 1992 *IEEE El. Dev. Lett.* **13** 206

Gruhle A, Kibbel H, Erben, U and Kasper E, 1993 *El. Lett.* **29** 415

Harame D L et al., 1990a *VLSI Symp. Technology* 47

Harame D L et al, 1990b *IEDM Techn. Dig.* 33

Harame, 1991 *VLSI Symp. Technology* 71

Hart P A H, 1981 *Handbook of semiconductors* **4** : Device Physics North-Holland Publishing Company (Ed. C. Hilsum) 88

Huang C H and Abdel-Motaleb I M, 1991 *IEEE Proceedings-G* **138**(2) 165

Iyer S S, Patton G L, Stork M C, Meyerson B S and Harame D L, 1989 *IEEE Trans. Elect. Dev.* 2043

Jain S C and Roulston D J, 1991 *Solid-State Electronics* **34** 453

Johannson H, Rudner S, Xu D.-X and Willander M, 1989 *Proceedings of the 5th international Workshop of Semiconductor Device*, New Delhi, India, (Ed. W. S. Khokle and S. C. Jain) 615

Jorke H and Herzog, H -J, 1985 *Proc. 1st. Int. Symp. Silicon MBE* (J. C. Bean, ed.), Proc. Vol. 85-7, Electrochemical Society Pennington, New Jersey352

Jorke H, 1988 *Surface Science* **193** 569

Kamins T I et al., 1989 *IEDM Techn. Dig.* 647

King C A, Hoyt J L, Gibbons F J, 1989 *IEEE Trans. El. Dev.* **36**, 2093

Kroemer, 1982 *IEE Proc.* 13

Kroemer H, 1985 *Solid-State Electronics* **28** 1101

Legoues F K, Rosenberg R and Meyerson B S, 1989a *Appl. Phys. Lett.* **54** 644

Legoues F K, Rosenberg R, Nguyen T, Himpsel F and Meyerson B S, 1989b *J. Appl. Phys.* **65** 1724

Leong W Y, Robbins D J, Young I M and Glasper J L, 1988 *Proc. UK. IT. Conf.* (INF. ENG. DIR, DTI, Victoria St. London, 1988) 382

Lu P-F, Comfort J H, Tang D D, Meyerson B S and Sun J Y-C, 1990 *IEEE Elect. Dev. Lett.* 336

Lundstrom M S, 1986 *Solid-State Electronics* **29** 1173

Manku T and Nathan A, 1991 *Journal of Applied Physics* **69** 8414

Matutinovic-Krstelj Z, Prinz E J, Schwarz P V and Sturm J C, 1991 *IEEE El. Dev. Lett.* **12** 163

McGregor J M, Manku T and Roulston D J, 1992 *Solid-State Electronics* **34** 421

Meister T F et al., 1992 *IEDM Techn. Dig.* 401

Meister T F, Stengl R, Felder A, Rein H-M and Treitinger L, 1993 *Proceedings of the 23rd European Solid State Device Research Conference* 203

Meyerson B S, Ganin E, Smith D A and Nguyen T N, 1986 *Journal of The Electrochemical Society* 1986

Meyerson B S, Uram K J and Legoues F K, 1988 *Appl. Phys. Lett.* **53** 2555

Middlebrook R D, 1963 *Solid-State Electron.* **6** 555

Moll J L and Ross, 1956 *I. M. , Proc. IRE* **1** 72

Narozny P, Kohlhoff D, Kibbel H and Kasper E, 1990 *Proceedings of the 20th ESSDERC*, Nottingham 477

Noel J P, Rowell N L, Houghton D C and Wang A, 1992 *Appl. Phys. Lett.* **61** 690

Patton G L et al., 1990a *IEEE El. Dev. Lett.* **11** 171

Patton G L, Stork J M C, Comfort J H, Crabbé E F, Meyerson B S, Harame D L and Sun J Y.-C, 1990 *IEDM Tech. Dig.* 13

Poortmans J, Jain S C, Caymax M, Van Ammel A, Nijs J, Mertens R P and Van Overstraeten R, 1992a *Microelectronic Engineering* **19** 443

Poortmans J, Caymax M, Van Ammel A, Libezny M and Nijs J, 1992b *MRS Symp. Proceedings Vol 281* (Ed. Charles W. Tu) 409

Poortmans J, 1993a Ph. D. thesis Katolieke Universiteit Leuven

Poortmans J, Jain S C, Totterdell D H J, Caymax M, Nijs J, Mertens R P and Van Overstraeten R, 1993b to be published in Solid-state Electronics

Prinz E J, Garone P M, Schwarz P V, Xiao X and Sturm J C, 1989 *IEDM Techn. Dig* 639

Prinz E J and Sturm J C, 1991a *IEEE El. Dev. Lett.* **12** 661

Prinz E J, Garone P M, Schwarz P V, Xiao X and Sturm J C, 1991b *IEEE El. Dev. Lett.* **12** 42

Pruymboom A, Slotboom J W, Gravesteijn D J, Fredriksz C W, Van Gorkum A A, Van De Heuvel J M, Van Rooij-Mulder J M L, Streutker G and Van de Walle G F A, *IEEE El. Dev. Lett.* **12** 357

Roulston D J, 1990 *IEEE El. Dev. Lett.* **11**

Roulston D J and McGregor J M, 1992 *Solid-State Electronics* **35** 1019

Sasaki K, Fukazawa T and Furukawa S, 1987 *IEDM Tech. Dig.*, 186-189

Sato F, Hashimoto T, Tatsumi T, Kitahata H and Tashiro T, 1992 *IEDM Techn. Dig.*

Shafi Z A, Martin A S R, Whitehurst J, Ashburn P, Godfrey D J, Gibbings C J, Post I R C, Tuppen C G, Booker G R and Jones M E, 1991a *Microelectronic Engineering* **15** 135

Shafi Z A, Gibbings P, Ashburn P, Post I R C, Tuppen C G and Godfrey D J, 1991b *IEEE Trans. El. Dev.* **38** 1973

Slotboom J W, Streutker G, Pruijmboom A and Gravesteijn D J 1991 *IEEE El. Dev. Lett.* **12** 486

Stork J M C, 1988 *IEDM Techn. Dig.* 550

Sturm J C, Prinz E J and Magee C W, 1991 *IEEE El. Dev. Lett.* **12**, 303

Sugii T, Ito T, Furumara Y, Doki M, Mieno F and Maeda M, 1988 *IEEE El. Dev. Lett.* **9** 87

Sugii T, Yamazaki T, Arimoto Y, Ito T, Furumura Y, Namura I, Goto H and Tahara A, 1992 *Microelectronic Engineering* **19** 335

Symoens J, Ghannam M, Neugroschel A, Nijs J and Mertens R, 1987 *Solid-State Electronics* **30** 1143

Suzuki K, 1991 *IEEE Trans. El. Dev.* **38** 2512

Taft R C, Hayden J D and Gunderson C D, 1991 *IEDM Tech. Dig.* 869

Takahashi M, Tabe M and Sakakibara Y 1987 *IEEE El. Dev. Lett.* **109** 475

Tang and Solomon, 1979, *IEEE J. Sol. State Circuits* 679

Tang, 1989, *IEEE Elect. Dev. Lett.* 67

Toh K Y, Chuang C T, Chen T C, Warnock P, Li G P, Chin K and Ning T H, 1989 *IEEE ISSCC* 224

Toh K Y, Warnock J, Cressler J, Jenkins K, Danner D and Chen T.-C, 1991 *Tech. Dig. of the Bipolar Circuits and Technology Meeting*, Minneapolis, Minnesota 136
Ugaijin M and Amemiya Y, 1991, *Solid-state Electron.* **34** 593
Willander M, Shen G.-D, Xu D.-X and Ni W.-X, 1988 *J. Appl. Phys.* **63**(10) 5036
Yano K, Nakazato K, Miyamoto M, Onai T, Aoki M and Shimohigashi K, 1991 *IEEE Trans. El. Dev.* **38** 555-565

7

Field-Effect Transistors, Infrared Detectors, and Resonant Tunneling Devices in Silicon/Silicon-Germanium and δ-Doped Silicon

Magnus Willander

7.1   INTRODUCTION

In the last ten years there has been a tremendous development in III-V semiconductor devices. The advances in the development of the modulation-doped field-effect transistors (MODFETs), quantum devices such as resonance devices, and quantum well based infrared (IR) detectors were possible with experience gained in epitaxial growth techniques during the last decade. MODFETs are already used for commercial purposes and quantum well IR detectors will probably be the second category of quantum devices (second to semiconductor lasers) to be put into industrial production. It also seems reasonable to expect that resonant tunneling (RT) devices will eventually be the third category of the quantum devices to be used widely in electronics.

Advances made in low temperature silicon epitaxial growth techniques using molecular beam epitaxy (MBE) and chemical vapour deposition (CVD), will allow similar progress outlined above for III-V devices will occur for silicon based heterostructures. The progress in silicon based heterostructure devices with improved performance and novel functionality realised at small dimensions will make silicon monolithic integration approach possible since conventional silicon is compatible. The ease of integration of silicon based heterostructure with conventional silicon process, and the potential of Si/SiGe and δ-doped devices offer significant advantages over previously proposed monolithic integration approaches using mixed technologies. This is specially true when hybrid solutions between silicon integrated chips and III-V devices is not suitable, and when improvements in conventional silicon technology is desirable.

In this chapter we will discuss both the physics and the improvements which have been achieved for Si/SiGe quantum well MOSFETs and MESFETs. We will then look at the advantages of quantum well and heterostructure IR-detectors, and explain their operation. Finally, we will briefly discuss some topics regarding resonant tunneling Si/SiGe diodes and transistors. The analysis of δ-doped structures will be provided with reference to both the FETs and RT devices.

7.2 FIELD-EFFECT TRANSISTORS

7.2.1 Silicon/Silicon-Germanium

The physics behind the improved hole mobility lies mainly in the influence of strain on the valence band (Fu et al, Fu et al 1991). For electrons the improved mobility is due to the splitting in the 2-fold and 4-fold states at the X point conduction band minimum. A detailed discussion on the electronic properties of strained SiGe layer is given in Chapter 1.

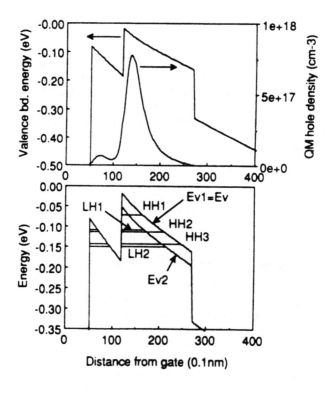

Figure 7.1 P-channel SiGe quantum-well MOSFET structure for gate voltage -1.25 V (Hoffmann 1991).

There is a need for devices with improved current drive capability as well as high frequency properties. These requirements, together with improved hole and electron mobilities in quantum wells, have lead to demonstrations in the last several years of Si/SiGe MOSFETs and MESFETs with high performance characteristics.

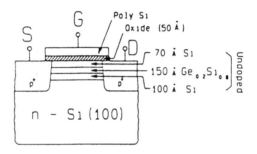

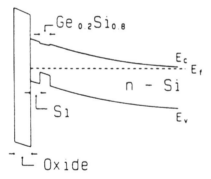

Figure 7.2 A schematic diagram of the p-type $Si_{0.8}Ge_{0.2}$ MOSFET and its energy band diagram (Na-yak et al 1991).

The first p-type MESFET (MODFET) based on a two dimensional hole gas in a buried quantum well was presented in 1985 (Pearsall et al 1985, Pearsall et al 1986). This transistor which had a Schottky gate, showed a maximum transconductance of 3 mS/mm for a gate length of around 2 µm. To eliminate the alloy scattering, a MODFET with a pure germanium channel was fabricated by Murakarim et al in 1991 possessing a hole mobility as high as 8600 cm^2/(Vs) and a saturation velocity of 1.5×10^7 cm/s at T=77K. The intrinsic transconductance measured for the gate length of 4 µm was 37 mS/mm. Since the occurence of a larger leakage current between the source and gate (at higher gate biases) can dramatically lower the transconductance (Chen and Willander 1992b), this type of transfer characteristics behaviour remains a major drawback in heterojunction FETs utilizing Schottky gate. Therefore, the modulation doped MOSFET (MOD-MOSFET) is much more promising than the Schottky gated MODFET. Before we describe the different experimental results obtained for MODMOSFETs, we should briefly discuss the results of modeling the charge control process in the channel from self-consistent numerical solution of the Schroedinger and the Poisson equations (Hoffmann 1991).

The charge control process is primarily influenced by the Ge concentration in the well and the thickness of the Si spacer between the oxide and the well (Figure 7.1) and is insensitive to the well thickness (down to 6nm). Generally, the properties of a p-channel SiGe/Si MODMOSFET will be improved if the Ge is increased and spacer thickness reduced. However, higher Ge concentration and a thinner spacer also leads to larger carrier scattering rates. As an example of the improved quantum well carrier density, Hoffman (1991) showed that a reduction of the spacer thickness from 7 nm to 2 nm (x=0.2) causes the quantum well channel carrier density to increase by a factor 6.

Enhancement-mode quantum well $Ge_{0.2}Si_{0.8}$ PMOS has been fabricated and characterized by the UCLA group (Nayak et al 1991) and a schematic drawing of their transistor is shown in Figure 7.2. The band discontinuity between Si and $Ge_{0.2}Si_{0.8}$ was 0.15 eV for the valence band and 0.02 eV for the conduction band. A thermally grown oxide was used as isolating gate material and all epitaxial semiconductor layers were grown by molecular beam epitaxy using rapid thermal processing in all high-temperature steps. For a gate length of 0.7 μm and a drain-to-source voltage of 2.5 V, a transconductance g_m of 64 mS/mm was obtained. (For a similar Si PMOS g_m =80 mS/mm was obtained.) The channel mobility was determined to 155 $cm^2/(Vs)$ for a long channel device as opposed to a similar Si PMOS with a channel mobility of 122 $cm^2/(Vs)$.

The low frequency noise behaviour of this transistor has also been investigated by Chang et al (1991). At room temperature the generation-recombination, (g-r) noise was dominate, while below 70K the 1/f noise becomes more important. The (g-r) noise originates from the interaction between holes and traps in the cap layer. The 1/f noise comes from tunneling of holes in the channel to the traps in the cap layer.

A similar structure was grown using a combination of rapid thermal processing and chemical vapour deposition by the Princeton group (Garone et al 1991a). The gate oxides were plasma deposited at 350 °C/30s, and Ge concentrations in the channel were 0.2, 0.3, and 0.4 for different transistors. Enhancement of 50 % and over 100 % in channel carrier mobility occured at room temperature and at T=90 K respectively compared with Si PMOS (Garone et al 1991b, Garone et al 1992).

The IBM group (Verdonckt-Vandebrock et al 1991) demonstrated a modulation-doped SiGe-channel p-MOSFET (grown by MBE), (Figure 7.3). By using an undoped graded $Si_{1-x}Ge_x$ channel with a boron-doped layer underneath, the thre-

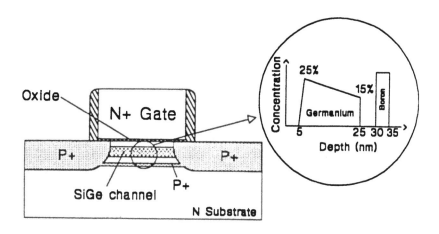

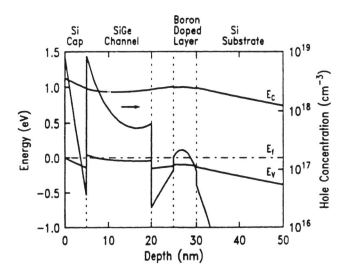

Figure 7.3 A schematic cross-section of the MODMOSFET and its energy band diagram for V_{gs}=-2V (Verdonckt-Vandebrock et al 1991).

shold voltage was tailored without degrading the mobility. Thus hole mobilities as high as 220 cm^2/(Vs) at T=300 K and 980 cm^2/(Vs) at T=82 K were obtained. These three demonstrations of p-channel SiGe/Si MOSFETs during the early

1990s showed that improvements of transistor functions can be expected from incorporation of a SiGe buried channel.

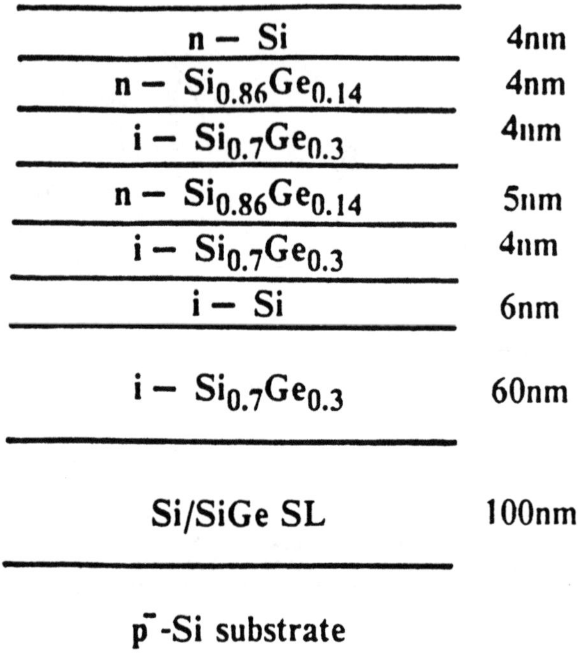

n — Si	4nm
n — Si$_{0.86}$Ge$_{0.14}$	4nm
i — Si$_{0.7}$Ge$_{0.3}$	4nm
n — Si$_{0.86}$Ge$_{0.14}$	5nm
i — Si$_{0.7}$Ge$_{0.3}$	4nm
i — Si	6nm
i — Si$_{0.7}$Ge$_{0.3}$	60nm
Si/SiGe SL	100nm
p$^-$-Si substrate	

Figure 7.4 A cross-section of the grown layers to obtain a high electron mobility channel. The 6 nm-thickness intrinsic Si layer contains the two-dimensional electron gas (Ismail et al 1991a).

N-type quantum-well FETs use the tensile strain in the silicon layers to cause a downshift of the two-fold conduction band valleys. The conduction band offset can now be much larger than kT (150 meV or larger). This also results in a larger in-plane effective mass (m*=0.9m$_0$) and a smaller effective mass (m*=0.2m$_0$) perpendicular to the layers. Hence large enough band offset can be obtained for confinement of electrons for n-type MODFET design, while it also opens the possibility for room temperature operation of RT devices (due to the low electron effective mass compared with the hole).

Recently a new kind of graded Si/SiGe buffer layer has been demonstrated with a very low dislocation density in the tensile Si layers (Ismail 1991a, Ismail 1991b, LeGoues 1991). At IBM, UHV/CVD was used to grow a n-type modulation-doped Si/SiGe structure on a Si/SiGe superlattice structure (see Figure 7.4). In this samples electron mobilities of 1800 cm^2/(Vs), 9000 cm^2/(Vs), and 19000 cm^2/(Vs) were measured at room temperature, 77 K, and 1.7 K. Similar high values of the electron mobilities were obtained in structures grown by the Daimler Benz (Schubert et al 1991b). Mii et al demonstrated even higher electron mobilities (Mii et al 1991). The best reported value is 173000 cm^2/(Vs) at 1.5 K (Schäffler et al 1992).

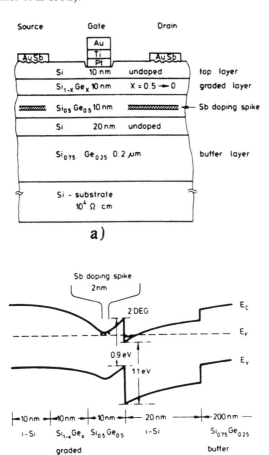

Figure 7.5 a) A schematic cross-section view of the first n-channel MODFET. The 20-nm thickness undoped Si layer contains the two-dimensional gas. b) The band-structure of n-MODFET. (Daembkes et al 1985).

The first n-type Si/SiGe MOS was fabricated by using high dose Ge ion implantation (Selvakumar and Hecht 1991). This transistor also showed higher transconductance than for the Si MOST. However, using these very promising results for n-type growth of Si/SiGe quantum wells on graded Si/SiGe superlattices, one could expect very good results for n-typ quantum-well FETs.

A series of breakthroughs for n-type HEMTs was done by the Daimler-Benz group (Daembkes et al 1985, Daembkes et al 1986). They demonstrated n-type Si/SiGe HEMT (see Figure 7.5). For the gate length of 1.6µm this device

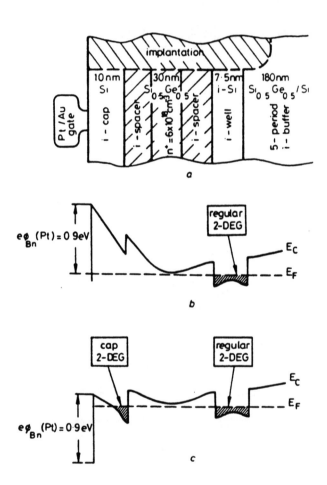

Figure 7.6 The structure of the two-electron channel MODFET (König et al 1991).

showed an extrinsic transconductance of 50 mS/mm. A few years later Daembkes (1988) also proposed a complementary MODFET structure. A relaxed $Si_{0.75}Ge_{0.25}$ alloy buffer layer was used on which a strained Si (n-channel) was grown. Above this layer a strained $Si_{0.5}Ge_{0.5}$ layer (p-channel) was grown and finally a Si cap layer (see Figure 7.5). The carriers in the channels are supplied from the source contacts. Using a graded SiGe buffer layer to improve the crystal quality of the channel, an extrinsic transconductance as high as 340 mS/mm and 670 mS/mm at 300 K and 77 K respectively were experimentally obtained for gate lengths of 1.4 μm (König et al 1992). (A large portion in the improvement

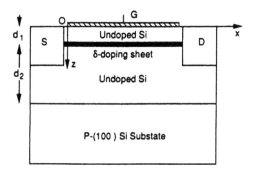

Figure 7.7 The schematic illustration for the structure of the silicon δ-doped FET in (Chen and Willander 1992a).

the transconductance resulted from the use of a recessed gate.) These values are of the same order as those for similar III-IV MODFETs and similar results were obtained by the IBM group, (i.e. g_m= 600 mS/mm at T=77 K for channel lengths of 0.25 μm (Ismail 1992).)

Finally, a twin electron channel MODFET (see Figure 7.6) was also realized by the Daimler-Benz group (König et al 1991) decreasing the gate current and simultaneously increasing the transconductance by moving the conducting channels near the gate.

7.2.2 δ-doped Field-Effect Transistors

Buried channels can also be created in Si by using δ-function-like doping profiles. A Schottky-gate δ-doped FET is shown in Figure 7.7. The characterization of the sublevels in the V-shaped structure as well as the mobility properties has been described in (Chen and Willander 1991). The advantages of the δ-doped Schottky-gate FETs are: (1) reduced short-channel effects, (2) improved transconductance, and thus, (3) higher speed operation as compared to homogeneously doped FETs.

When the gate length becomes comparable with the gate-channel distance, short channel effects become important. Since the distance between gate-channel can be made short by using δ-doping, the short channel effect is reduced.

The improved transconductance is a result from the short distance between gate and channel. If we approximate f_T by

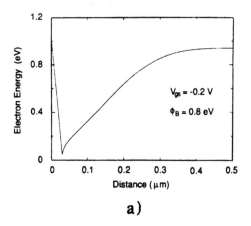

a)

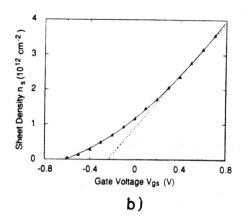

b)

Figure 7.8 a) The conduction band profile for the Si δ-doped FET. ($N_{2d}=4\times10^{12}$ cm^{-2}, $d_1=30$ nm, $T_{2d}=2$ nm, $d_2=150$ nm, $N_{buf}=10^{15}$ cm,$^{-3}$ $P_{sub}=10^{16}$ cm^{-3}). The Fermi level has been set to zero as an energy reference. b) The sheet density of the channel electrons n_S vs the gate voltage V_{gs} in a δ-FET with the same parameters as those in Figure 7.2 but $d_1=20$ nm. Numerical results (•), qudratic fit (—), and linear fit (---) (Chen and Willander 1992a).

$$f_T = g_m/(2\pi C_{gs}),\qquad(7.1)$$

we note that f_T is increased mainly due to increase the transconductance g_m, and the gate-channel capacitance C_{gs}, can be held constant by decreaseing gate length to compensate for any increase in C_{gs} due to decrease in gate-channel distance.

The self-consistent solution of the Schroedinger equation for the 2DEG and the Poisson equation for the electric potential distribution gives the numerical relationship of sheet density of the channel electrons, n_s, vs. the applied gate voltage, V_{gs}. The difference between a linear relation of the numerical results is shown in Figure 7.8. A relation between n_s and V_T can be written :

$$n_s = a + bV_{gs} + cV_{gs}^2.\qquad(7.2)$$

Using this expression for the charge control model and the two-piece velocity field model for silicon, the effect of δ-doping on f_T was determined from equation (7.2) (Chen and Willander 1992a). The results are shown in Tables 7.1 and 7.2. The influence of the non-linear charge control model compared to the linear charge control model on the d.c. output characteristics is shown in Figure 7.9.

The δ-doped MOST was investigated theoretically (Yamaguchi et al 1983). Here we should also mention that hot carrier effects are suppressed as well as the problem with increased threshold voltage (as a consequence of increase in substrate doping required to avoid punch-through) when scaling the transistor to smaller dimensions. If a thin p-channel is grown below the conducting δ-doped n-channel, it can be shown from numerical simulations that the punch-through is suppressed by introduction of an extra thin p-layer. The threshold voltage is found to be more stable when the gate length is decreased (Yamaguchi et al 1988). These facts have been demonstrated experimentally for δ-doped MOS tran-

Table 7.1 The effect of δ-doping sheet density N_{2d} on the performances of a Si δ-FET (d_1=30 nm, L=0.3 μm, V_{gs}=0 V, V_{ds}=2 V) (Chen and Willander 1992a).

N_{2d} (10^{12} cm^{-2})	I_{ds} (mA mm^{-1})	g_m (mA mm^{-1})	C_{gs} (10^{-2} pF mm^{-1})	f_T (GHz)
3.0	30.8	103.8	78.8	17.2
4.0	80.9	135.1	82.8	21.3
5.0	140.5	150.4	84.4	23.2
6.0	190.2	147.2	84.8	22.4
7.0	227.2	140.5	83.2	20.1

Table 7.2 The influence of the top layer thickness d_1 on the performances of an Si δ-FET ($N_{2d}=4\times 10^{12}$ cm^{-2}, L=0.3 μm, $V_{gs}=0.2$ V, $V_{ds}=2$ V) (Chen and Willander 1992a).

d_1 (100 Å)	I_{ds} (mA mm^{-1})	g_m (mA mm^{-1})	C_{g} (10^{-2} pF mm^{-1})	f_T (GHz)
2.0	57.2	145.3	106.7	18.5
2.5	80.0	151.4	97.1	21.2
3.0	109.7	154.4	88.0	23.0
3.5	129.2	149.2	81.2	23.6
4.0	155.7	140.9	72.8	24.1

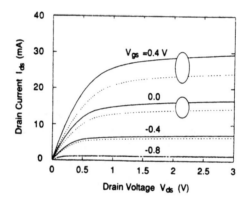

Figure 7.9 The effect of the square term of the charge control model on the d.c. output characteristics (solid line: c≠0, dashed line: c=0) for the intrinsic device of Figure 7.8.a with gate length L=0.3 μm and width W=200 μm (Chen and Willander 1992a).

sistors with 2-μm channels, using Atomic Layer Deposition (ALD) (Van Gorkum et al 1982, Nakagawa et al 1989).

N-channel δ-doped MOS transistors with gate lengths down to 0.8 μm were recently demonstrated (O'Neill et al 1992). They were grown by MBE and the post-processing steps were similar to conventional MOS transistors without any high temperature steps. A transition in the I-V characteristic shows that a parasitic channel between the gate oxide and the quantum well was formed. The channel mobility was determined to be 155 cm^2/(Vs).

Finally, the work function of poly-$Si_{1-x}Ge_x$ is found to vary with the content x, thus it is possible to find an optimum value for p^+-poly-$Si_{1-x}Ge_x$ which is suitable as a gate material for both the NMOS and PMOS transistors.

7.3   INFRARED DETECTORS

Two types of detectors will be discussed in this chapter. 1) IR detectors based on creation of the photocurrent by excitation of carriers from the valence band to the conduction band (interband absorption). These detectors (p-i-n photodetectors) can detect wavelengths up to around 1.5 μm when Ge is introduced into the silicon crystal, with a small extension of the cut off wavelength when an electric field is applied to the detector. 2) IR detectors based on creation of photocurrent by excitation of carriers over a Si/SiGe hetero-interface. These detectors can detect wavelengths up to 18 μm or more.

For long wavelength infrared (LWIR) detectors it is also possible to work with δ-doped induced wells and use the intersubband absorption as an operation mechanism. In silicon this concept has not yet been demonstrated for detectors. However, fundamental studies of intersubband absorption in Sb and B δ-doped quantum well structures have been performed (Temple et al 1990, Lee and Wang 1992a). The advantage with the silicon based detectors is the possibility for monolithic integration with silicon electronics. Even if the silicon detectors will not obtain better figures of merit than the best values for other materials, this is of less importance. For example, the fiber amplification in fiber-optic communication will decrease the requirements of the detectors.

7.3.1 p-i-n Detectors

For fiber-optic communications, hybrid solutions are usually used (e.g. InGaAsP photodetectors mounted on Si chips). The first attempt to make an integrated solution was made by Lury et al (1984). A Ge p-i-n photodiode was grown on a $Si_{1-x}Ge_x$ buffer layer on a Si substrate, reducing the number of threading dislocations and internal quantum efficiency of 41 % at 1.45 μm and 300 K was obtained. The disadvantages of this were structure with a Ge terminated surface and high noise (a problem for avalanche photodetection) due to a low ratio of the ionization coefficients for holes and electrons.

These problems were solved by using a wave-guide structure (Temkin et al 1986, Pearsall et al 1986) (see Figure 7.10). The different internal efficiencies for different wavelengths and Ge concentrations in the active layer are shown in Figure 7.10. Additionally, good results from high-frequency-response measurements were obtained (e.g. a rise time of 50ps). This initial work showed that Si/SiGe superlattice avalanche photodiodes could compete with InGaAs p-i-n detectors for wavelengths around 1.3 μm.

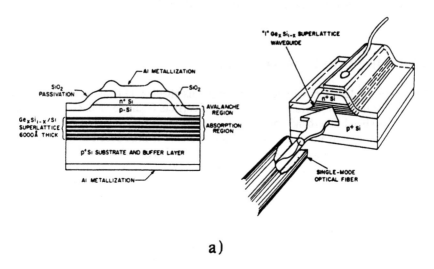

a)

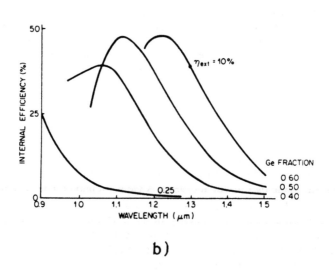

b)

Figure 7.10 a) Schematic diagram of the epitaxial structure of the separated avalanche and absorption $Si_{1-x}Ge_x/Si$ infrared waveguide detector (Pearsall et al 1986). b) Room-temperature spectral response of the detector (Temkin et al 1986).

The absorption edge for excitons is sensitive to the electric field which means that one can move the absorption edge into the IR region for Si/SiGe superlattice structure to wavelengths longer than 1.5 μm. Hence, self-electrooptic effect devices (SEEDs) can be used as modulators and realized in Si/SiGe structures.

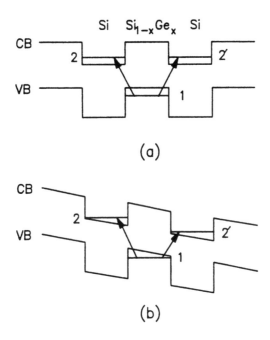

(a)

(b)

Figure 7.11 A type II $Si_{1-x}Ge_x$ multiple quantum well. a) Non-biased. b) Biased (Park et al 1990).

If a quantum well structure is grown on a relaxed $Si_{1-x}Ge_x$ buffer layer, a type 2 bandstructure is created (see Figure 7.11). Park et al (1990) experimentally showed the influence of an electric field on the quantum well structure in the wavelength range 1—2 μm. The experimental results for a type 2 p-i-n diode-structure consisting of 150 periods of 7-nm thick quantum wells separated by 7-nm Si barriers gave a redshift in the absorption edges (see Figure 7.12).

7.3.2 Heterojunction Internal Photoemission Detectors

Let us now go over to discuss detectors based on creation of the photocurrent by excitation of the carriers over a Si/SiGe heterojunction barrier. First we will analyse the development of $Si_{1-x}Ge_x$/Si heterojunction internal photoemission (HIP) detector for detections of IR light (in the range 8—16 μm). The primary purpose in introducing a HIP detector was to extend wave-length range compared with the Schottky barrier diodes, which are only usuable to 5 μm and to obtain an increase of the quantum efficiency for Schottky barrier diode detectors (as Pt/Si and Ir/Si). A solution to the low quantum efficiency was obtained by using

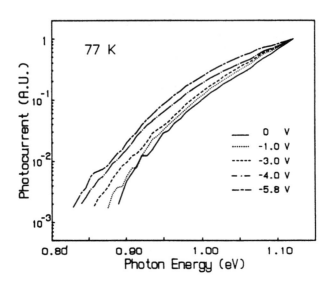

Figure 7.12 Photocurrent spectra vs. photon energy at T=77K as function of reverse bias across the p-i-n diode (Park et al 1990).

internal photoemission over a heterostructure as indicated by Sheperd et al (1971) and first implemented in a Si/SiGe heterostructures during 1990 by the Jet Propulsion Laboratory (Lin et al 1990a, Lin et al 1990b). They used a structure as shown in Figure 7.13. By using a SiGe/Si heterostructure internal photo-emission (HIP) operation, they obtained 3—5 % internal quantum efficiency Y_i, several times higher than for Schottky barrier diodes. One reason for this im-provement is that, for Schottky detectors, photons can excite carriers far below the Fermi level and yet not have enough energy to overcome the barrier. On the other hand, HIP detectors have a narrow band of absorbing states (Figure 7.13) which leads to a sharper turn-on and higher sensitivity near the cutoff frequency. Mathematically, the difference in Y_i can be seen from a comparision of the inter-nal quantum efficiency for a Schottky-barrier IR detector (equation 7.3a) and a heterojunction IR detector (equation 7.3b) (neglecting reflections of photoexcited carriers at the interface)

$$Y_i = (h\nu - \psi)^2/(8E_f h\nu) \tag{7.3a}$$

$$Y_i = (h\nu - \psi)^2/((8E_f h\nu)^{1/2}(E_f + \psi)^{1/2}). \tag{7.3b}$$

where E_f is the Fermi energy, ψ is the barrier height, and $h\nu$ is the photon ener-gy. Since E_f is usually much smaller for semiconductors than for metals, the in-ternal quantum efficiency is larger for HIP detectors than for Schottky-barrier de-tectors. However, the absorptance A can be larger for Schottky-barrier detectors,

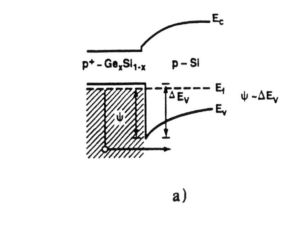

a)

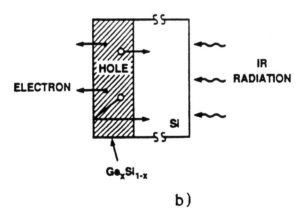

b)

Figure 7.13 a). Energy-band diagram and structure of the p^+-$Si_{1-x}Ge_x$/p-Si HIP de-tector. b). Structure of the p^+-$Si_{1-x}Ge_x$/p-Si HIP detector (Lin et al 1990a).

and this fact increases the external quantum efficiency Y ($Y=AY_i$) for Schottky-barrier diodes. Another reason for improvement of the external quantum efficien-cy is that the reflection of the photoexcited carrier at the interface is lower for the semiconductor heterointerface than for the metal-semiconductor interface.

The external quantum efficiency Y is controlled by the absorptance (A) and the internal quantum efficiency Y_i. The principal implementation of the absorptance is by increasing the boron doping in the $Si_{1-x}Ge_x$ layer (forming an extra absorbing metallic layer on top of the $Si_{1-x}Ge_x$ layer) or by incorporating an extra optical cavity (Tsauer et al 1991a). The Lincoln Laboratory, MIT, (Tsauer et al 1991a) showed experimentally that Y increased (Figure 7.14) when an absorbing Pt layer was deposited on the SiGe layer. They also suggested that a double-heterojunction structure should increase the internal quantum efficiency.

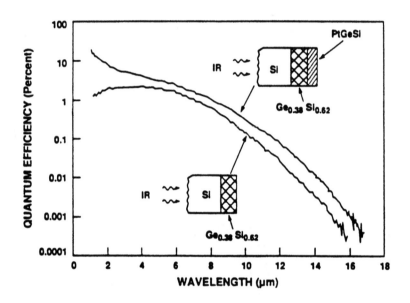

Figure 7.14 External quantum efficiency for a SiGe/Si HIP detector with and without a Pt-Ge-Si overlayer (Tsauer et al 1991a).

Carrier scattering decreases ψ_i because the scattering events can reduce the energy of the photoexcited carriers. As a result of this effect for a Si/SiGe HIP detector, the thickness of the SiGe layer doped to 2×10^{20} cm^{-3} should, in practice, be around 20 nm.

The controlling of the cutoff frequency is made by the choice of germanium concentration, and also the boron concentration in the $Si_{1-x}Ge_x$ layer due to heavy doping effects on the bandgap narrowing. Jain et al (1992) showed this theoretically for a boron concentration of 10^{20} cm^{-3} at 100K for different germanium concentrations. In an improved analyses the influence of the existing energy sublevels in the V-shaped potential at the Si/SiGe interface, and the nonparabolicity of the valence band for higher energies should be taken into account. Hence, a complete theoretical explanation of the photoresponse of Si/SiGe HIP detectors is still missing.

A practical implication was demonstrated by the MIT group (Tsauer et al 1991b) who fabricated 400×400-element arrays which had $Si/Si_{0.56}Ge_{0.44}$ HIP detectors (λ=9.3 μm, T=53 K, dark current $\approx 10^{-8}$ A/cm^2). They also demonstrated a 320×244-element arrays which had $Si/Si_{0.60}Ge_{0.40}$ HIP detectors (λ=10.5 μm, T=48 K). The SiGe arrays had excellent uniformity (Tsauer et al 1991a).

7.3.3 Intersubband absorption Detectors

Long-wave (8—16 μm) IR detectors have also recently been demonstrated in p-type Si/SiGe intersubband absorption multiple quantum wells at UCLA (Karunasiri et al 1991, Park et al 1992a). They made a multiple quantum well IR detector which consisted of 50 periods of 3-nm-thickness $Si_{0.85}Ge_{0.15}$ wells (p-doped to $\approx 10^{19}$ cm^{-3}) and separated by 50-nm-thickness undoped Si barriers. Subsequently, primarily free carrier absorption and intersubband absorption were the dominating processes in this detector.

The photocurrent, I_p, can be obtained through to number of photogenerated carriers at a given quantum well times the probability of their reaching the contact (Karunasiri et al 1991)

$$I_p(E) \approx 2(hv)^{-1} e P_0 \, \alpha d \sum_{n=1}^{\infty} exp(-nL/(v(E)t)) \qquad (7.4)$$

where P_0 is the incident interface power, hv is the photon energy, α is the absorption coefficient and d is the well width. The period of the multiple quantum well is given by L, n is the number of quantum wells, $v(E)$ is the drift velocity, and t is the excited carrier lifetime. This equation can be resolved assuming that the absorption depth is much larger than the well width. In moderate electric fields the responsivity $R(E)=I_p(E)/P_0$ can be written (Karunasiri et al 1991)

$$R(E) \approx (hv)^{-1} 2e\alpha\tau (d/L)\mu E/(1+\mu E/v_s) \qquad (7.5)$$

where v_s is the saturation velocity and $v(E)=\mu E/(1+\mu E/v_s)$. The measured and simulated results (by using equation (7.5), $\mu \approx 1000$ cm^2/(Vs) and $\tau = 15$ ps) are shown in Figure 7.15. $R(E)$ can be increased by decreasing the barrier width ($\approx 1/2$). However if one goes to far, the detectivity $D*$ could be decreased due to a decrease in the carrier lifetime. D^* can be written (Ka-runasiri et al 1991) as:

$$D^* = \eta(eA\cos\Theta/I_d)^{1/2}(2hv)^{-1} \qquad (7.6)$$

where $\eta=1-exp(-2\alpha d)$ is the quantum efficiency, A is the area of the device, Θ is the incident angle, and I_d is the dark current. If L^{-1} is too small, I_d will increase and then D^* will decrease. D^* was around 10^9 cm(Hz)$^{1/2}$/W at T=77 K for λ=9.5 μm. This was obtained for 45° incidence illumination to the quantum wells.

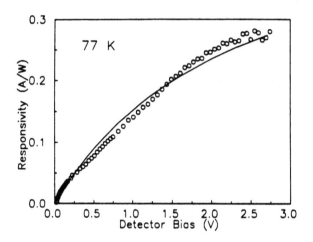

Figure 7.15 Responsivity of the $Si_{0.85}Ge_{0.15}$/Si multiple quantum well infrared detector as function of applied bias across the device at 77 K. Solid curve is from equation (7.5) (Karunasiri et al 1991).

The ATT group has published results from a p-type $Ge_{0.25}Si_{0.75}$/Si strained layer quantum well IR detector (People et al 1992). They obtained D^*=3.3×10^9 cm(Hz)$^{1/2}$/W for normal incidence illumination at 77 K and λ=10.8 μm.

Normal incidence electron intersubband transistors have been experimentaly demonstrated in SiGe/Si multiple quantum wells grown on Si (110) substrates by the UCLA group (Lee and Wang 1992b) although it should be noted that the most common substrates for VLSI is Si (100) substrates. They also demonstrated strong absorption peaks from 8.3 μm to 5 μm in $Ge/Si_{1-x}Ge_x$ (0.4≤x≤0.8) multiple quantum well structures grown on Si (100) substrate. Therefore normal incidence detection of light has been demonstrated in QW (quantum well) structures based on the silicon technology for both n- and p-types. These results offer the opportunity for integration of QW detectors operating with normal incident light with silicon VLSI circuits. The main advantages of the GeSi QW detectors is that they can detect the normal incident infrared radiation but normal incident detection mechanisms for p- and n-type GeSi QW are very different.

For p-type $Si_{1-x}Ge_x$/Si QW the bandgap at the Γ point between the valence band and the conduction band is so narrow that the band mixing can be significant if the x value is high (e.g., more than 0.15 (Park et al 1992b)). Thus, the electron transitions between different valence bands are induced in response to the normal incident radiation. The strain of QW and non-parabolicity of the energy band increases this mixing such that the normal incident radiation absorption is also increased.

For n-type GeSi QW, the off-diagonal components of the unisotropic effective mass tensor provide the coupling between electron and the normal incident. By using some special methods of choosing coordinate system and matrix calculation, a general analytical formulae for the oscillator strength and absorption coefficient as functions of the sample growth direction, and anisotropic effective masses for normal incident radiation were derived by the Device Physics Group in Linköping (Xu and Willander, Xu et al 1993, Xu et al). Using the concept of invariable quantities of ellipsoidal constant energy surface under the the coordinate transformation, the limit of the normal incident absorption was investigated. The conclusions are: the greater the difference between the longitudinal and transverse effective masses, the better the normal incident radiation absorption limit is reached. Compared with a QW which has silicon as a well material, QWs with germanium as a well material therefore have higher normal incident radiation absorption coefficient. Adjusting the growth direction of QW, one can obtain the maximum limit of absorption coefficient for the normal incident radiation. For a QW whose well material is silicon, however, the absorption coefficient for the normal incident radiation is always smaller than that for parallel incident radiation regardless of how the growth direction is selected.

Finally, it should also be mentioned that if carriers are not injected from the metal contact, but instead are injected into the wells via interband tunneling from adjacent layers, a dramatically increase in the IR photocurrent response, photoconductance and photocurrent gain will result. In addition, low noise and high response speed operation of the detectors will be obtained (Shen et al).

7.4 RESONANT TUNNELING DEVICES

7.4.1 Resonant Tunneling Diodes

Resonant tunneling (RT) has been investigated for more than 20 years. At the end of the 80s the first negative differential resistance (NDR) was observed in a double barrier $Si/Si_{1-x}Ge_x$ RT (Liu et al 1988, Ree et al 1988) and in transistor structures (Ismail et al 1992). In the early 90s RT in triple barrier structures (Xu et al 1992) and in δ-doped induced double barrier NDRs were also observed (Ni et al 1992). All of these three structures were designed for hole tunneling, since it is easier to get large valence band offsets and, at the same time, get high quality interfaces (see Section 7.2). However, the large hole mass also resulted in bad temperature performance for these devices with a peak-to-valley ratio (PVR) of 3 at 10 K. When raising the temperature, PVR was decreased to around 1.3 at

T=77 K and was smeared out at higher temperatures. This behavior is similar for p-type III-IV RT devices.

For electron tunneling, excellent results have been obtained for III-V (PVR≈103 at T=300 K). It is possible for Si/SiGe heterostructure to make the conduction band offset large (ΔE_c≈150 meV can be obtained), and in combination with the small electron mass (compared with the hole mass), it should be possible also to observe NDR in Si/SiGe RTD at room temperature. Recently IBM has also measured room temperature NDR in electron-tunneling RTDs (Ismail et al 1991b) (Figure 7.16). The key point was to grow a graded Si/SiGe superlattice (Si strained and SiGe unstrained) buffer layer with very low dislocation densities near the RTD device (see Figure 7.16). It should be mentioned that the measured

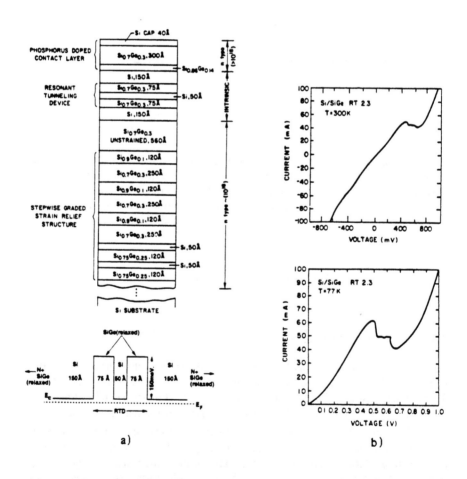

a)   b)

Figure 7.16 a). The electron resonant tunneling diode, its cross-section, and band diagram. b) The IV characteristic of the device (Ismail et al 1991b).

IV curve showed large barrier heights for the metal-semiconductor contacts. Still, the results for room temperature operation obtained by Ismail et al (1991b) has not been reproduced by other groups.

However, most of the works up until today deal with p-type RTD devices. It is therefore important to consider the temperature behaviour and the magnetic field influence on a p-type RTD.

The current transport through the DBRT can be written as:

$$J(T)=J_{res}(T)+J_{thermionic}+J_0 \qquad (7.7)$$

where J_{res} is the resonant tunneling current density which is approximated to:

$$J_{res}=A_\gamma Tln((1+exp(-(E_\gamma E_{F.C}-\alpha eV)/(kT))/(1+exp(-(E_\gamma E_{F.C}-(1-\alpha)eV)/(kT)))) \qquad (7.8)$$

A_γ depends on the transmission coefficient and effective mass, T is the temperature, and E_γ is the resonant energy level. $E_{F.C}$ is the Fermi energy level of the emitter contact, V is the total external applied bias, and αeV is the energy difference between the emitter valence band edge and the bottom quantum well. If the activation energy $\Phi_\gamma=E_\gamma E_{F.C}-\alpha eV \gg kT$, we can write

$$J_{res}=A_\gamma Texp(-\Phi_\gamma/(kT)). \qquad (7.9)$$

$J_{thermionic}$ is the thermionic current density and expressed as:

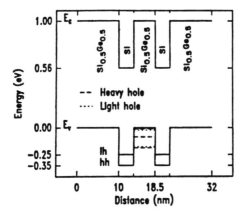

Figure 7.17 Band diagram and structure of the Si/SiGe resonant tunneling diode used for Table 7.3 (Gennser et al 1991a).

$$J_{thermionic} = A^* T^2 exp(-\Phi_{barrier}/(kT)) \qquad (7.10)$$

where A^* is the effective Richardson constant. J_0 is a temperature-insensitive term involving pure tunneling current.

In this way, Gennser et al (1991a) investigated the temperature behaviour of the structure in Figure 7.20 and measured the different resonant energies (see Table 7.3). As a comparision, the temperature dependence of the current in Si/SiGe double and triple barrier resonant tunneling structures grown by the Daimler-Benz group and measured by the Device Physics Group in Linköping (Xu et al 1991, Xu et al 1992) are also shown (Figure 7.18). From the temperature investigations it was concluded that hole RT is not possible for higher temperatures and that temperature investigations can be a tool for investigating the resonance energies as well as the transport properties of light and heavy holes in the tunneling process.

Table 7.3 Comparison between calculated and measured energy level using current temperature dependent measurement on the structure in Figure 7.17 (Gennser et al 1991a).

	Calculated resonant energies (meV)	Measured activation energies (meV)	Measured resonant energies (meV)
H.H.0	25 ± 3.5	16 ± 5	37 ± 5
L.H.0	57 ± 7	29 ± 5	50 ± 5
H.H.1	100 ± 12.5	60 ± 10	81 ± 10
L.H.1	220 ± 35	170 ± 20	191 ± 20

Magnetotunneling experiments are of great interest, as they provide a probe into the physical nature of the tunneling process. The dispersion relations of the quantum well states, and thus band-mixing and non-parabolicity of the SiGe layers, can be investigated using magnetotunneling spectroscopy (Hayden et al 1991). By monitoring the shifts of the resonances with magnetic fields both in the plane and transverse to the quantum well, Schubert et al (Schubert et al 1991a) could verify the heavy hole and light hole nature of the first two resonances. Using angle-resolved magnetotunneling spectroscopy, in which the magnetic field is rotated in the plane of the quantum well, a much larger in-plane anisotropy of the valence bands in Si/SiGe has been demonstrated (Gennser et al 1991b) as compared with GaAs/AlGaAs.

With the magnetic field transverse to the quantum well, tunneling occurs through Landau levels. In Figure 7.19 we illustrate an ideal magnetotunneling

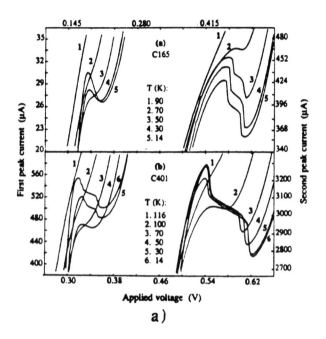

a)

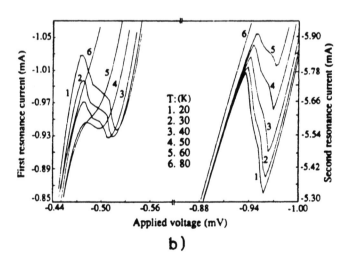

b)

Figure 7.18 a). Temperature dependence of a double barrier hole resonant tunneling structure, and in b) for triple barrier hole resonant tunneling structure (Xu et al 1991, Xu et al 1992).

situation (scattering and thickness fluctuations will smear out the steps in the IV characteristics). The steps appear (and some of them have been found in p-type RTDs (Gennser et al 1992)) when the Fermi level in the emitter is aligned with the Landau levels. As the two dimensional density of states is sharply peaked, the current will increase in steps as tunneling occurs through the higher Landau levels.

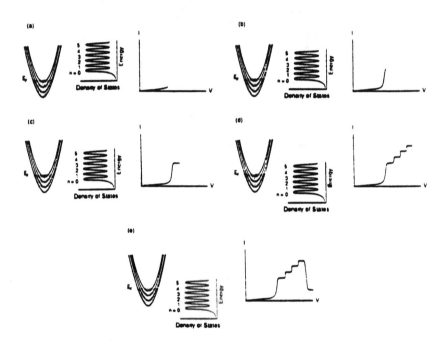

Figure 7.19 Schematic depiction of Landau level tunneling. The emitter Landau levels and quantum well density of states are shown for different bias points. The emitter states are filled below the Fermi level, as shown by the shaded area. Tunneling can only occur through Landau levels with the same index, resulting in steps in the I-V characteristics, as shown next to each respective tunneling situation (Gennser et al 1992).

Noise is both of practical importance and of importance in understanding fundamental physics. For Si/SiGe RTDs, few experimental results have been published (Okada et al 1988). The noise from the diode begins to dominate over the system noise at temperatures below 77 K. The flicker noise has a frequency dependence as $1/f^2$ and shot noise contributes little to the noise spectra at lower frequencies. Analysis of the spectra is a complex task due to the fact that the tunneling process is still not fully understood. In principle, shot noise should dominate if there is no inelastic scattering at the interfaces, i.e. if there is a coherent tunneling process. However more sophisticated phenomena such as a fluctuation of the Fermi energy during the transport process should be taken into account.

Fluctuations of the stored charges in the well will also make the interpretation more difficult.

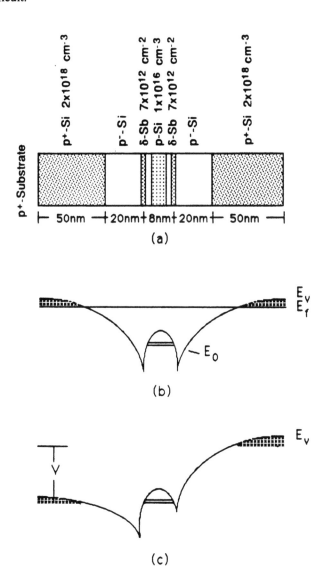

Figure 7.20 Schematic presentations of a) a δ-doped layer structure for RTD, b) the valence band at zero bias and c) with bias (Ni et al 1992).

Another approach to obtain negative differential conductance is to use δ-doped-induced double barrier structures in silicon (Ni et al 1992). NDR was observed from p^+-p^--d(n^+)-d(p)-d(n^+)-p^--p^+-doped Si (001) multilayer structures (Figure

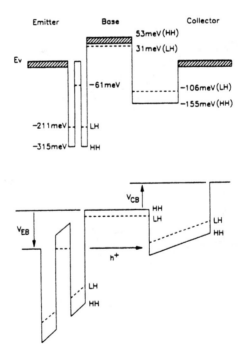

Figure 7.21 Cross-sectional structure of a SiGe RHET (upper), the valence-band diagram in equilibrium (middle) and under biased condition (lower) (Rhee et al 1990).

7.20). The widths of the δ-doped layers were smaller than 2 nm. NDR was observed up to 145 K with a PVR value of 1.23 at T=50 K. As illustrated in Figure 7.20, high barrier heights (≈1 eV) can be obtained outside the well. Problems, however, arise with increasing barrier heights inside the well since the well can not be doped too high in order to avoid scattering.

7.4.2 Resonant Tunneling Transistors

A resonant hot electron transistor was also demonstrated in 1989 (Rhee et al 1990). Hot holes were injected from the emitter and after few scattering events in the base. This was primarily due to a thin base and the low temperature with which the holes enter the collector (see Figure 7.21). Even though this first demonstration showed low NDR, it should be possible to improve its performance such as by using prewell.

7.5   SUMMARY

Untill today, Schottky gated and isolated gate FETs with buried strained SiGe and Si channels have been demonstrated. These transistors have shown improved transconductances and current drive capabilities in comparision with conventional FETs. However, there are still insufficient experimental investigations at high frequencies. The δ-doped silicon FETs have also been as well as theoretically predicted and demonstrated to have a good frequency performance.

Long-infrared wavelength detectors using internal photoemission in Si/SiGe heterostructures show higher quantum efficiency than their counterparts in Pt/Si and Ir/Si. The HIP detectors have also been integrated as 400×400-element arrays into silicon chips. P-type and n-type SiGe/Si multiple quantum wells for 8—14 μm photoresponse have recently been demonstrated using normal incidence light.

There have been demonstrations of n-type Si/SiGe resonant tunneling diodes operating at room temperature although problems to repeat the experiments exist. Hole resonant tunneling Si/SiGe double and triple barrier diodes are however limited to only low temperature operation, while hole resonant tunneling δ-doped Si structures have recently shown better temperature behaviour than the p-type Si/SiGe resonant tunneling diodes.

Acknowledgement

It is a pleasure to thank Drs U. Gennser and Y. Mamontov for reading the manuscript.

REFERENCES

Chang J, Nayak D K, Ramman V K, Woo J C S, Park J S, Wang K L, and Vismanathan C R 1991 *Microelectronic Engineering* **15** 19
Chen Q and Willander M 1991 *J. Appl. Phys.* **69** 8233
Chen Q and Willander M 1992a *Solid-State Electronics* **35** 687
Chen Q and Willander M 1992b *Solid-State Electronics* **35** 1493
Daembkes H, Herzog H, Jorke H J, Kibbel H, and Kasper E 1985 *IEDM Technical Digest* 768
Daembkes H, Herzog H, Jorke H J, Kibbel H, and Kasper E 1986 *IEEE Trans. Electron Devices* **ED-33** 633
Daembkes H 1988 *Proc. 2nd Int. Symp. Silicon MBE* ed J C Bean and Showalber (New Jersey: Pennington) p 15
Fu Y, Chen Q, and Willander M 1991 *J. Appl. Phys.* **70** 7468
Fu Y, Grahn K, and Willander M *IEEE Trans. Electron Devices* to be published
Garone P M, Venkataraman V, and Sturm J C 1991a *IEEE Electron Device Lett.* **12** 230
Garone P M, Venkataraman V, and Sturm J C 1991b *IEDM Proc.* 29
Garone P M, Venkataraman V, and Sturm J C 1992 *IEEE Electron Device Lett.* **13** 56
Gennser U, Kesan V P, Iyer S S, Bucelot T J, and Yang E S 1991a *J. Vac. Sci. Technology B* **9** 2059
Gennser U, Kesan V P, Syphers D A, Smith T P, III, Iyer, S S, and Yang E S 1991b *Phys. Lett.* **67** 3828
Gennser U 1992 *Ph. D. thesis* (Coloumbia Univ.)
Hayden R K, Maude D K, Eaves L, Valadares E C, Menini M, Sheard F W, Huges O H, Portal J C, and Curz L 1991 *Phys. Rev. Lett.* **66** 1749
Hoffmann K R 1991 *Mat. Res. Soc. Proc.* **220**
Ismail K, Meyerson B S, and Wang P J 1991a *Appl. Phys. Lett.* **58** 2117
Ismail K, Meyerson B S, and Wang P J 1991b *Appl. Phys. Lett.* **58** 973
Ismail K, Meyerson B S, Rishton S, Chu J, Nelson S, and Nocera J 1992 *IEEE Electron Device Lett.* **13** 229
Jain S C, Poortmans J, Nijs J, Van Mieghem P, Mertens R P, and Van Overstraeten R 1992 *Microelectronic Engineering* **19** 439
Karunasiri R P G, Park J S, and Wang K L 1991 *Appl. Phys. Lett.* **59** 2588
King T J, Pfiester J R, and Saraswat K C 1991 *IEEE Electron Device Lett.* **12** 533
König U and Schäffler F 1991 *Electronics Lett.* **27** 1405
König U, Boers A J, Schäffler F, and Kasper E 1992 *Electronics Lett.* **28** 160
Lee C and Wang K L 1992a *J. Vac. Sci. Technology B* **10** 992

Lee C and Wang K L 1992b *Appl. Phys. Lett.* **60** 2264

LeGoues F K, Meyerson B S, and Morar J F 1991 *Phys. Rev. Lett.* **66** 2903

Lin T L and Maserjian J 1990a *Appl. Phys. Lett.* **57** 1422

Lin T L, Eric E W, Ksendzav A, Dejewski S M, Fathauer R W, Krabach T N, and Maserjian J 1990b *IEDM Technical Digest* 614

Liu H C, Landheer D, Buchanan N, and Houghton D C 1988 *Appl. Phys. Lett.* **52** 1809

Lury S, Kastalsky A, and Bean J C 1984 *IEEE Trans. Electron Devices* **ED-31** 1135

Mii Y J et al 1991 *Appl. Phys. Lett.* **59** 1611

Murakarim E, Nagawa K, Nishida A, and Migao M, *IEEE Electron Device Lett.* **EDL-12** 71

Nakagawa K, Van Gorkum A A, and Shiraki Y 1989 *Appl. Phys. Lett.* **54** 1869

Nayak D K, Woo J C S, Park J S, and Wang K L 1991 *IEEE Electron Device Lett.* **12** 154

Okada Y, Xu J, Liu H C, Landheer D, Buchanan M, and Houghton D C 1989 *Solid State Electronics* **32** 797

O'Neill A G, Wood A C G, Phillips P, Whall T E, Parker E H C, Dundlach A, and Taylor S 1992 *Proc. ESSDERC'92* eds H E Maes, R R Mertens, and R J Van Overstraeten (Elsevier) 743

Ni W X, Fu Y, Willander M, and Hansson G V 1992 *Proc. 21st Int. Conf. on the Physics and Semiconductors* (Beijing)

Park J S, Karunasiri R P G, and Wang K L 1990 *J. Vac. Sci. Technology* **B8** 217

Park J S, Karunasiri R P G, and Wang K L 1992a *Appl. Phys. Lett.* **60** 103

Park J S, Karunasiri R P G, and Wang K L 1992b *Appl. Phys. Lett.* **61** 681

Pearsall T P, Bean J C, People R, and Fiory A T 1985 *Proc. 1st Int. Symp. Silicon MBE* ed J C Bean (New Jersey: Pennington) p 400

Pearsall T P, Temkin H, Bean J C, and Lury S 1986 *IEEE Electron Device Lett.* **EDL-7** 330

People R, Bean J C, Bethea C G, Sput S K, and Peticolas L J 1992 *Appl. Phys. Lett.* **61** 1122

Ree S S, Park J S, Karunasiri R P G, Ye A, and Wang K L 1988 *Appl. Phys. Lett.* **53** 204

Rhee S S, Chang G K, Carns T K, and Wang K L 1990 *Proc. IEDM* 651; 1990 *Appl. Phys. Lett.* **59** 1061

Schubert G, Abstreiter G, Gornik E, Schäffler F, and Luy J F 1991a *Phys. Rev.* **B43** 2280

Schubert G, Schäffler F, Berson M, Abstreiter G, and Gornik E 1991b *Appl. Phys. Lett.* **59** 3318

Schäffler F, Többen D, Herzog H J, Abstreiter G, and Holländer B 1992 *Semicond. Sci. Technol.* **7** 260

Selvakumar C R and Hecht B 1991 *IEEE Electron Device Lett.* **12** 444

Shen G D, Xu D X, Willander M, and Hansson G V unpublished

Sheperd F D, Vickers V E, and Yang A C 1971 *U. S. Patent No. 3,603,847, September 7*

Temkin H, Pearsall T P, Bean J C, Logan R A, and Luryi S 1986 *Appl. Phys.*

Lett. **48** 963

Temple G, Schwarz N, Muller F, and Koch F 1990 *Thin Solid Films* **184** 171

Tsauer B Y, Chen C K, and Marino S A 1991a *Proc. SPIE* (San Diego)

Tsauer B Y, Chen C K, and Marino S A 1991b *IEEE Electron Device Lett.* **12** 293

Verdonckt-Vandebrock S, Crabbe E F, Meyerson B S, Harame D L, Restle P J, Stork J M C, Megdanis A C, Stanis C L, Bright A A, Kroesen G M W, and Warren A C 1991 *IEEE Electron Device Lett.* **12** 447

Yamaguchi K, Shiraki Y, Kabayama Y, and Murayama Y 1983 *Jpn. J. Appl. Phys. Suppl.* **22** 267

Yamaguchi K, Nakagawa K, and Shiraki Y 1988 *20th Conf. on Solid-State Devices and Materials* (Tokyo) p 17

Van Gorkum A A, Nakagawa K, and Shiraki Y 1982 *Jpn. J. Appl. Phys.* **26** L1933

Xu D X, Shen G D, Willander M, Hansson G V, Luy J F, and Schäffler F 1991 *Appl. Phys. Lett.* **58** 2500

Xu D X, Shen G D, Willander M, Luy J F, and Schäffler F 1992 *Solid-State Electronics* **35**

Xu W, Fu Y, and Willander M 1993 *Phys. Rev.* October 15

Xu W and Willander M unpublished

Xu W, Fu Y, and Willander M unpublished

8

Crystalline Silicon-Carbide and Its Applications
Toshihiro Sugii

8.1 INTRODUCTION

Because of its mechanically hard, refractory, and chemically inactive properties, SiC has been used for coating components such as susceptors, carbon heaters, and tubes used under high temperatures. SiC has also been used for abrasives and firebricks. However, one of the principal driving forces for the current interest in this material has resulted from its potential for high-power, high-speed, high-temperature, light emitting, and radiation-hard microelectronic devices based on the properties shown in Table 8.1. From the viewpoint of discrete devices, Johnson analyzed the high-frequency and high-power capabilities of various semiconductors, and used the following equation to relate breakdown field, E_B, and electron saturation drift velocity, V_{sat}, with his figures of merit, FM1[1].

$$FM1 = (E_B \cdot V_{sat}) / 2 \pi \qquad (8.1)$$

Baliga based his figures of merit, FM2, for power devices[2]. Considering ON-resistance, FM2 is given by the following equation for field-effect transistors.

$$FM2 = \mu \cdot E_B^2 \cdot V_G^{0.5} / (2 \cdot V_D^{1.5}) \qquad (8.2)$$

Here, μ is the carrier mobility, E_B is the breakdown field, and V_G and V_D are the gate and drain voltages. In Johnson's and Baliga's figures of merit for selected semiconductor materials (Table 8.2), 3C and 6H-SiC have characteristics superior to the other materials with the exception of diamond. Diamond has, however, no shallow donors or acceptors and, at present, is difficult to dope and not available in large sizes. A perusal of these figures of merit coupled with device requirements for high-speed, high-temperature microprocessors and radiation-hard application makes clear the present interest in SiC growth, characterization, and device applications.

The sections that follow review crystalline SiC, including polycrystalline SiC, from the viewpoints of growth technique and device application.

Table 8.1 Electric properties of 3C and 6H-SiC

Properties	3C	6H
Energy gap at 300 K (eV)	2.3	3.0
Electron mobility at 300 K (cm²/V•s)	1000	300
Electron saturation drift velocity (cm/s)	2.5×10^7	——
Breakdown field (V/cm)	——	3×10^6
Dielectric constant	9.7	10

Table 8.2 Johnson's and Baliga's figures of merit (ratios to silicon)

	Johnson	Baliga
Silicon	1.0	1.0
GaAs	6.9	9.5
InP	16.0	—
Diamond (n-type)	—	453.7
Diamond (p-type)	—	357.2
6H-SiC	—	13.1
3C-SiC	1138	—

8.2   PHYSICAL PROPERTIES

In crystalline SiC, the smallest distance between Si and C atoms is 1.89 Å. Each Si (or C) atom is surrounded tetrahedrally by four C (or Si) atoms with sp^3 hybrid bonds (covalency is about 88%). The SiC structure has many polytypes.

One can imagine them as different stacking sequences of the same silicon, carbon layer. This consists of a close hexagonal packed layer of silicon which has immediately above a layer of carbon atoms. There are three possible situations, A, B, and C in the stacking of the SiC layers. The next layer must not have the same position. In such an arrangement, every Si atom is surrounded tetrahedrally by four carbon atoms and vice versa. Thus, numerous modifications are possible. If layers are stacked periodically ABC ABC patterns, 3C-SiC (ß-SiC) results. Stacking the layers in ABCACB ABCACB patterns, results in 6H-SiC. Figure 8.1 shows the arrangement of atoms for 3C-SiC and 6H-SiC.

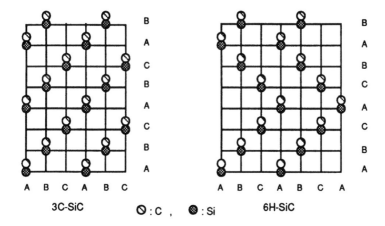

Figure 8.1 Arrangement of atoms for 3C-SiC and 6H-SiC

Table 8.3 Lattice constants and energy gaps of common SiC polytypes

Polytype	Lattice constant (Å)		Energy gap (eV at 4.2 K)
	a	c	
2H	3.076	5.048	—
4H	3.076	10.046	3.77
6H	3.081	15.117	3.02
8H	3.079	20.147	2.80
15R	3.073	37.30	2.99
21R	3.073	52.78	2.85
33R	—	82.5	3.00
3C	4.360	—	2.30

Table 8.3 gives the most frequent SiC polytypes and their lattice constants, and energy gaps[3]. The first number in the polytype expressions shows the repetion of periodicity and the latter indicates the crystal symmetry such as cubic (C), hexagonal (H), and rhombohedral (R). 3C-SiC is also known as ß-SiC. Hexagonal and rhombohedral structures are often called α-SiC.

8.3   CRYSTALLINE SiC GROWTH

As mentioned previously, SiC is thermally and mechanically stable. These characteristics make it difficult to obtain high-quality, large crystalline SiC at temperatures reasonable for device application. Many studies are ongoing to obtain high-quality crystals at lower temperatures. Since SiC has many polytypes, investigation naturally turns to controlling the polytypes. Unlike Si, large ingots are not available at this time. Thus, the study of heteroepitaxy on foreign substrates results. These perspectives have led to the development of a variety of growth methods, including reduced or atmospheric pressure chemical vapor deposition (CVD), solid-source or gas-source molecular beam epitaxy (MBE), liquid phase epitaxy (LPE), reactive deposition involving the volatilization of Si in C_2H_2, reactive sputtering of Si target in a mixture of Ar and CH_4, and carbon ion implantation into Si. In addition to growth methods, a variety of substrates have been used for homoepitaxial and heteroepitaxial growth. It is difficult to cover all of these mentioned growth methods in a single chapter. This section reviews a couple of typical growth methods. Due to their importance in electronic devices, polytypes are restricted to 3C-, 4H-, and 6H-SiC.

8.3.1   3C-SiC Growth

3C-SiC has not been used in electronic device application because growth has been limited to the mm diameters possible with sublimation above 2000°C or solution growth using a Si melt in a graphite crucible[4,5]. To obtain larger quantities of single-crystal SiC, SiC is commonly grown heteroepitaxially on foreign substrates. Single-crystal Si has been almost universally used as the substrate because of the availability of large high-quality substrates, though a lattice mismatch of about 20% and a difference in thermal expansion coefficients of about 8% have serious effects on heteroepitaxial growth. Although SiC devices have been studied for many years, practical devices are hard to fabricate because of the lattice mismatch. One problem has been the difficulty in the reproducible growth of high-quality single-crystal SiC over a large area. This difficulty is drastically reduced by adopting an initial reaction of a Si substrate surface with a carbon-containing gas, and successive epitaxial growth on the carbonized layer of relatively thick, crack-free 3C-SiC films using individual gases containing C and Si. Nishino et al. first reported the growth of high-quality 3C-SiC on a Si substrate at 1350°C using this carbonized layer[6]. Similar results and development followed, with the several reports on crystallinity and electrical properties[7-12]. Figure 8.2 gives an example of the temperature program for reproducible growth of 3C-SiC on Si[11]. Heteroepitaxial growth was at atmospheric pressure using SiH_4, C_3H_8, and H_2. A Si substrate was put on a SiC-coated graphite susceptor after etching in HF and rinsing in deionized water. The substrate was inductively heated in a H_2 atmosphere with 200 kHz. Single-crystal 3C-SiC was heteroepitaxially grown on (100) Si by etching, carbonization, and CVD. Carbonization was performed at 1360°C for 2 minutes in a C_3H_8 and H_2 flow.

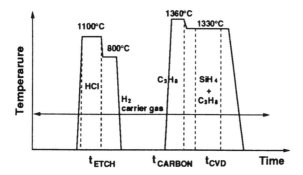

Figure 8.2 Temperature program for SiC single-crystal growth: t_{ETCH}, t_{CARBON}, and t_{CVD} mean the time for etching, carbonization, and single-crystal growth[11].

The cleanness of the Si substrate before carbonization is important for the carbonized layer formation. If the Si surface after HCl gas etching is not mirror-like, the reflection electron diffraction (RED) pattern of the carbonized layer contains faint rings, indicating the existence of a polycrystalline layer. The environment during heating affects the quality of the carbonized layer. Before the substrate is heated for carbonization, the reaction tube should be saturated with C_3H_8 and the substrate temperature quickly raised to 1360°C. The carbonization time is also a key parameter in layer quality. The optimum thickness for a Si(100) substrate is about 30 nm. The surface region (8 nm) of the layer had a SiC stoichiometry, (Si/C) of 1, but the remaining region changed compositionally with depth from SiC to Si. 3C-SiC was grown by subsequent CVD at 1330°C with SiH_4, C_3H_8, and H_2. The surface of the layer grown was smooth and mirror-like. Increasing thickness decreased the fullwidth of half maximum (FWHM) of the X-ray rocking curve, indicating that the crystallinity becomes better with thickness. The crystallinity of the epitaxial layer was improved by reducing the growth rate. The narrowest FWHM was about 0.3°. The RED pattern showed a streaky spot pattern indexed as 3C-SiC.

Before 3C-SiC can be applied to electronic devices, a lower growth temperature is required to reduce dopant redistribution, contamination, and thermal stress in the wafer. Furumura et al. grew 3C-SiC on a 4-inch Si substrate at 850°C and 1000°C without an intentional carbonized layer[13-15]. Figure 8.3 shows the experimental apparatus. The conventional cold-wall reactor was evacuated by mechanical booster and rotary pumps. The substrates were 4-inch Si wafers with a (111) orientation tilted 4° toward the $[\bar{2}11]$ direction. The carbon source gas was C_3H_8 for growth at 1000°C and C_2H_2 for growth at 850°C. The Si source gas was $SiHCl_3$. In the temperature program for 3C-SiC growth at 1000°C (Figure 8.4), Si substrates were preheated in the reactor with a H_2 flow at 1000°C to eliminate the native oxide on the Si substrates. Source gases were introduced immediately after preheating. The pressure during preheating and growth was 400 Pa.

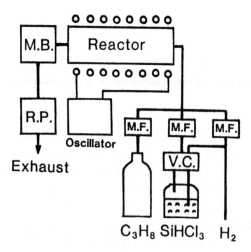

Figure 8.3 Schematic diagram of the experimental apparatus

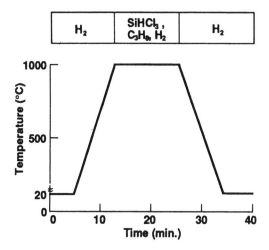

Figure 8.4 Temperature program for SiC single-crystal growth at 1000°C

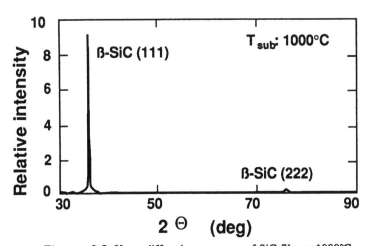

Figure 8.5 X-ray diffraction spectrum of SiC film at 1000°C

The double-crystal X-ray diffraction spectrum of a 400-nm-thick specimen is shown in Figure 8.5. Only the diffraction peaks from the (111)-oriented 3C-SiC and (111)-oriented Si substrate are visible. The SiC exhibited a mirror surface over the entire substrate. No surface defects, such as cracks or slip lines, were observed. The Furumura group also discussed the growth mechanism for heteroepitaxy using $SiHCl_3$-C_3H_8-H_2 gas systems[14]. Si and C nucleation must continuously take place in the system for SiC growth in the <111> direction. The first step in the growth process is the carbonization of an area on the Si substrate. The next step for SiC growth requires Si nucleation. However, according to their another experiments, this rarely occurs. They proposed that the Si atoms or undissociated $SiHCl_3$ molecules migrating on the carbonized surface are promptly carbonized again and that the SiC monolayers are easily formed. Given this assumption, the top surface of the grown SiC is always terminated by carbon atoms. This coincides fairly well with their experimental finding that Si does not grow well on SiC surfaces if the SiC surface has been somehow terminated.

Gas source MBE (GSMBE) seems to lower the growth temperature effectively because ultra-high vacuum environments prevent contamination of the sample surface, and the substrate surface can be cleaned at relatively low temperatures. Another benefit originates from the relatively high vacuum during growth. Various techniques, including RHEED and AES, can be used to evaluate growth surfaces. 3C-SiC GSMBE has been reported from some laboratories. Fuyuki et al. have been studying 3C-SiC atomic layer epitaxy by GSMBE using a surface superstructure[16,17]. They used Si_2H_6 and C_2H_2 as source gases, since these gases decompose at relatively low temperatures. The source gases were introduced alternately. (100) 3C-SiC grown on Si by CVD was used as a substrate. During the deposition, the working pressure ranged from 5×10^{-6} to 5×10^{-5} torr. Flux was controlled by a variable leak valve and monitored by a quadruple mass spectroscope and an ion gage. 3C-SiC was epitaxially grown at 1000°C. When the Si_2H_6 and C_2H_2 source gases were introduced, the surface superstructure was found to change periodically. When Si_2H_6 was introduced, the surface superstructure of the initial (1 x 1) pattern changes to (3 x 2), which maintained after the Si_2H_6 supply is stopped. When C_2H_2 begins to flow, the surface superstructure returned to the initial (1 x 1) pattern. After an appropriate number of cycles, single-crystal 3C-SiC with a smooth, mirror-like surface was obtained. The growth rate was determined by the number of Si atoms forming the surface superstructure, which gives atomic level control in SiC growth. Motoyama et al. used $SiHCl_3$ and C_2H_4 for GSMBE[18]. Single-crystal 3C-SiC films were obtained at 1000°C. Prior to SiC growth, carbonization was done with C_2H_4 gas alone. The crystal growth was monitored continuously by RHEED. The RHEED patterns revealed that carbonization occurred at 750°C and the lattice mismatch between Si and SiC crystal was satisfactorily relaxed. All crystal growth processes were at lower temperatures than those of CVD growth.

We investigated 3C-SiC growth on Si(111) substrates using GSMBE with C_2H_2 gas and Si solid sources[19]. To lower the growth temperature, we studied the effect of cracking C_2H_2 molecules before they reach the substrate surface. The GSMBE system (Figure 8.6) consisted of an E-gun to produce a Si molecular beam, a substrate heating system, a RHEED apparatus, and a quadruple mass analyzer (QMA). The chamber was evacuated with an ion-pump to 1 x 10^{-8} Pa.

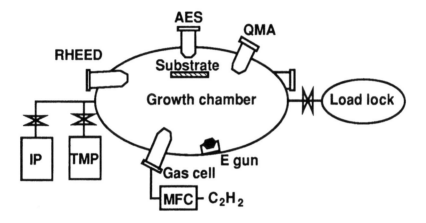

Figure 8.6 Gas-source MBE system for SiC growth

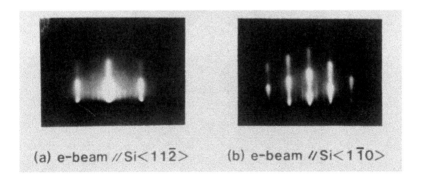

Figure 8.7 RHEED patterns of the carbonized layer at 900 °C. The substrate was Si(111) with 0° off-angle. The primary beams were in <11$\bar{2}$> azimuth for (a) and <1$\bar{1}$0> azimuth for (b).

SiC was grown in two steps, that is, Si surface carbonization and SiC deposition. Cracking C_2H_2 molecules lowers the carbonization temperature to 820°C. The RHEED pattern revealed that SiC (6 x 3) reconstruction of the carbonized layer occurred (Figure 8.7). The SiC (6 x 3) reconstruction disappears when the substrate temperature exceeds 920°C, and reappears when the temperature drops below 920°C. Once the reconstructed surface was exposed to air, none of the extra streaks were observed again by RHEED. The carbonized layer was found to be 1-2 nm thick as measured with an ellipsometer and X-ray photoelectron spectroscopy (XPS). The layer composition of $Si_1C_{0.82}$ was obtained by XPS. Single-crystal 3C-SiC was grown on the carbonized layer homoepitaxially at 900°C with C_2H_2 and Si. The cross-sectional TEM image of the layer grown (Figure 8.8) reveals that the crystalline quality is nearly the same as that grown by conventional CVD at higher temperatures. Auger electron spectroscopy (AES) showed the carbon concentration to be 52%. The ratio of Si to C atoms was nearly one. Despite a Si molecular beam intensity of about one tenth the C_2H_2 intensity, the grown film composition was nearly stoichiometric. Therefore, desorption or reflection of C_2H_2 must occur during epitaxy. C_2H_2 molecules or their decomposed products seem predisposed to react with Si atoms.

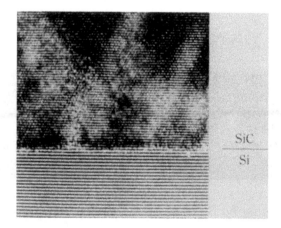

Figure 8.8 Cross-sectional TEM image of the SiC/Si interface. The carbonization temperature and following SiC growth temperature with C_2H_2 and Si molecules were at 900°C.

The low diffusion constant of SiC requires doping temperature of 1800°C or more and long process times[20-22]. At these temperatures, it is not possible to use passivating oxide layers to mask against diffusion. This makes the precise fabrication of SiC planar devices very difficult. Doping during SiC growth or by ion implantation has been investigated. Bartlett et al.[23], Long et al.[24], and Nishino et al.[25] produced p-type 3C-SiC by adding B_2H_6 or $AlCl_3$ during the SiC growth. Kim et al.[26] have been studying P and N, and B and Al for n- and p-type dopants. Furumura et al. studied n-type doping of relatively low-temperature-grown 3C-SiC[14,27]. Figure 8.9 shows secondary ion mass spectroscopy (SIMS) of phosphorus-doped 3C-SiC grown at 1000°C from $SiHCl_3$, C_3H_8, and PH_3. The phosphorus was uniformly doped in the SiC and its concentration sharply decreased in the Si substrate. Even at a growth temperature of 1000°C, phosphorus diffused into the Si was below the SIMS detection limit. This is due to the low diffusion constant of the impurity in SiC. Once the Si surface is covered with SiC, the impurity cannot diffuse through the SiC layer. Figure 8.10 shows the relationship between the sheet resistance and PH_3 flow rate. The SiC thickness was 400 nm. With the PH_3 flow rate below 1 sccm, the phosphorus concentration increased and sheet resistance decreased. Hall measurement showed the carrier concentration to be about 10^{18} /cm^3, and the active fraction of phosphorus is several percent.

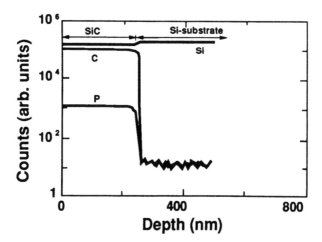

Figure 8.9 SIMS depth profile of phosphorus-doped SiC grown at 1000°C from $SiHCl_3$, C_3H_8, and PH_3

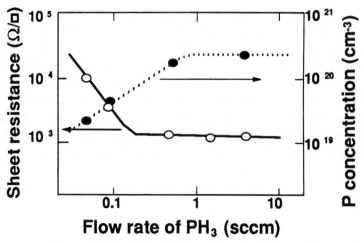

Figure 8.10 Sheet resistance and phosphorus concentration as a function of PH_3 flow rate

Doping by ion implantation is an alternative if the associated radiation damage is annealable at a reasonable temperature and the implanted atoms become electrically active. Several studies have been reported on ion implantation[28-32]. Matsunami et al. investigated P^+- and N_2^+- implantation into highly-resistive B-doped 3C-SiC grown layers at room temperature[28]. Their implantation conditions are listed in Table 8.4. Annealing after implantation was in an Ar atmosphere for 30 minutes.

Table 8.4 Ion implantation conditions

No.	Implanted ion	1st dose(cm²) 2nd	1st energy (eV) 2nd	total dose (cm⁻²)
1	P^+	1×10^{15} 3×10^{14}	100K 25K	1.3×10^{15}
2	P^+	3×10^{14} 9×10^{13}	100K 25K	3.9×10^{14}
3	N_2^+	1.5×10^{14} 4.5×10^{13}	100K 25K	1.95×10^{14}

Figure 8.11 shows electrical activation as the ratios of sheet carrier concentration to total dose. Electrical activation increased with annealing temperature. For P$^+$-implantation, the layer with a total dose of 3.9 x 10^{14} /cm^2 showed higher electrical activation than that with a 1.3 x 10^{15} /cm^2 dose. The N$_2$$^+$-implanted layer showed higher electrical activation than the P$^+$-implanted layer.

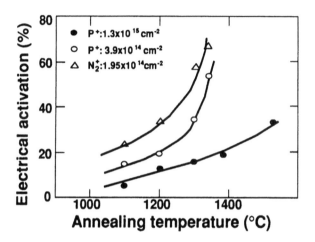

Figure 8.11 Electrical activation rate as a function of annealing temperature[28]

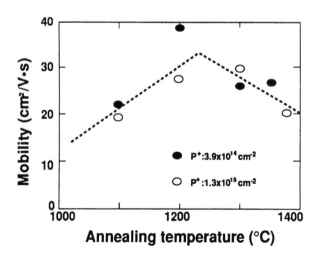

Figure 8.12 Relation between electron mobility and annealing temperature[28]

The relationship between electron mobility and annealing temperature (Figure 8.12) shows the mobility was restricted by residual damage below 1200°C. Above 1200°C, the mobility decreased due to increased amounts of ionized impurities. Arsenic ions were also used as donor impurities for 3C-SiC[29,30]. To provide a uniform arsenic atom density, arsenic was doubly implanted to a depth of 200 nm with doses of 5×10^{15} /cm^2 at 60 and 150 keV. The substrate was at room temperature and not intentionally heated. Subsequent furnace annealing was at 1100°C for 30 minutes in a nitrogen ambient. The SIMS impurity depth profiles (Figure 8.13) shows that the double implantation produced relatively uniform impurity distributions. Figure 8.14 shows the relationship between the sheet resistance and annealing temperature for arsenic and phosphorus. To provide near-equivalent projection ranges, ion implantation was done at an acceleration energy of 60 keV for arsenic and 30 keV for phosphorus. The annealing time was 30 minutes. Figure 8.14 shows two features. The first is that SiC layers grown on (111) Si exhibit a much lower sheet resistance than those grown on (100) Si. This is attributed to the differences in carrier concentration. That is, the active fraction of implanted atoms is more than an order larger for (111) Si. By measuring the carrier depth profile with step etching and sheet resistance measurement, active fractions were found to be 0.1 for SiC grown on (111) Si and 0.005 for that grown on (100) Si. A large number of defects in SiC grown on (100) Si act as sinks for impurities due to impurity segregation. The second feature is the differences in the dopant atoms. The degrees of damage caused by the implanted atoms seems to be responsible for this result.

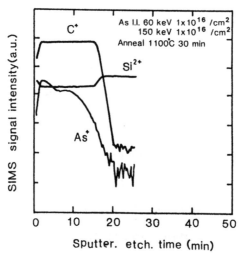

Figure 8.13 Implanted As atoms depth profile after annealing at 1100°C

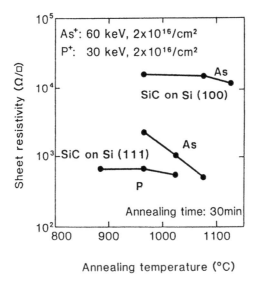

Figure 8.14 Sheet resistivity vs annealing temperature for two wafer orientation. The implanted atoms were arsenic and phosphorus. Annealing was 30 min.

8.3.2 α-SiC growth

α-SiC, especially 4H- and 6H-SiC show promise for use in developing light-emitting diodes (LEDs). 6H-SiC has been mass-produced using the Acheson method. Crystals fabricated by this way cannot be used as substrates for semiconductor devices because the crystals grown are too small for practical use. So, there needs a crystal growth to obtain much larger ingots. Tairov developed a 6H-SiC single-crystal fabrication technique using seed crystals in a modified Lely method in a vacuum[33]. Ziegler et al. used this method to produce crystal diameters up to 20 mm[34]. Inch-size 6H-SiC crystals were fabricated by Koga et al., using a water-cooled quartz tube as the reaction chamber[35,36]. The seed of the 6H-SiC platelets with {0001} faces and SiC powder were placed into a graphite crucible heated by RF induction. 6H-SiC growth was done at temperatures from 2000 to 2400°C. The temperature gradient between the source and seed was set for 15-25°C/cm, and the Ar gas was controlled to keep reaction chamber pressure low. X-ray oscillation photography, Raman, photoluminescence, and transmission spectra showed that the as-grown samples were single-crystals with 6H modifications. Table 8.5 shows the electrical properties.

Table 8.5. Electrical properties of grown crystals

Type	Carrier concentration (cm^{-3})	Resistivity (Ω·cm)	Hall mobility (Max.) (cm^2/V·s)
undoped (N)	9×10^{15}	6.4	112
N	$2 \times 10^{17} \sim 2 \times 10^{18}$	$4 \times 10^{-2} \sim 2 \times 10^{-1}$	100
P	$1 \times 10^{15} \sim 3 \times 10^{18}$	$5 \times 10^{-1} \sim 1 \times 10^{3}$	32

To fabricate LEDs, it is necessary to grow doped SiC epitaxially. Liquid-phase epitaxial growth (LPE) using Si melt has been widely adopted. The dipping technique, one of the LPE methods, made it possible to grow epitaxial multilayers under independent doping conditions. A crucible filled with Si and a small amount of dopant was heated to and held at 1700°C under a purified Ar gas flow[37,38]. Substrates obtained by the modified Lely or Acheson method were fixed to a holder and dipped in the Si melt. Carrier concentrations in the epitaxial layers were controlled from 5×10^{16} to 3×10^{18} for n-type samples and 5×10^{16} to 3×10^{17} /cm^3 for p-type. Their maximum Hall mobilities were 252 cm^2/V·s for an n-type at thickness of 3 to 5 μm and 54 cm^2/V·s for a p-type of 6 to 13 μm.

Due to the high temperature in LPE growth of around 1700°C, precise control of impurity doping and a pn junction is rather difficult. As an alternative, CVD of 6H-SiC on 6H-SiC substrates has been studied. Single crystals of 6H-SiC were epitaxially grown at 0.1-0.15 μm/min at 1600°C using SiCl$_4$-C$_3$H$_8$-H$_2$[39,40]. Epitaxial growth was confirmed by X-ray and RED analysis. Impurity doping was also investigated with CVD growth of 6H-SiC on 6H-SiC substrates[41,42]. The conductivity type and carrier concentration of the grown layer were controlled using B$_2$H$_6$ as a p-type dopant and N$_2$, AsH$_3$, or PH$_3$ as n-type dopants. Doping was most effective with B$_2$H$_6$, and least with N$_2$.

The remarkable progress in homoepitaxial growth of 6H-SiC is step-controlled epitaxy proposed by Matsunami et al.[43,44]. High-quality single-crystals of 6H-SiC can be grown on angled-lapped 6H-SiC {0001} at 1500°C, which is much lower compared with liquid-phase epitaxial growth temperature or conventional CVD growth temperature. The density of surface steps is increased by angled lapping. Utilizing step-flow growth (lateral growth from the steps), single crystalline 6H-SiC with a very smooth surface is grown at lower temperature. Single crystals of 6H-SiC grown by the Acheson method were used as substrates. The low-index basal planes of the crystals are either {0001}Si or {000$\bar{1}$}C faces. To control the step density, the basal planes were angle-lapped with carborundum and polished with diamond paste. Crystal growth was carried out mainly at 1500°C under atmospheric pressure using SiH_4, C_3H_8, and H_2.

Since flat 4H-SiC substrates are not available, 4H-SiC can be obtained heteroepitaxially on 6H-SiC substrates by vacuum sublimation [45,46]. The seeds of a 6H-SiC {0001} substrate and poly-crystalline SiC materials were placed in a graphite crucible, with a temperature gradient between them. Growth was from 2000 to 2400°C under pressures of 2 to 60 mbar. Table 8.6 shows the relationship between growth conditions and the resultant polytypes. 4H-SiC grew on the C-face of {0001} 6H-SiC at temperatures of 2300°C or more. 6H-SiC grew predominantly on both the Si- and C-faces at 2200°C or less. N-type 4H-SiC ingots were obtained by introducing purified nitrogen gas. Since 4H-SiC has a larger bandgap (3.3 eV) than 6H-SiC, attempts have been made to fabricate violet or bluish-purple LEDs.

Table 8.6 Relation between growth conditions and resultant polytypes grown by sublimation method

Temperature (°C)	Polarity	Pressure (mbar)				
		~2	5	7	10	~60
2000	Si - face					
	C - face		6H		6H	
2100	Si - face	6H*				
	C - face	6H*				
2200	Si - face	6H*		6H	6H	
	C - face	6H*		6H		
2300	Si - face	6H*			6H	6H*
	C - face	4H+6H*		4H		4H
2400	Si - face	Poly*		6H		
	C - face	4H+Poly		4H+6H	4H	

*Time constant for evacuating pressure is 20 minutes.

8.3.3   Polycrystalline SiC growth

Polycrystalline SiC has been studied for electronic devices because it can be prepared at lower temperatures than single-crystal SiC. Low-temperature growth effectively suppresses impurity redistribution, which is indispensable for high-performance (high-density and high-speed) electronic devices. However, there seems to be a tradeoff between the low growth temperature and film qualities such as carrier mobility, resistivity, and interface properties. Polycrystalline SiC has been prepared by CVD[47,48] and electron beam evaporation[49,50]. In the CVD growth example, Si_2H_6 and C_2H_2 were used for the source gases. Growth was at atmospheric pressure. The crystal quality depended on the growth temperature. This is demonstrated in Figure 8.15, which shows the RHEED patterns of 50-nm films grown at various substrate temperatures. The pattern prepared at 800°C has a mixture of halos and weak Debye rings. At 850°C, Debye rings which indicate a polycrystalline structure can be seen clearly. Spotty patterns appears at 900°C. From the above results, polycrystalline SiC can be obtained at temperatures above 850°C. Figure 8.16 shows the Fourier transform infrared spectroscopy (FTIR) spectrum of SiC grown at 900°C with a C_2H_2 to Si_2H_6 gas flow rate of 6. Only one absorption line is seen due to stretching vibration between Si and C at around 800 cm^{-1}.

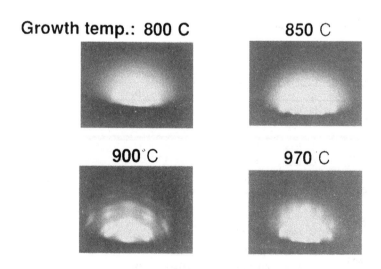

Growth temp.: 800 C   850 C

900°C   970 C

Figure 8.15 RHEED patterns of SiC layers with various growth temperatures. The film was 50 nm thick.

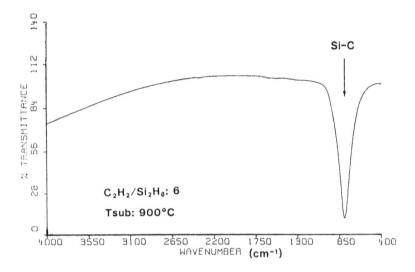

Figure 8.16 FTIR transmittance spectrum of the SiC film grown at 900°C. The gas flow ratio of C_2H_2 and Si_2H_6 was 6.

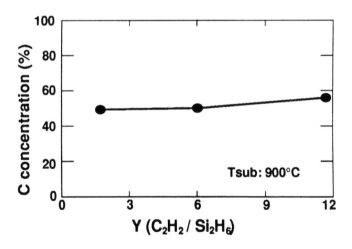

Figure 8.17 Carbon concentration of film grown with various source gas ratios

Single-crystal SiC grows with a Si-to-C ratio constant of 1. Polycrystalline SiC, however, does not. Figure 8.17 shows the carbon concentration, obtained by Auger electron spectroscopy (AES), of films grown with various source gas ratios ($Y = C_2H_2 / Si_2H_6$). Carbon concentration increases very slightly with Y. Considering the accuracy of AES, carbon concentration is around 50% at all source gas ratios investigated. Growth of the SiC film proceeds as the Si-to-C ratio remains 1. On the surface, C_2H_2 may react only with Si_2H_6 or its decomposed products, without carbon being deposited by the C_2H_2 molecules reaction. The polycrystalline SiC bandgap was around 2 eV, which is nearly the same as that of 3C-SiC. Polycrystalline SiC can be obtained by electron beam evaporation of a SiC source[49,50]. X-ray diffraction was used to analyze the film. Diffraction peaks were observed for samples prepared at substrate temperatures above 600 °C. The film appeared to be polycrystalline in this temperature range, and as substrate temperature increased, crystallinity improved. N-type doping was studied for the film grown by CVD. Phosphine (PH_3) was used as a doping gas. Figure 8.18 shows resistivity as a function of the PH_3 flow rate. Initially, the resistivity rapidly decreased but, when the flow rate exceeded 1 sccm, the resistivity was saturated. Minimum resistivity was 4.3 x 10^{-3} Ω·cm. Hall mobility was around 14 cm²/V·s when the carrier concentration was less than 5 x 10^{19} /cm³. On films prepared by electron beam evaporation without intentional doping, the film showed n-type conduction. Carrier concentrations were 5 x 10^{16} to 5 x 10^{17} /cm³. Hall mobilities depended on the growth temperature, reaching 20 cm²/V·s at 900 °C.

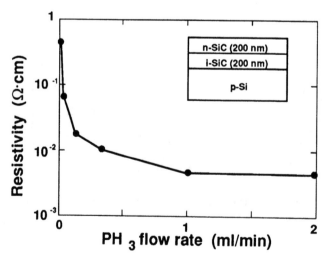

Figure 8.18 Resistivity as a function of the PH_3 flow rate

8.4. DEVICE APPLICATION

As the operational limits of electronic and optical devices made of conventional Si and GaAs semiconductors become clearer, the need for SiC devices increases. The following three sub-sections review device application of SiC in the last several years. SiC application in devices is generally classified into three. First is bipolar devices which use 3C-SiC/Si heterojunctions. Since the bandgap of 3C-SiC is wider than that of Si and since it can be produced as single-crystals, 3C-SiC shows promise for use in heterojunction bipolar transistors (HBTs). The second group of applications is field-effect transistors (FETs). SiC's great thermal and chemical stability, coupled with its wide bandgap, are primary reasons why SiC has been studied for use in high-temperature devices. 3C-SiC has a high breakdown field and a high saturation electron drift velocity (2.5×10^7 cm/s) which enable high-power and high-frequency operation. The third group of applications is LEDs fabricated using α-SiC which emit blue to bluish-purple light corresponding to the polytype bandgap.

8.4.1 3C-SiC/Si heterojunction bipolar transistor (HBT)

Advanced Si bipolar transistor techniques such as self-alignment and trench isolation have drastically downsized bipolar devices, and the accompanying parasitic resistances and capacitances. The per-gate delay of the high-performance circuits has been pushed around 20 ps. One way to further reduce gate delay is to decrease the base width and base resistance with a widegap emitter. The principal benefit of using a widegap emitter is a major reduction in the intrinsic base resistance while retaining sufficient current gain. Among the numerous candidates for widegap material including GaP, a-Si, a-SiC, and SiOx, 3C-SiC has been extensively studied because of its thermal stability, low resistivity, and high electron saturation velocity[27,30,51,52].

To obtain the expected performance, the band structure of the 3C-SiC/Si heterojunction must function as a base current barrier. We studied the discontinuity in the 3C-SiC/Si energy band using photoemission spectroscopy and determined the band structure in the heterojunction[53,54]. The schematic band structure for the 3C-SiC/Si heterojunction is shown in Figure 8.19. $\Delta E1$ and $\Delta E2$ are the binding energy differences between the valence band edges and the Si2p core levels for Si and 3C-SiC. The binding energy difference of Si2p core levels across the interface is ΔE. The valence band discontinuity, ΔEv, is given by

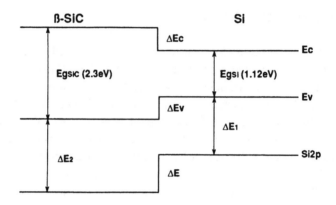

Figure 8.19 Schematic band structure for 3C-SiC/Si heterojunction

$$\Delta Ev = \Delta E1 - \Delta E2 + \Delta E. \qquad (8.3)$$

The conduction band discontinuity, ΔEc, is derived from

$$\Delta Ev + \Delta Ec = \Delta Eg, \qquad (8.4)$$

where ΔEg is the difference between the 3C-SiC and Si bandgaps. Once $\Delta E1$, $\Delta E2$, and ΔE are known, ΔEv and ΔEc can be obtained. $\Delta E1$ was obtained from a Si substrate and $\Delta E2$ from the thick 3C-SiC film (a few hundred nanometers) on a Si substrate. ΔE was measured from the thin 3C-SiC film (a few nanometers) on Si. Figure 8.20 shows the resulting band structure in equilibrium for an n-type 3C-SiC/p-type Si (emitter/base) heterojunction. The potential barrier, ΔEb, for holes in the Si is about 1 eV higher than that for electrons in SiC. With this SiC as a widegap emitter in an HBT, the injection of holes from the Si base to the SiC emitter is smaller than the injection of electrons from the SiC to the Si base by a factor of $\exp(\Delta Eb / k \cdot T)$, where k is Boltzmann's constant and T is temperature.

The current flow mechanism of the SiC/Si heterojunction is essential in determining the heterojunction system's applicability to HBTs. To investigate the current flow mechanism, we fabricated a heterojunction with n-SiC/p-Si[29,30]. n-type doping of the SiC was done with As ion-implantation. A mesa-type structure was fabricated by reactive ion etching with CF_4 and O_2.

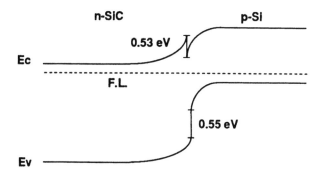

Figure 8.20 Resulted band lineup in equilibrium when a heterojunction of n-type 3C-SiC/p-type Si was formed.

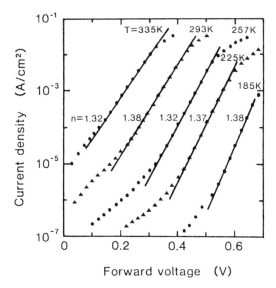

Figure 8.21 Forward current density-voltage characteristics of the n-type 3C-SiC/p-type Si heterojunction diode at various temperature.

Figure 8.21 shows the forward current density-voltage characteristics for various temperatures between 185 and 350 K. For all the experimental temperatures, there is a region where the current depends exponentially on the applied voltage. An analysis of the slope of the straight lines shows that the value of the diode ideality factor n is almost constant for various temperatures. This means that the current density-voltage characteristics can be described as

$$J = J_0 \cdot exp(q \cdot V / n \cdot k \cdot T). \qquad (8.5)$$

Since the n value is around 1.35, the current flow mechanism of the n-SiC/p-Si heterojunction diode is a combination of diffusion and recombination currents.

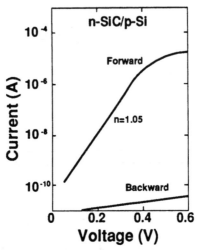

Figure 8.22 Current-voltage characteristics of the n-SiC/p-Si heterojunction

N-type SiC doping during SiC growth was also studied for the current-voltage characteristics of the n-SiC/p-Si heterojunction (Figure 8.22). The n-SiC layer was 60 nm thick and had a resistivity of 3×10^{-2} Ω•cm. The diode ideality factor is 1.05, much better than that of the As-implanted SiC/p-Si heterojunction. Damage in the As-implanted junction apparently is not completely removed during annealing.

A heterojunction bipolar transistor with a single-crystal 3C-SiC emitter was fabricated as shown in Figure 8.23. Undoped SiC is first grown epitaxially on an n-type Si(111) substrate followed by phosphorus ion implantation. The epitaxial growth was done using a $SiHCl_3$-C_3H_8-H_2 system at 1000 °C and 200 Pa. The epitaxial film was 200 nm thick. Boron ions were implanted at 70 keV at doses from 5×10^{12} to 5×10^{14} /cm^2 through the SiC to form the base region. Boron redistribution was suppressed by ion implantation through previously grown SiC. Reactive ion etching was used to etch away the 3C-SiC, except in the emitter area. The exposed Si was then oxidized slightly to form SiO_2. Ni was sputtered to form the emitter and base electrodes and aluminum was evaporated for the collector. As shown by the simulated impurity depth profile at the intrinsic transistor region (Figure 8.24), the base is about 200 nm wide.

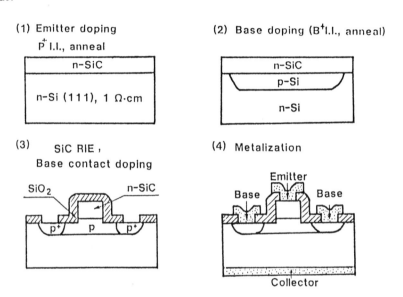

Figure 8.23 Transistor fabrication steps

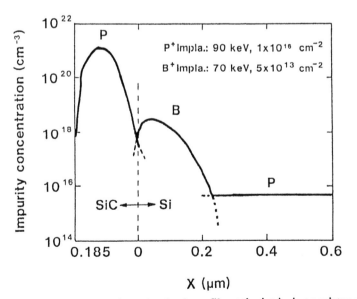

Figure 8.24 Simulated impurity depth profile at the intrinsic transistor region

Figure 8.25 shows the current gain as a function of the collector current for the different boron doses used in base fabrication. The maximum current gain of about 800 is obtained with an emitter carrier concentration of 1×10^{19} /cm^3, about two orders smaller concentration than for conventional poly-Si emitters. The relationship between the base current, I_b, and the emitter-base voltage, V_{eb}, is described by $I_b = I_0 \cdot \exp((q \cdot V_{eb}) / n \cdot k \cdot T)$, and the ideality factor n was around 1.1, suggesting that the diffusion current dominates.

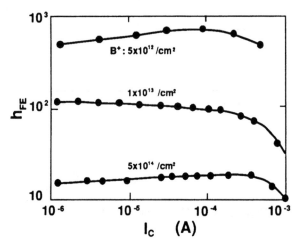

Figure 8.25 Current gain as a function of the collector current for different boron doses used in base fabrication

According to calculations, the large lattice mismatch of about 20% between 3C-SiC and Si should produce 2.4×10^{14} dangling bonds per cm^2 for a Si(111) substrate. The existence of a large number of misfit dislocations at the heterojunction and in the SiC epitaxial layer is confirmed by a lattice image obtained using a cross-sectional transmission electron microscope. Therefore, such misfit dislocations are not effective recombination centers, so the base current flows mainly by diffusion. This current flow mechanism is the main reason for the high current gain.

8.4.2   Pn junctions, Schottky diodes, and FETs

High-temperature semiconductor devices are useful for electronic control systems used under harsh environments such as those for automobiles, airplanes, and spacecrafts. SiC is promising for such devices because of its wide bandgap and physical stability.

Suzuki et al. have fabricated 3C-SiC pn junction diodes with mesa structures. They used CVD growth on Si substrates with impurity doping during crystal growth[55,56]. An undoped n-type layer was grown on an n-type Si substrate (conc. 1 x 10^{17} /cm³) followed by an aluminum-doped p-type layer (2 x 10^{17} /cm³). Diode characteristics were not adversely affected even during operation at 500°C. At room temperature, relatively good rectification is observed. With a junction area of 3.1 x 10^{-2} mm², ON voltage was about 1.2 V and the reverse leakage current was 5 µA at 5 V. Unfortunately, high-temperature rectification is significantly degraded with increased reverse leakage current and decreased built-in potential.

Pn junction diodes in α-SiC have been investigated vigorously at Cree Research, Inc., for use as high-temperature rectifiers, UV photodiodes, and blue LEDs. Rectifiers with blocking voltage from 15-1400 V and a forward current rating of 400 mA at about 3.0 V have been fabricated, packed for operation at 350°C, and tested[57]. For a 710 V rectifier, the reverse bias leakage current density at 650 V was shown to increase from 10^{-5} to 10^{-4} A/cm² for 300 to 673 K. The reverse recovery time of these devices is 15 to 25 ns, making them useful for switching power supplies.

Many FETs fabricated using SiC have been reported. In an early stage of device development, drain current was not saturated because of the leakage current through the SiC pn junction between the active n-layer and the buried p-layer. Advances in epitaxial growth techniques have made possible to fabricate FETs with suppressed leakage. Figure 8.26 shows room temperature drain current vs drain voltage characteristics of a 3C-SiC MESFET[58,59]. It was fabricated with an n-type 3C-SiC thin film epitaxially deposited on a 7-µm buried p-type 3C-SiC layer which was previously grown by CVD on a Si(100) substrate. The gate voltage, Vg, was varied from 0.6 V to -1.5 V. Very good drain current saturation is achieved as the drain voltage increases. The maximum transconductance in the saturated region was 0.64 mS/mm. For Vg < -2 V, the drain current was almost independent of the gate voltage, preventing the device from being fully turned off. This indicates some leakage current between the source and drain, possibly caused by the pn junction underneath the thin n-type layer, by the leakage current between the gate and the source, and/or by defects in the 3C-SiC film. The defect density, including stacking faults and antiphase domain boundaries in the bulk of the film, is very high.

Defect concentrations have been greatly reduced by growing the 3C-SiC in the (111) orientation on 6H-SiC {0001} substrates. Higher quality FETs have been fabricated with these films. The structure of the concentric ring device developed by Palmour et al. is shown in Figure 8.27[60]. A 100-μm-diameter circular dot, which functions as the drain, is entirely surrounded by a concentric gate ring which has a 20-μm-wide connecting strip extending to a 100 x 100 μm contact pad. The MOSFETs were fabricated in 1.2 μm n-type layers (n = 1-3 x 10^{15} /cm^3) on 0.5 μm p-type 3C-SiC thin films grown by CVD at 1773 K on the Si{0001} face of 6H-SiC crystals. The MOSFET's drain current-voltage characteristics at 296 K (Figure 8.28) show very stable drain current saturation. The leakage current, with a V_{DS} of 25 V in the off state (Vg = -15 V) was 3.75 μA. With a Vg of 2.5 V, the maximum room temperature transconductance with V_{DS} fixed at 20 V was 5.32 mS/mm. The transistor's operation was demonstrated up to 923 K. For temperatures up to 573 K, transconductance increased. It decreased with temperatures beyond 673 K. The maximum transconductance observed was 11.9 mS/mm on a 2.4-μm channel device at 673 K.

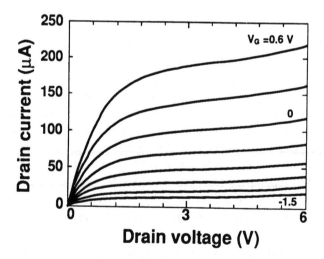

Figure 8.26 Drain current-voltage characteristics at room temperature of a MESFET with a gate 3.5 μm long[58]

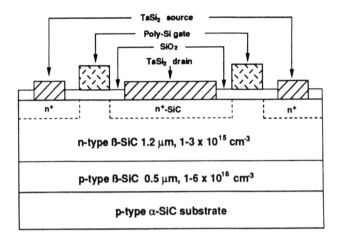

Figure 8.27 Cross section of the depletion-mode MOSFET structure[60]

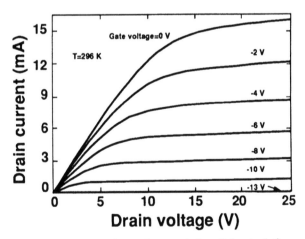

Figure 8.28 Drain current-voltage characteristics of the depletion-mode n-channel MOSFET in 3C-SiC(111) thin film at 296 K. The gate was 7.2 μm long and 390 μm wide[60].

An inversion-type MOSFET was fabricated using a 3C-SiC layer grown on a Si(100) substrate[61,62]. A substrate 2° off (100) toward (011) was used to eliminate antiphase boundaries and obtain a flat surface for reducing pn junction leakage. Epitaxial 3C-SiC growth was carried out in a horizontal quartz reactor tube at atmospheric pressure. Carbonization was done using C_3H_8 followed by CVD growth using a SiH_4-C_3H_8-H_2 system. Crystal growth was at 1350°C. Figure 8.29 is the cross-sectional view of the MOSFET. A channel layer of B-doped p-SiC with a thickness of 2 μm was grown on undoped n-SiC with a thickness of 7 μm. Because of the lattice mismatch between Si and SiC, a layer of over several microns was needed to reduce crystal defects. Using B_2H_6 during CVD growth, B-doped p-type SiC was obtained. Hole measurement of samples grown under similar doping conditions showed hole concentration and mobility of almost 1 x 10^{15} /cm^3 and 17 cm^2/V•s. The source and drain were formed in two steps by P$^+$ ion implantation. The first implantation was at a dose of 1 x 10^{15} /cm^2 with an ion energy of 100 keV to form an n$^+$ layer for the source and drain. To increase carrier concentration near the surface and reduce contact resistance, the second implantation was at 3 x 10^{14} /cm^2 with 25 keV. Annealing was in an IR radiative heating furnace in an Ar atmosphere at 1080°C for 1 hour. The sheet resistance of the implanted layer was 1.5 x 10^3 Ω/square. To form a gate insulator, SiC thermal oxidation was done at 1050°C with dry oxygen for 6 hours. The gate oxide was about 60 nm thick. The gate was 20 nm long and 500 μm wide.

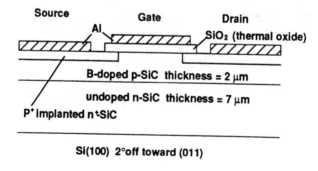

Source   Gate   Drain

Al   SiO$_2$ (thermal oxide)

B-doped p-SiC thickness = 2 μm

undoped n-SiC thickness = 7 μm

P$^+$ implanted n$^+$SiC

Si(100) 2°off toward (011)

Figure 8.29 Cross section of the inversion-type MOSFET[61]

Figure 8.30 shows the current-voltage characteristics of the FET at 310 K. This
was the first report on the characteristics of the inversion-type MOSFET. For a
gate bias of 10 V, the leakage current between the source and gate was less than
1 μA. A tendency of saturation of drain current can be seen. The threshold
voltage was 2.8 V and the effective mobility was about 100 cm^2 /V•s.
Inversion-type MOSFETs using α-SiC reported by Davis et al.[63] had a
maximum transconductance of 0.27 mS/mm at a gate voltage of 24 V. The
threshold voltage was high, about 9 V. Increasing the temperature to 673 K
increased the maximum transconductance to 1.03 mS/mm and decreased threshold
voltage from 9 to 0.25 V. Because of their higher transconductance, lower
threshold voltage, and relatively low leakage current, the devices appear
promising for high- temperature operation.

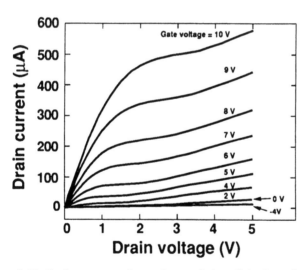

Figure 8.30 Drain current-voltage characteristics of the inversion-type
MOSFET[61]

Kelner et al. showed improved performance in a buried-gate junction field
transistor (JFET)[64] which used an n-type 3C-SiC (111) thin film grown on the
Si {0001} face of a p-type 6H-SiC substrate. Devices with 4 μm gate length
had maximum room temperature transconductance of 20 mS/mm. Using a
charge control model, experimental data showed that the average field-effect
mobility (~565 cm^2/V•s) was close to the measured Hall mobility (~470
cm^2/V•s) and that the electron saturation velocity in the channel is close to the

predicted theoretical value of 2 x 10^7 cm/s. The model suggests that transconductance values exceeding 100 mS/mm and channel currents higher than 400 mA/mm can be achieved, making this type of SiC FET attractive for high-power, high-frequency application.

Yoshida et al. fabricated Schottky-barrier diodes on unintentionally doped 3C-SiC films grown on (100) Si with nickel and gold as the ohmic and rectifying contacts[65]. The barrier height was 1.15±0.15 eV according to capacitance measurements and 1.11±0.03 eV from photoresponse results. Fujii et al. studied the dependence of Schottky characteristics by evaporation of Au on n-type SiC's on the different Si substrate orientation ((n11) Si, n=1,3,4,5,6, and (100) Si)[66]. The Schottky-barrier diodes of the 3C-SiC films on (611) Si, (411) Si, and (111) Si show excellent characteristics compared with those using Si(100) substrates, i.e., reverse leakage currents are small, ideality factors are close to unity, and barrier heights are large. These dependences on substrate orientation are probably from the differences in the surface crystal imperfections such as stacking faults or antiphase boundaries. Furukawa et al. fabricated Schottky-barrier diodes using crack-free 3C-SiC on Si(111) substrates with Au as the Schottky metal[67], that showed excellent rectification characteristics with a reverse leakage current of about 1.1 x 10^{-3} A/cm^2 at -5 V. The ideality factor is 1.4 to 1.6 and the barrier height is between 0.9 to 1.1 eV.

Schottky-barrier FETs were fabricated in 3C-SiC[68-70]. First, boron-doped high-resistivity layers of about 5 μm were epitaxially grown at 1350°C on p-type Si (100) substrates. Next, undoped 3C-SiC layers were grown to about 650 nm. The undoped 3C-SiC has n-type conduction, a carrier concentration of 3 x 10^{16} /cm^3, and a room temperature Hall mobility of 770 cm^2/V•s. The on voltage was about 1.1 V and saturation of the drain current was observed. A maximum transconductance of 1.1 to 1.7 mS/mm in the saturation region and a threshold voltage of about -4 V were obtained. As temperature increased, the soft-breakdown voltage of the 3C-SiC Schottky diodes decreased. Transconductances of 0.5, 0.32, and 0.15 mS/mm were obtained for temperatures of 200, 300, and 400°C. The room temperature drain current-voltage characteristics remained unchanged after heating to 400°C in air, and then cooling the devices, confirming the stability of the undoped and boron-doped 3C-SiC epitaxial layers, as well as the Au Schottky-barrier contacts.

MESFETs fabricated in α-SiC reported by Davis et al. appear promising for high temperature operation[63]. The device had a gate 13 μm long, and 1 mm wide. The maximum transconductance was 1.5 mS/mm with a threshold voltage of 4.4 V. I-V characteristics actually improved when the device was heated to 373 K. The maximum transconductance increased to 1.65 mS/mm, and the current saturation was very good to the drain voltage of 30 V. The high-speed capabilities of MESFETs were shown by Palmour et al.[71]. Submicron-geometry MESFETs fabricated in 6H-SiC show 7 dB power gain at 1 GHz. The highest f_T measured was 2.9 GHz. Based on these results and device modeling, 6H-SiC MESFETs should eventually operate at frequencies as high as 10 GHz with output power levels 5 times higher than equivalent GaAs devices. Such high power devices would be a major advantage in radar communications.

Although SiC has not yet been applied to high-voltage active devices, it is a strong candidate. High-voltage Si-based semiconductor devices have been used extensively. But, as their blocking voltage exceeds 200 volts, the ON-state voltage drop of these devices increases rapidly leading to unacceptably high power dissipations. It has been theoretically shown[72] that SiC-based Schottky rectifiers and power MOSFETs have lower ON-state voltage drops than even the best 5000-volt Si bipolar devices.

8.4.3   Light emitting diodes (LEDs) and detectors

Blue diodes using 6H-SiC and fabricated by LPE are now commercially available[57]. The devices emit light with a peak wavelength of 470 nm and a full width at half maximum of 70 nm. The optical power output is typically between 12 and 17 μW at a forward current of 20 mA and 3 V biasing. This represents a power efficiency of ~0.02 to 0.028%, the highest, to our knowledge, for any commercially available blue LEDs. Blue diodes fabricated by CVD have also been reported[73-76], but the power efficiency was, at best, 0.002%. To obtain higher efficiency, the crystal quality of substrates and the controllability in epitaxial growth must be improved. Edmond et al reported UV photodetectors made with 6H-SiC[57]. The 2.9 eV energy band allows for inherently low dark currents and a high quantum efficiency, even at high temperatures. Preliminary devices exhibit quantum efficiency of about 70% with a peak response at 270 nm. The wavelength cut-off is at ~420 nm corresponding to the bandgap. The dark current density at -1.0 V and 300 K is about 10^{-14} A/cm². The large, 3.27 eV, bandgap of 4H-SiC among α-SiC polytypes corresponds to a light range from blue to ultraviolet. Since 4H-SiC substrates became available by a sublimation method, LEDs using the substrates were reported[77,78]. An epitaxial layer was grown on the substrate by LPE at about 1700 °C, and Al (acceptor) and N (donor) doping was used to form donor-acceptor pairs. A peak wavelength of electroluminescence was about 420 nm, and the full width at half maximum was 47 nm. The maximum light intensity at 20 mA was 2.2 mcd .

8.5.   SUMMARY

This chapter reviewed physical properties, recent developments in growth techniques, and device applications for SiC. Electronic devices with encouraging characteristics have been reported in the last decade, mainly due to developments in SiC growth technique. We expect that in the next decade, many of the new electronic devices will have developed characteristics and features originating in the promising physical properties of SiC.

ACKNOWLEDGMENT

The author is pleased to thank many colleagues for discussions and suggestions: T. Aoyama, A. Fukuroda, M. Doki, F. Mieno, T. Eshita, M. Maeda, Y. Furumura, and T. Ito

REFERENCES

[1]   E. O. Johnson, RCA Review, 2 6, 163 (1965)
[2]   B. J. Baliga, IEEE Electron Device Lett., 1 0, 455 (1989)
[3]   R. C. Marshall. J. W. Faust, Jr., and C. E. Ryan, eds., Silicon Carbide, Proc. Int. Conf., 3rd, 1973, University of South Carolina Press, Columbia, South Carolina, 1974
[4]   W. E. Nelson, F. A. Halden, and A. Rosengreen, J. Appl. Phys., 3 7, 333 (1966)
[5]   J. A. Lely, Ber. Deut. Keram. Ges., 3 2, 229 (1970)
[6]   S. Nishino, J. A. Powell, and H. A. Will, Appl. Phys. Lett., 4 2, 460 (1983)
[7]   P. Liaw and R. F. Davis, J. Electrochem. Soc., 1 3 2, 642 (1985)
[8]   A. Addamiano and J. A. Sprague, Appl. Phys. Lett., 4 4, 525 (1984)
[9]   S. Nishino, H. Suhara, and H. Matsunami, Ext. Abstr. of 15th Conf. on SSDM, 317 (1983)
[10] S. Nishino, K. Shibahara, S. Dohmae, and H. Matsunami, Final program and Late News Abstr. of 16th Conf. on SSDM, 8, (1984)
[11] S. Nishino, H. Suhara, H. Ono, and H. Matsunami, J. Appl. Phys. 6 1, 4889 (1987)
[12] K. Sasaki, E. Sakuma, S. Misawa, and S. Gonda, Appl. Phys. Lett., 4 5, 72 (1984)

[13] Y. Furumura, M. Doki, F. Mieno, and M. Maeda, Electronics and Comunications in Japan, Part 2, **7 0**, 705 (1987)

[14] Y. Furumura, M. Doki, F. Mieno, T. Eshita, T. Suzuki, and M. Maeda, 10th CVD Symp. Proc., 435 (1987)

[15] T. Eshita, T. Suzuki, T. Hara, F. Mieno, Y. Furumura, M. Maeda, T. Sugii, and T. Ito, Mat. Res. Soc. Symp. Proc. **1 1 6**, 357 (1988)

[16] T. Fuyuki, M. Nakayama, T. Yoshinobu, and H. Matsunami, Proc. Molecular Beam Epitaxy, 104 (1988)

[17] T. Fuyuki, M. Nakayama, T. Yoshinobu, H. Shiomi, and H. Matsunami, J. Cryst. Growth, **9 5**, 461 (1989)

[18] S. Motoyama and S. Kaneda, Appl. Phys. Lett., **5 4**, 242 (1989)

[19] T. Sugii, T. Aoyama, and T. Ito, J. Electrochem. Soc., **1 3 7**, 989 (1990)

[20] L. J. Kroko and A. G. Milnes, Solid-State Electron., **9**, 1125 (1966)

[21] R. B. Campbell and H. S. Berman, Mat. Res. Bull., **4**, S211 (1969)

[22] O. J. Marsh and H. L. Dunlap, Rad. Effects, **6**, 301 (1970)

[23] R. W. Bartlett and R. A. Muller, Mater. Res. Bull., **4**, 5341 (1969)

[24] N. N. Long, D. S. Nedzvetski, N. K. Prokofera, and M. B. Reifman, Opt. i. Spektroskopiya, **2 9**, 388 (1970)

[25] S. Nishino, H. Suhara, and H. Matsunami, Ext. Abstr. of 15th Conf. on SSDM, 317 (1983)

[26] H. J. Kim and R. F. Davis, J. Electrochem. Soc., **1 3 3**, 2350 (1986)

[27] T. Sugii, T. Aoyama, Y. Furumura, and T. Ito, Proc. 37th American Vacuum Soc. Symp., TC3-WeM4 (1990)

[28] T. Takeuchi, K. Shibahara, and H. Matsunami, SSD86-177, 51 (1986)

[29] T. Sugii, T. Ito, Y. Furumura, M. Doki, F. Mieno, and M. Maeda, Tech. Dig. 1986 Symp. on VLSI Technol., 45 (1986)

[30] T. Sugii, T. Ito, Y. Furumura, M. Doki, F. Mieno, and M. Maeda, J. Electrochem. Soc., **1 3 4**, 2545 (1987)

[31] R. E. Avila, J. J. Kopanski, and C. D. Fung, J. Appl. Phys., **6 2**, 3469 (1987)

[32] J. A. Edmond, K. Das, and R. F. Davis, J. Appl. Phys., **6 3**, 922 (1988)

[33] Yu. M. Tairov and V. F. Tsvetkov, J. Cryst. Growth, **5 2**, 146 (1981)

[34] G. Ziegler, P. Lanig, D. Theis, and C. Weyrich, IEEE Trans. Electron Device, **E D - 3 0**, 277 (1983)

[35] K. Koga, T. Nakata, and T. Niina, Ext. Abstr. 17th Conf. on SSDM, 249 (1985)

[36] T. Tanaka, K. Koga, Y. Matsushita, Y. Ueda, and T. Niina, Springer Proceed. in Physics, **4 3**, 26 (1988)

[37] A. Suzuki, M. Ikeda, N. Nagao, H. Matsunami, and T. Tanaka, J. Appl. Phys., **4 7**, 4546 (1976)

[38] H. Matsunami, M. Ikeda, A. Suzuki, and T. Tanaka, IEEE Trans. Electron Devices, **E D - 2 4**, 958 (1977)

[39] H. Matsunami, S. Nishino, M. Odaka, and T. Tanaka, J. Cryst. Growth, **3 1**, 72 (1975)

[40] S. Nishino, H. Matsunami, and T. Tanaka, J. Cryst. Growth, **4 5**, 144 (1978)

[41] G. Gramberg and M. Koniger, Solid State-Electron., **1 5**, 285 (1972)

[42] W. von Munch and I. Pfaffeneder, Thin Solid Films, **3 1**, 39 (1979)

[43] H. Matsunami, Springer Proceed. in Physics, **7 1**, 3 (1992)

[44] W. S. Yoo, S. Nishino, and H. Matsunami, J. Cryst. Growth, **9 9**, 278 (1990)

[45] K. Koga, T. Nakata, Y. Ueda, Y. Matsushita, Y. Fujikawa, T. Uetani, and T. Niina, Ext. Abstr. 176th Electrochem. Soc. Meet., **8 9**, 689 (1989)

[46] Y. Ueda, T. Nakata, K. Koga, Y. Matsushita, Y. Fujikawa, T. Uetani, and T. Niina, Mat. Res. Soc. Proc. 162, 427 (1990)

[47] T. Sugii, T. Aoyama, and T. Ito, J. Electrochem. Soc., **1 3 6**, 3111 (1989)

[48] T. Sugii, T. Yamazaki, K. Suzuki, T. Fukano, and T. Ito, Tech. Dig. IEDM, 659 (1989)

[49] K. Kamimura, Y. Nishibe, and Y. Onuma, Springer Proceed., in Physics, **4 3**, 207 (1989)

[50] Y. Onuma, S. Miyashita, Y. Nishibe, K. Kamimura, and K. Tezuka, Springer Proceed. in Physics, **4 3**, 212 (1989)

[51] T. Sugii, T. Ito, Y. Furumura, M. Doki, F. Mieno, and M. Maeda, IEEE Electron Device Lett., **EDL-9**, 87 (1988)

[52] T. Sugii, T. Yamazaki, Y. Furumura, and T. Ito, Abstr. 4th Int. Conf. Amorphous and Cryst. SiC and other IV-IV Mats., V-1, (1991)

[53] T. Aoyama, T. Sugii, and T. Ito, Proc. 2nd Int. Conf. Formation of Semicon. Surf., 143 (1989)

[54] T. Aoyama, T. Sugii, and T. Ito, Appl. Surf. Sci. **4 1 / 4 2**, 584 (1989)

[55] A. Suzuki, A. Uemoto, M. Shigeta, K. Furukawa, and S. Nakajima, Ext. Abstr. 18th Conf. on SSDM, 101 (1986)

[56] K. Kurusawa, A. Uemoto, M. Shigeta, A. Suzuki, and S. Nakajima, Appl. Phys. Lett., **4 8**, 1536 (1986)

[57] J. A. Edmond, H. S. Kong, and C. H. Carter, Jr., Abstr. 4th Int. Conf. Amorph. and Cryst. SiC and Other IV-IV Mats., III.4 (1991)

[58] H. S. Kong, J. W. Palmour, J. T. Glass, and R. F. Davis, Appl. Phys. Lett., **5 1**, 442 (1987)

[59] J. T. Glass, Y. C. Wang, H. S. Kong, and R. F. Davis, Mat. Res. Soc. Symp. Proc. **1 1 6**, 337 (1988)

[60] J. W. Palmour, H. S. Kong, and R. F. Davis, Appl. Phys. Lett., **5 1**, 2028 (1987)

[61] K. Shibahara, T. Saito, S. Nishino, and H. Matsunami, IEEE Electron Device Lett., **EDL-7**, 692 (1986)

[62] K. Shibahara, T Saito, S. Nishino, and H. Matsunami, Ext. Abstr. 18th Conf. on SSDM, 717 (1986)

[63] R. F. Davis, J. W. Palmour, and J. A. Edmond, Mat. Res. Soc. Symp. Proc. **1 6 2**, 463 (1990)

[64] G. Kelner, M. Shur, S. Binari, K. Sleger, and H. Kong, Springer Proceedings in Physics, **4 3**, 184 (1988)

[65] S. Yoshida, K. Sasaki, E. Sakuma, S. Misawa, and S. Gonda, Appl. Phys. Lett., **4 6**, 766 (1985)

[66] Y. Fujii, A. Ogura, K. Furukawa, M. Shigeta, A. Suzuki, and S. Nakajima, Mat. Res. Soc. Symp. Proc., **1 1 6**, 351 (1988)

[67] K. Furukawa, A. Uemoto, Y. Fujii, M. Shigeta, A. Suzuki, and S. Nakajima, Ext. Abstr. of 19th Conf. on SSDM, 231 (1987)

[68] S. Yoshida, H. Daimon, M. Yamanaka, E. Sakuma, S. Misawa, and K. Endo, J. Appl. Phys., 6 0, 2985 (1986)

[69] H. Daimon, M. Yamanaka, E. Sakuma, S. Misawa, K. Endo, and S. Yoshida, Jpn. J. Appl. Phys., 2 5, L592 (1986)

[70] H. Daimon, M. Yamanaka, M. Shinohara, E. Sakuma, S. Misawa, K. Endo, and S. Yoshida, Appl. Phys. Lett, 5 1, 2106 (1987)

[71] J. W. Palmour, J. A. Edmond, H. S. Kong, and C. H. Carter, Jr., Absr. 4th Int. Conf. Amorph. and Cryst. SiC and Other IV-IV Mats., III.1 (1991)

[72] B. J. Baliga, Absr. 4th Int. Conf. Amorph. and Cryst. SiC and Other IV-IV Mats., VI.1 (1991)

[73] W. V. Munch, et al., Solid-State Electron., 1 9, 871 (1976)

[74] W. V. Munch, J. Electron. Mater., 6, 449 (1977)

[75] S. Nishino, A. Ibaraki, H. Matsunami, and T. Tanaka, Jpn. J. Appl. Phys., 1 9, L353 (1980)

[76] K. Shibahara, N. Kuroda, S. Nishino, H. Matsunami, Jpn. J. Appl. Phys., 2 6, L1815 (1987)

[77] K. Koga, T. Nakata, Y. Ueda, Y. Matsushita, Y. Fujikawa, T. Uetani, and T. Niina, Ext. Abstr. 176th Electrochem. Soc. Meeting, 8 9, 689 (1989)

[78] V. A. Dmitriev, L. M. Kogan, Ta. V. Morozenko, B. V. Tsarenkov, and A. E. Cherenkov, Sov. Phys. Semicond., 23 (1), 23 (1989)

PART TWO:

Polycrystalline silicon

9

Large Grain Polysilicon Substrates
for Solar Cells

S. Martinuzzi and S. Pizzini

9.1   INTRODUCTION

Since the very early age of semiconductor silicon, all the efforts of the world silicon manufacturers were devoted to avoid the presence of structural defects, like dislocations, twins, grains and subgrain boundaries, which were proven to be extremely harmful for the lifetime of minority carriers.

So, polycrystalline silicon, in any form or shape produced, was only a scientific curiosity until the oil crisis of 1972, when the western countries were forced to consider alternative energy sources, and within them, to photovoltaics.

Solar energy, infact, albeit very cheap and practically infinite requires vast surfaces and very sophisticated devices to be captured in sufficient amount, due to its low density of power at ground level.

For this and for many other reasons (availability, low cost, negligible interaction with the environment) silicon was identified very soon as the ideal candidate material for terrestrial photocells.

After more than 30 years of in-field experience with silicon solar cells, this original idea has been demonstrated to be well founded. It was also recognized that polycrystalline silicon solar cells offer the challenge to compete in efficiency with those made with single crystal silicon.

This conclusion, which appears at a first glance somewhat arbitrary, is supported by the impressive progress achieved in polycrystalline silicon manufacturing processes.

As we will show in the next Sections, this material is grown in square section molds to reduce the shaping losses and presents a typical microstructure characterized by columnar grains, which could be several cm in diameter. As this grain size is exceedingly larger than the diffusion length of minority carriers, the recombination losses at grain boundaries (GBs) are negligible with respect to other losses (i.e. deep level recombination, dislocations, etc.).

Furthermore, this material is currently produced at an industrial scale by several manufacturers. The lower cost and the square shape of the slices which can be obtained by shaping and sawing the ingots with relative minor amounts of kerf losses when compared to CZ silicon, make indeed the cast silicon materials particularly suitable for solar cell manufacturing purposes. Moreover the progresses in the growth techniques were specifically aimed at the improvement of the solar cell efficiency (Herzer 1991).

The silicon ribbon route, as the alternative to the casting process, is still on an intermediate step of development, except for the single case of the EFG (Edge Film Fed Growth) Mobil Solar process (Kalejs 1987). Here, high quality about 300 µm thick ribbons are obtained by laser cutting meters-long octagon shaped "tubes".

Both routes offer significative advantages, at least in principle, with respect to the single crystal route, but present also several important drawbacks.

As an example, the electrical and structural properties of polycrystalline silicon are entirely depending on the particular growth process used for its production. Therefore, many "qualities" of polycrystalline silicon exist as many production processes are currently used.

The aim of this chapter is to discuss this topic presenting, however, more our personal views on the matter than a critical review of the existing technical and scientific literature on this topic.

9.2   GROWTH OF POLYCRYSTALLINE (MULTICRYS-TALLINE) SILICON

A list of the processes which are today at an industrial scale, with their most relevant features, is reported in Table 9.1. Some of these processes, are discussed in details in this Section, for their relevance in the economics of the European solar cell market. Polycrystalline silicon seemed in fact the natural candidate as a low cost substrate for solar cells, as it was already shown that the diffusion

Table 9.1 Polycrystalline silicon growth techniques

Producer	Technique/Trade name	Remarks	Year	Ref.
Wacker	Casting/SILSO	The charge is premelted and poured in the casting mold where a frozen Si layer prevents sticking and impurity contamination. Ingot weight (IW) = 100 kg. Kyocera has been licenced to produce SILSO in Japan.	1975 1987	Authier Helmreich
Solarex	DS/SEMIX	The charge is molten and crystallized in one single mold. Proprietary crucible features. IW = 30 kg.	1976 1988	Lindmayer Brenneman
Crystal Systems	DS/HEM	Helium is used as the coolant. Highly directional solidification, with very large grains. Vacuum processing. IW = 50 kg - 80 kg.	1972 1987	Schmid Chattak
Photowatt	DS/POLIX	The Si charge is encapsulated by a liquid, to prevent impurity contamination from the graphite mold. The mold is reusable. IW = 60 - 100 kg.	1980 1987	Fally Fally
Eurosolare Crystallox	DS	Highly directional solidification in a quartz crucible. Silicon sticking is prevented by a thick silicon nitride coating. IW = 60 - 100 kg.	1981 1991	Pizzini Dorriy
NEC/OTC	DS/MRC	Directional solidification in quartz crucibles with silicon nitride as releasing agent.	1981	Saito
Mobil Solar	EFG	Ribbon growth on octagon and nonagon shape. The material is large grained, generally largely supersaturated in carbon. The output is about 9000 cm^2/h in the octagon version of the puller.	1987 1991	Kalejs Wald
Bayer	DS/SOPLIN	A planar solidification interface is adopted for improving the impurity rejection and for better electrical homogeneity.	1992	Koch
Osaka Titanium	Cold crucible/EM Casting	Silicon is fused in electromagnetic field where it leviates with graphite susceptor walls. Possibility of continuous casting.	1990	Kaneko

lenght of minority carriers and the conversion efficiency approached that of single crystal silicon when the grain size was larger than a fraction of a cm (Dugas 1987) (see figure 9.1 and 9.2).

Furthermore, the casting technology is a process widely used in the metallurgy of light metal alloys and of superalloys and it was selected in alternative to the conventional Czochralski process in view of its being, at least potentially, a viable, low cost technology.

Within the many problem which could be indeed forsecast for silicon, the major consisted in the choice of the proper crucible material.

In fact, silicon not only has a large volume expansion when it solidifies but it is also known to wet and react in the liquid state with most of the materials which could withstand the high process temperatures (about 1450-1500°C).

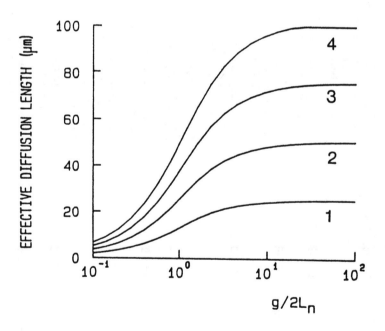

Figure 9.1 Effective diffusion length vs the ratio grain size/bulk diffusion length. $N_A = 10^{16} cm^{-3}$; Grain boundary recombination velocity : 10^6 cm/s.
1) $L_n = 25 \mu m$; 2) $L_n = 50 \mu m$; 3) $L_n = 75 \mu m$; 4) $L_n = 100 \mu m$ [from Dugas (1987)].
(Reproduced by permission of Ed. de Physique)

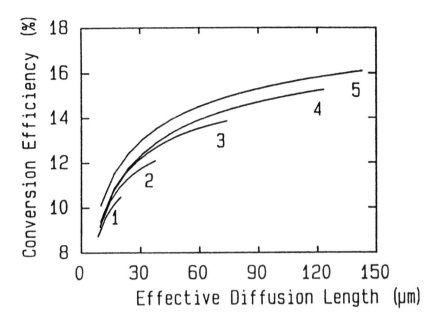

Figure 9.2 Variation of the conversion efficiency of solar cells as a function of the effective diffusion length, for different grain sizes : 1) g = 100 μm ; 2) g = 200 μm ; 3) g = 500 μm ; 4) g = 2 mm ; 5) equivalent monocrystalline cell.

Base : p-type, $N_A = 10^{16}$ cm^{-3} ; base thickness = 500 μm.

Emitter : n-type , $N_D = 5 \times 10^{19}$ cm^{-3} ; emitter thickness = 0.2 μm.

Grain boundary recombination velocity S = 10^6 cm/s [from Dugas (1987)]. (Reproduced by permission of Ed. de Physique)

This property not only implies that impurity contamination would be an immediate consequence of the use of any crucible not explicitly designed for this use, but that wetting would cause the set-up of interface stresses if the expansion coefficients of silicon and of the crucible material would not match properly, which is the most general case.

The Wacker process (Authier 1975, Helmreich 1985a , Helmreich 1987b).

The early considerations upon the need to reduce dramatically the cost of solar cells in order to compete with conventional energy sources were the hint for Wacker Chemitronic to start with a research program aiming to develop a low cost casting technique for large polycrystalline silicon ingots.

The key issue of the Wacker process, succesfully developed at the Wacker Heliotronic Laboratories, was to use graphite as the mold and to carry out the initial solidification of the charge quickly enough to create in correspondance of the mold walls a thick silicon crust which behaves as a liner or as the true working crucible.

In this way silicon crystallizes without or with minor contaminations from the mold.

In order to do it, the charge is previously molten in a quartz crucible and then poured in a graphite crucible in which the charge is allowed to solidify.

This is apparently a variant of the cold crucible technique, with the advantage that the silicon "crucible" could be held at any temperature below the melting point of silicon, in order to avoid thermal stresses, which would degrade the material quality. A further advantage claimed by the manufacturer is that fast diffusing impurities are gettered by the chilled layer.

Finally, the mold is reusable, as fast freezing avoids silicon sticking and it can be built in detachable elements.

To provide a proper solidification of the charge, which is intended to be truly unidirectional, the process is carried out in a controlled vertical temperature gradient, and the heat of solidification is extracted predominantly from the bottom by an insulated water-cooled shaft. To prevent the freezing of the upper portion of the molten charge (which would cause the set-up of stresses associated to the difference in the molar volume of the solid and of the fused silicon and, possibly, the fracture of the charge) a radiation heater is mounted above the mold. In spite of these precautions, the molten charge solidifies starting also from the walls, originating a characteristic grossly columnar structure which is typical of this process (see figure 9.3a).

By this process 100 kg ingots grown with a growth rate of about 0.5 kg/h are currently obtained.

The material produced with this technology is known with the trade name Silso. It is relatively large grained, and coherent grain boundaries, twins, intragrain dislocations are the most common extended defects present.

A deeper insight on the microstructure of this material shows that in some portions of the ingot a characteristic microcrystalline (called "grit") structure appears, where the local values of the minority carrier diffusion lenght are particularly low, and that dislocations with a density and a distribution which vary locally are the most dangerous defects as lifetime killers.

This "grit" structure is however totally absent in the wafers now currently available on the market. At this time, the Wacker polycrystalline silicon together with the Crystal System one is the only commercial material freely available on the western market, as all the other materials are for custom use.

The Crystal Systems Heat Exchanger Process (HEM) (Schmid 1972a , Schmid 1975b , Chattak 1987)..

The HEM process is based on a technology which was developed by Crystal Systems for the industrial production of large, high quality sapphire crystals weighing up to 48 kg and which is now used also for other materials.

The process, which has been demonstrated to yield high quality polycrystalline and single crystalline materials (depending whether a seed is used or not), is based on the use of a exceptionally well designed system for extracting the heat of solidification on the bottom of the crucible by an heat exchanger, using helium as the cooling fluid.

In turn, the vertical gradient is controlled carefully by the furnace heaters, in such a way that, ideally, heat is extracted only from the bottom and true directional solidification (DS) is carried out.

By this technique, which does allow to maintain a convex-shaped interface along the entire duration of the solidification, not only single crystalline ingots could be grown without spurious nucleations from the walls, but impurity rejection by segregation occurs in such a condition that never an impurity excess capable to induce constitutional supercooling is accumulated in the stagnant layer, due to the increasing size of the solid-liquid interface.

As the crucible, high purity graded silica is used, which prevents the ingot from cracking during cooling. The HEM process, which is routinely operated under a vacuum of 0.1 mbar, offers definitive advantages over the other directional solidification technologies, in that no argon as cover gas is needed, no moving parts are necessary and the heat extraction can be carried out in a very controlled and simple fashion.

Vacuum processing, however, presents the definitive disadvantage that some carbon transport from the graphite heaters to the melt occurs (see Section 1.2 for details), without appreciable influence on the electrical properties of the material, provided SiC formation is prevented. Single crystalline or polycrystalline (multicrystalline) ingots of a maximum weight of about 100 kg can be grown by this technique.

The Photowatt Polix Process (Fally 1980a , Fally 1987b).

Also the Polix process is a true directional solidification process, which has been originally developed in the Marcoussis CGE Laboratories and which has been then scaled up by Photowatt to an industrial process.

The distinctive feature of this process is a reusable graphite crucible, where the silicon charge is molten and then crystallized. A direct contact between silicon and graphite, which would induce saturation of the charge with carbon and, possibly, also strong impurity contamination unless using very pure graphite, is avoided by the use of a liquid encapsulation technique, where the encapsulant is a material which is stable in the presence of both silicon and graphite and which is immiscible with molten silicon.

The heat extraction is carried out with a proprietary technique, essentially based on a controlled radiative heat loss from the bottom of the crucible.

A disadvantage of this technique is that during the cooling of the silicon ingot toward room temperature after complete solidification of the charge, the encapsulant solidifies and makes difficult the extraction of the charge from the crucible.

The ingot is therefore extracted in a successive step, via RF heating up to the melting point of the encapsulating material.

The material produced with this technology is columnar like the HEM one, with large grains (see figure 9.3b) where dislocations and twins are normally present at variable densities. Ingots up to a weight of 80 kg are currently casted and the quality of the material is fairly good as the mean diffusion length of minority carriers is around 90 μm.

The Heliosil-Italsolar (now Eurosolare) Process (Pizzini 1981a , Margadonna 1989 , Dorrity 1991).

This process was originally developed at a pilot plant stage at the Heliosil Laboratories and then scaled up to an industrial process by Italsolar.

The distinctive feature of this process, which is also a true DS process, is the use of a silica crucible which is protected from wetting by molten silicon by a thick layer of silicon nitride, which is applied by a proprietary technology.

The crucible is supported by a pedestal, which allows the crucible to be lowered at constant speed in the furnace assembly. Joule heating by a graphite heater was used until the last developments (see below).

Heat extraction is carried out by an especially designed water cooled heat exchanger which couples the advantages of a highly efficient radiative loss at high temperature with a conventional liquid fluid cooled low temperature heat exchanger.

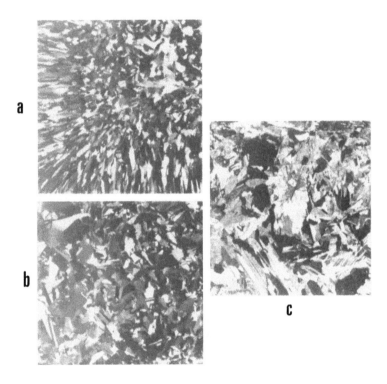

Figure 9.3 Optical micrograph of three 10 x 10 cm^2 multicrystalline silicon wafers after chemical etching : (a) Silso ; (b) Polix ; (c) Eurosolare.

Like in the case of the HEM technique, this process is carried out under a vertical temperature gradient which is carefully controlled by the furnace heaters and heat is extracted only from the bottom. This ensures a convex-shaped solidification interface, very large grains (see figure 9.3c), no spurious nucleations from the walls, very efficient segregation of impurities and practical absence of constitutional supercooling effects.

In its last version, jointly developed with Crystalox (Dorrity 1991), low frequency RF heating is used and the pedestal can be rotated in the range of 1-10 rpm.

At this time, multicrystalline ingots up to a weight of 100 kg, can be currently manufactured by this process. The quality of the material is in general very good, as 80 % of the ingot presents lifetime values which are comparable with those of CZ silicon, with substantial degradation effects close to the ingot walls and bottom.

9.3   THE ROLE OF OXYGEN, CARBON AND POINT DEFECTS IN POLYCRYSTALLINE SILICON

Oxygen and carbon in various amounts [1] which depend on the type of process used to cast it, are the major contaminants of polycrystalline silicon.
Carbon comes primarily from the graphite crucible, from the susceptor or from any other hot graphite part of the casting furnace, via a direct carbon dissolution process or a gas transportation process. In this last case, CO is the intermediate gaseous species which is formed by reaction of graphite with residual oxygen in the casting chamber

$$1/2\ O_{2,g} + C_s \rightleftharpoons CO_g \tag{3.1}$$

and which then reacts with liquid silicon to bring it in equilibrium with the gaseous atmosphere

$$CO_g + Si_l \rightleftharpoons C_{Si,l} + O_{Si,l} \tag{3.2}$$

In turn, oxygen is stripped off the melt as SiO vapour until stationary conditions are achieved, which depend also on the rate of reaction of the liquid charge with any oxidic material in contact with molten silicon, as is the case of processes employing silica (or oxide ceramics) as the crucible materials.

In this case, silica dissolves irreversibly in liquid silicon, until a dynamic equilibrium is established among the silica dissolution and the SiO sublimation reactions.

---

[1] As it is well known, the oxygen and carbon concentrations in silicon are evaluated from the infrared absorptions associated to interstitial oxygen (1106 cm^{-1}), and to substitutional carbon (607 cm^{-1}), using suitable conversion factors.

It is expected therefore that the oxygen and carbon content in the melt, and then, in the solid material, should feature some kind of equilibrium property which should be defined by a quasi-equilibrium constant

$$K = N_O \times N_C \tag{3.3}$$

where N_O and N_C are the actual concentrations of oxygen and carbon in the solid silicon.

Apparently, K accounts not only for the equilibrium under equation (3.2) but also for the repartition coefficients of carbon and oxygen within the solid and liquid phase (Pizzini 1988b).

In fact, it has been shown (see figure 9.4) that in the case of relatively small (few kg) silicon ingots grown by the directional solidification technique, the oxygen and carbon contents in the as-grown material seem to be really an equilibrium property of the solid.

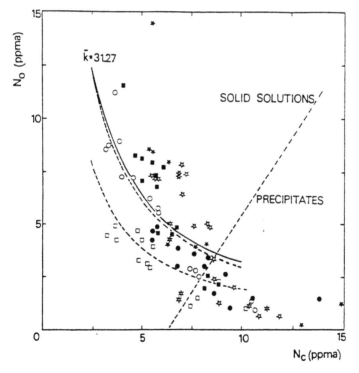

Figure 9.4 Experimental relationship between oxygen and carbon concentration in DS silicon [from Pizzini (1988b)].
(Reproduced by permission of Electrochemical Society)

This feature is a consequence of the fact that when the solid ingot is cooled at room temperature, at the end of the growth cycle, the solid solutions of oxygen and carbon remain in a metastable state which reflects the high temperature conditions (see below).

It is furthermore quite evident that the accomplishment of this pseudo-equilibrium property depends on some specific growth process parameters, like the free liquid to gas surface area or/and on the crucible surface to silicon charge volume ratio, if the rate of oxygen and carbon transfer from the gas phase to the liquid phase or the rate of silica dissolution are determining the rate of the overall process. On large ingots, therefore, it is no more expected that equation (3.3) shall be satisfied.

This concept is or it might be of practical use in many circumstances. As an example, silicon ribbons grown from a silicon melt through a graphite die can be enriched in oxygen in a relatively wide range of concentrations (up to some ppma) by equilibrating the freezing liquid on the top of the graphite die by CO at various partial pressures (Kalejs 1987).

As a consequence of carbon and oxygen dissolution processes during the growth cycle, in most practical cases a key feature of silicon ingots for solar cell uses is, however, that either oxygen or carbon, or both, are present as supersaturated solid solutions.

Therefore, any appropriate thermal treatment carried out in a later time will cause a shift towards the true equilibrium conditions.

It is however well known that oxygen and carbon segregation from their supersaturated solid solutions[2] is a slow process ruled by volume constraints (Hu 1991a , 1992b), being $V_{Si} = 2 V_{SiC}$ and $V_{Si} = 1/2 SiO_2$, where V_i is the molar volume of a species i. Here, "volume constraint" means that being the volume conservation law unsatisfied, any second phase formation requires the creation or the disappearence of new volume, or the second phase segregates in a very strained configuration.

Alternatively, stress release might occur via dislocation emission. As an example, the segregation of oxygen as an oxide phase is known to occur readily only at high temperatures (T > 900°C), when the yield strength of silicon is sufficiently low to release the stress associated to volume expansion via dislocation emission.

---

[2]segregation means here any homogeneous or heterogeneous process which results in the nucleation or in the precipitation of a second phase, both processes being easily followed by IR spectroscopy measurements.

The case of carbon is well represented in figure 9.5(Pizzzini 1993c) which reports the segregation behaviour of carbon in EFG silicon ribbons. This material is a good example of an oxygen poor, highly carbon supersaturated solid solution. Apparently, segregation of carbon occurs only at T > 1100°C and is associated to the contemporaneous segregation of oxygen.

At lower temperatures, oxygen and carbon segregation is shown to occur via interstitials Si_i or vacancies V_{Si} emission (and absorptions), possibly associated to the formation of intermediate species which behave as the precursors of the oxide or carbide phase, as it is schematically illustrated by the following equations (Hu 1991a), (Pizzini 1993c).

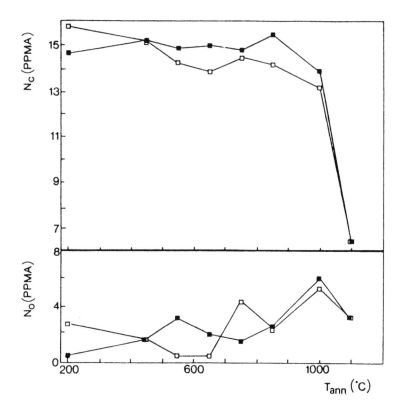

Figure 9.5 Evolution of the carbon and oxygen concentration with the annealing temperature in a polycrystalline silicon ribbon. Full squares, high dislocation density regions ; open squares, low dislocation density region.
(Reproduced by permission of Academie Verlag)

$$(1 + x) Si_{si} + 2O_i \qquad \rightarrow \qquad SiO_2 + xSi_i + \varepsilon' \qquad (3.4)$$

and

$$(1 - x) Si_{si} + C_{si} + xSi_i \quad \rightarrow \quad SiC + \varepsilon'' \qquad (3.5)$$

where ε is a strain energy term which can be made negligible if the cavity of the silicon matrix in which SiO_2 or SiC are contained fits exactly their volume.

In general, more complex processes account for the oxide and carbide segretation processes, as in either case both vacancies and interstitials do participate, at least in the initial phase of segregation.

The occuring of these processes is also the very reason why strong correlations among carbon and oxygen segregation processes are found, if both impurities are present in silicon (Taylor 1991), (Pizzini 1993c).

Apparently, they run at best when they could couple, as not only the growing carbide phase is a sink for the silicon selfinterstitials injected by the oxide phase during its growth, see equations (3.4) and (3.5), but the presence of any point defect supersaturation is suppressed by the occuring of the Frenkel equilibrium

$$Si_i + V_{Si} \quad \leftrightharpoons \quad Si_{Si} \qquad (3.6)$$

This form of stress compensation, typical of the high temperatures range, is associated, in an intermediate temperature range (between 450 and 850°C) to a more subtle kind of compensation which involves the direct formation of [C-O] complexes (Pizzini 1988b)

$$C_{Si} + O_i \quad \leftrightharpoons \quad [C\text{-}O] \qquad (3.7)$$

where carbon is in a substitutional position at low temperatures and in a interstitial position at high temperatures.

The coupling of oxygen and carbon segregation processes via point and extended defects could also explain why the carbon to oxygen ratio is shown to have some influence on the dislocation generation processes (Pizzini 1989d) and why there is some evidence that the oxygen and carbon segregation processes are favoured or inhibited by the presence of extended defects.

The effect of the oxygen and carbon content of a series of as-grown samples on their dislocation density, reported in figure 9.6, might be in fact understood as soon as we recognize that a sort of compensation among the stress field of carbon and oxygen occurs, when they segregate simultaneously, via [C-O] complexes formation.

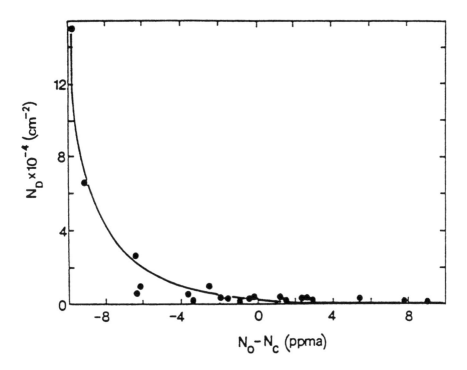

Figure 9.6 Experimental dependence of the dislocation density on the excess oxygen or carbon concentration [from Pizzini (1989d)].
(Reproduced by permission of Plenum Press)

Then, also the amount of tensile stress which must be released by dislocation emission in order to segregate the carbide or oxide phase depends on the excess of carbon over oxygen. The fact that the dislocation density increases very sharply with relatively small excesses of carbon depends on the solubility of carbon which is smaller that of oxygen.

It must be remarked, however, that the features of figure 9.6 might not necessarily represent a typical character of polycrystalline silicon.

In fact, carbon an oxygen might be very inhomogeneously distributed in the ingot and dislocations might be generated by different mechanisms. As an example, mechanical stresses associated to the different expansion coefficients of the crucible material and the solid silicon ingot or thermal stresses induced by an inhomogeneous thermal field, are as well causes of dislocation emission in the plastic range of silicon.

Also the effect of extended defects (grain boundaries and dislocations) on the point defect-assisted segregation of oxygen and carbon might be understood, if we assume that they behave as sources or sinks of point defects.

Actually, defects of either type (vacancies or interstitials) might be independently generated at internal surfaces, via a Schottky mechanism, in order to mantain or to reach the equilibrium concentration, with a free energy expenditure which equals

$$\Delta G_i = RT \ln \frac{a_*}{a_o} \tag{3.8}$$

where ΔG_i is the measure of the chemical work needed for the formation of defects in thermal equilibrium, and a_* and a_o are the actual and equilibrium activities of the i-type of defect.

Instead, dislocations are known to behave as sinks for selfinterstitials from their role in the kick-out mechanism of gold (Pichaud 1992).

We expect, therefore, an enhancement of the oxygen segregation and an hindering of carbon segregation in the presence of dislocations. In fact, in highly dislocated carbon-rich materials (see figure 9.5) the segregation of carbon is inhibited, the effect being larger as larger is the dislocation density (Pizzini 1993c).

Eventually, as local strain conditions, dislocations and point defects are known to induce the presence of band tails and gap states, we might expect that oxygen and carbon would have a non negligible effect on the electrical properties of polycrystalline silicon.

9.4 ELECTRICAL PROPERTIES OF MULTICRYSTALLINE SILICON WAFERS

9.4.1 Influence of defects and impurities

As we have shown in the previous Sections, polycrystalline (multicrystalline) silicon is a very complex system, and it is characterized by the simultaneous presence of point defects (vacancies and interstitials), various amounts of metallic and non metallic (oxygen, carbon and, possibly, also nitrogen) impurities and variable densities of extended defects (dislocations and grain boundaries).

We have also shown that the effective concentration of point defects, which are known to be electrically active, depends not only on intrinsic generation sources by a Schottky mechanism at external and internal surfaces (and by a Frenkel type of process in the bulk) but also on extrinsic sources or sinks like oxygen precipitates. As an example, oxygen precipitates have been shown in Section 1.2 to behave as self-interstitial sources and dislocations are known to behave as sinks for self interstitials.
Incidentally, a similar behaviour might be exibited by metallic impurity precipitates, if segregation occurs with a non negligible volume change.

Furthermore, although extended defects and isolated oxygen and carbon atoms in solid solution are known to be electrically inactive, interaction among them as well as with metallic impurities might induce electrical activity.

Eventually, also dopant impurities are known to participate to a number of equilibria which involve metallic (i.e. in B-Fe complexes) and non metallic impurities (i.e. B-H complexes).

The very complex pattern of the most important equilibria occuring in multicrystalline silicon among point defects and impurities, as well as of the interactions occurring among point defects, extended defects and metallic and non metallic impurities is reported in figure 9.7, just to show how complicated could be a thorough description of the effect of impurities and defects on the electrical properties of polycrystalline silicon.

Eventually, as these homogeneous and heterogeneous reactions are driven by chemical and mechanical potential gradients, and at least oxygen and carbon are always present in supersaturation conditions, the rate of these reactions increases at increasing temperatures and therefore they are strongly influenced by high temperature process steps during solar cell manufacturing, thus affecting the final throughput in terms of efficiency.

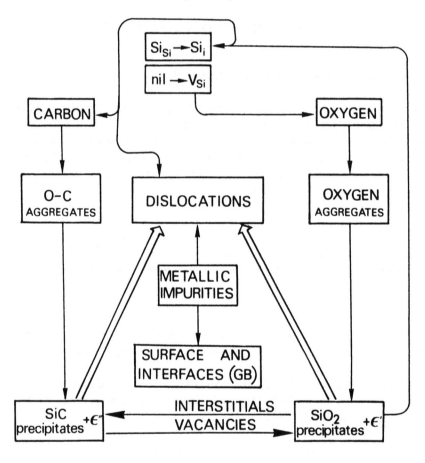

Figure **9.7** Schematic illustration of the correlations found among processes involving extended defects, point defects, oxygen, carbon and metallic impurities in silicon ; ε' and ε'' indicate the presence of residual stresses.

The parameter which is used systematically to qualify a multicrystalline silicon wafer in the as grown conditions or after proper thermal treatments is the diffusion length $L_{n,p}$,where the subscript indicates the type of minority carriers.

The measured L values, independently of the method used, are however only "effective" values, unless determined by a very local technique in a single crystalline region of the wafer which is far enough from any extended defect (GBs, dislocations, precipitates). Only in this last case we can consider that a diffusion length is measured.

Deviations from the "effective" value can be used to evaluate the homogeneity of the sample, which is better for low deviations.

Figure 9.8 shows as an example the very large variations of L in a given polycrystalline silicon samples.

Apparently, this is a typical feature of the very complex microstructure of this material, but very often these kinds of features can be found in other polycrystalline materials containing variable amounts of active GBs and dislocations.

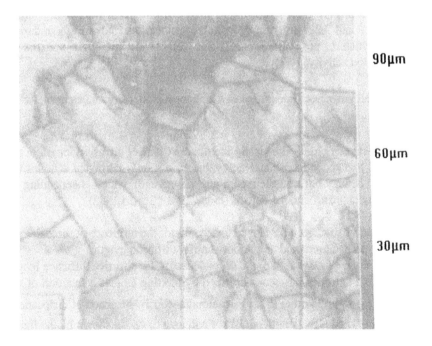

90μm

60μm

30μm

Figure   9.8 Local variations of the minority carrier diffusion length L_n in a Polix wafer.

High recombination rates are often a typical attribute of GBs after heat treatments at temperatures larger than 650°C, albeit in several circumstances GBs are found electrically active just in the as-grown conditions.

Segregation of impurities, like oxygen, carbon (and heavy metals) is generally observed at GBs by SIMS and by other spectroscopic techniques (Kazmerski 1982a , Maurice 1991) after annealing at sufficiently high temperatures. The increase of electrical activity might be attributed to segregated impurities and it is evaluated, like for external surfaces, by an effective interfacial recombination velocity, where the interface is located at the edge of the space charge region of these 2D defects (Card 1977 , Seager 1981 , Inoue 1981).

An enhanced activity is also observed when GBs are decorated with dislocations. In this case, the direct cause of enhanced activity migth be the local disorder increase, or a local excess of impurities, which are very efficiently captured by dislocations.

As in large grained multicrystalline materials the influence of GBs is very limited provided the grain size is higher than few mm, as it was already shown in figure 9.1 and 9.2, intragrain defects like dislocations, subgrain boundaries, impurity clusters, precipitates and point defects are the main sources of recombination activity.

Indeed, in all polycrystalline materials the dislocation density ranges from 10^3 to 10^7 cm^{-3}.

These defects can reduce drastically the minority carrier diffusion length in the grains and this reduction should be evaluated not only taking in account the dislocation density but also the recombination strength, which is depending on the impurity cloud around the defects (Inoue 1981 , Sopori 1988).

Several models have been developed to account for the influence of these defects on the cell photocurrent and on minority carrier diffusion length.
As an example, (Pizzini 1988b) the dependence of the effective diffusion length on the defect densities has been described by plotting L_{eff} as a function of the product $N_{dis} \cdot L_{GB}$, in order to consider simultaneously the action of dislocations (density N_{dis}) and GBs (length of GB per unit area L_{GB}), as shown by the figure 9.9.

The fits are acceptable, but the large dispersion of the experimental results is probably due to the different recombination strength of individual dislocations and GBs.

The model of (El Ghitani 1989a, b) computed instead the dependence of the effective diffusion length L_{eff} on the density of dislocation for different values of the true diffusion length L_n in the undislocated regions and for different values of the recombination strength S_d. S_d is evaluated by means of an effective interfacial recombination velocity at the edge of the cylindrical space charge region associated to the dislocation core.

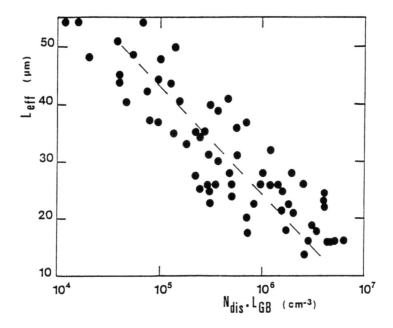

Figure 9.9 Dependence of L_{eff} as a function of the product length of GB per unit area and dislocation density [from Pizzini (1988b)].

The solid curves of the figure 9.10 describe the computed dependence of the effective diffusion length (L_{eff}) on N_{dis} for different S_d, when the value of L_n is about 90 μm.

Figure 9.11 shows the same dependence, but for one value of S_d ($S_d = 10^3 cms^{-1}$) and for different L_n.

It appears that the values of L_{eff} are dependent on N_{dis} and S_d, and decrease strongly when N_{dis} and S_d are higher than 10^3 cm^{-2} and 10^4 cms^{-1}, respectively. This dependence is more marked when L_n increases.

Experimental results fit very well the computed curves as it is shown in figure 9.10. Consequently L_{eff} could be expressed as :

$$L_{eff} = f(N_{dis}, S_d, L_n) \qquad (4.1)$$

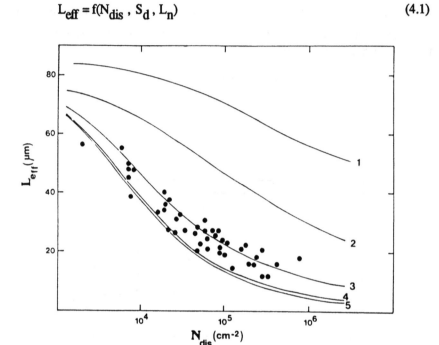

Figure 9.10 Calculated and experimental variations of the effective diffusion length (L_{eff}) as a function of the dislocation density (N_{dis}). Solid curves have been calculated for different values of dislocation recombination strength (S_d) :

(1) 5×10^2 cm s^{-1}; (2) 10^3 cm s^{-1}; (3) 10^4 cm s^{-1}; (4) 10^5 cm s^{-1} (5) 10^6 cm s^{-1}, full circles, experimental results [from El Ghitani (1989 a, b)]. (Reproduced by permission of Mat. Sci. Engineering)

These results show first that interaction among 2D and 3D defects plays a paramount role and that the improvement of the material (i.e. an increase of its effective diffusion length) could be obtained either by inducing a decrease of the recombination activity of dislocations (S_d) or an increase of the diffusion length.

Alternatively, the decrease of the density of the extended defects by a proper improvement of the growth technique would lead to the same result.

These conclusions are of crucial importance, as there is the indication that any improvement passes through suitable manipulations of the material, adressed to reduce the bulk impurity contaminations, and to the passivation of extended defects, when the density of these defects can not be reduced by the growth.

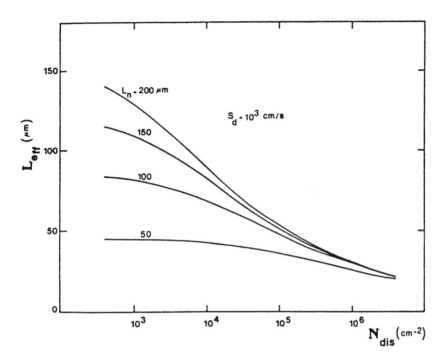

Figure 9.11 Plot of L_{eff} as a function of N_{dis} for different L_n values in the range between 50 and 200 μm and for $S_d = 10^3$ cm s^{-1} [from El Ghitani (1989c)]. (Reproduced by permission of Mat. Sci. Engineering)

9.4.2 Improvement techniques

Obviously the improvement of the wafers can result from the reduction of the defect density, and it can be obtained by a more careful control of the cooling rate and of the heat transfer during crystal growth.

An example of how effective could be the control of growth parameters is given by the cast ingots prepared by Solarex, the wafers of which present a regular grain size and few etch pits after etching. As a consequence, the effective diffusion length is greater than 100 μm (Doolittle 1990).

Similar results could be achieved with other multicrystalline materials (Silso, Polix, Eurosolare) after submitting them to hydrogen passivation or/and external gettering.

Hydrogen passivation

It is well known that frontside hydrogen ion implantations at low energy (< 2 keV) is the most efficient method to obtain in few minutes drastic increases of L_{eff}, photocurrent (J_{sc}) and photovoltage (V_{oc}) in N^+P cells (Seager 1982b). Hydrogen ion implantation is carried out by means of a Kaufman source, the voltage and the ion current of which are about 1 kV and 1 mA/cm^2 respectively. The temperature during implantation of the implanted samples could be kept in the range between 300°C and 400°C.

The observed improvements result from bulk effects, as it was reported by several authors that GBs and dislocations are passivated (dislocations are passivated up to a depth of hundred microns below the implanted surface) and it was shown that the larger increases of L_{eff} are obtained with materials having initial values in the range between 20 and 50 μm (Dube(1984 , Panitz 1984 , Martinuzzi 1987a , Ammor 1986 , Muller 1989). In addition, also deep levels associated to metallic impurities could be neutralized (Pearton 1990).

However, frontside implantation has a poor influence on the material when L_{eff} is higher than 100 μm as shown by the figure 9.12. In addition, the implanted surface of solar cells could be damaged and the collecting grid and the emitter surface altered (Panitz 1984 , Martinuzzi 1987a , Muller 1989).

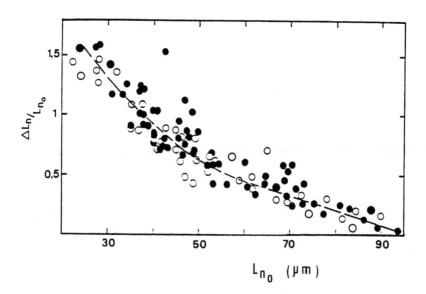

Figure 9.12. Relative variation of effective diffusion length versus initial values for hydrogen implanted samples. (o) N+P junctions ; (●) Al-Si diodes.

Consequently, high quality material cannot be improved unless long treatments are applied. During such long treatments, implantation damage increases the series resistance which in turn degrades the fill factor of the cells.

It was also suggested that the surface defects can getter fast diffuser (Cu ; Ni ; Au) because the samples are heated at about 400 to 500°C during the implantations, but their unavoidable presence changes the optical surface properties and minimizes the expected increase of efficiency (Panitz 1984).

In order to avoid these secondary effect, backside hydrogenation (Martinuzzi 1987b) by means of ion implantation and also by annealing in H_2 gas flow (provided the gas is adsorbed by a p-type surface) can be used. This is possible because 200 μm thick wafers are actually produced thanks to the use of wire saws .

Hydrogenations could be carried out also by other several different methods using, as an example, an hydrogen   plasma and or simply a flow of pure hydrogen.

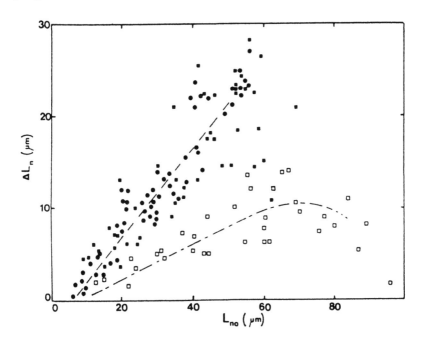

Figure   9.13 Increases of diffusion lengths versus initial value (backside anneals in H_2). (□) 400 μm thick cells , (●, ■) 200 μm thick cells [from Martinuzzi (1987b)]. (Reproduced by permission of IEEE Pu. Services)

In this last case the samples were annealed at 400°C for 30 mn in a silica tube flown with 99,99 % pure hydrogen gas.

The results reported in figure 9.13 show the variation of the increase ΔL_n of diffusion lengths as a function of the initial value L_{no}, for 400 and 200 μm thick cells.

The beneficial effect of hydrogen is clearly observed also with solar cells, as I_{sc} and V_{oc} are increased while the dark reverse current I_o is decreased. The existence of a correlation between the variation of these three parameters after the hydrogen treatment is illustrated by the figure 9.14. Indeed, if the classical relation

$$V_{oc} = \frac{AkT}{q} \log \left(\frac{I_{sc}}{I_o} - 1 \right)$$

is verified, a linear relation must be obtained if we plot

$$\frac{V_{oc(H)}}{A_{(H)}} - \frac{V_{oc}}{A} \quad \text{versus} \quad \log \frac{I_o}{I_{o(H)}} \cdot \frac{I_{sc(H)}}{I_{sc}}$$

where (H) indicates the value after hydrogenation, A is the ideality factor of the junction, and the slope of the straight line must be kT/q as it can be deduced from figure 9.14.

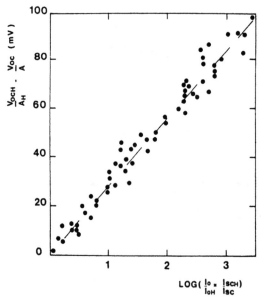

Figure 9.14 The slope of the broken line which fit the experimental results is 0.026 V, very close to kT/q as expected [from Martinuzzi (1987b)]. (Reproduced by permission of IEEE Pub. Services)

Similar results are obtained by the different hydrogenation methods indicating that they differ by the mode of introduction of hydrogen only. Better results are obtained by implantation, but they are essentially due to the high dose of hydrogen introduced in the material. During the annealing in gas flow it seems that adsorption of gas occurs at the emergences of GBs and dislocations, where chemical reactions between segregated impurities and hydrogen favour the dissociation of the hydrogen molecules. Actually, adsorption of hydrogen gas on single crystals containing dislocations has been observed by (Kveder 1985). It has been also demonstrated that annealing in hydrogen can influence the electrical properties of microcrystalline silicon layers (Lam 1982) and polycrystalline solar cells (Agarwal 1985).

The in-diffusion of hydrogen, and consequently the passivation of the bulk recombination centers, depends on the presence of dislocations, of impurities, mainly oxygen, and also on the surface and bulk concentration of hydrogen itself. It should be pointed out that under the same conditions of implantation, (Sopori 1991b) have observed that samples having lower concentration of oxygen exhibit deeper penetration at defects. These authors have also observed using XTEM studies that hydrogen segregates at dislocations nodes, this segregation being more pronounced at "clean" dislocations, and no segregation is observed if the hydrogen concentration is below 10^{16} cm^{-3}.

A comparison of hydrogen diffusion in different silicon wafers shows that the depth of diffusion within large grains was found to be the same and even higher than along GBs and dislocations. This effect could be associated to a strong interaction of hydrogen with point defects which could also explain the enhancement of hydrogen diffusivity. Oxygen instead appears to retard the diffusion of hydrogen , probably because hydrogen interact with oxygen as well as precipitates, as was already shown by (Kazmerski 1985b).

Simultaneous desorption of hydrogen and in some cases, a worsening of the L values is observed at temperatures higher than 400°C, suggesting that hydrogen has mostly reacted with segregated impurities during passivation, forming hydrides which are less stable than silicon hydrides.

In spite of intensive investigations, the role, the state and the diffusion mechanism of hydrogen in silicon is not yet well understood. Nevertheless, the preceding results are sufficiently indicative of the fact that the interaction of impurities with the extended crystallographic defects governs the electrical properties of multicrystalline silicon wafers.

Gettering of metallic impurities

In order to improve the electrical properties of the material it might be necessary to reduce the impurity concentration. Gettering is a mean of reducing or eliminating metallic impurities by removing them or localizing them in inactive regions of the wafers. As in solar cells the photovoltaic effect is developed in the bulk of the device, external gettering must be used only and the region in which impurities are accumulated must be chemically etched at the end of the treatment.

During the gettering, all impurities which could be removed are interstitial atoms which diffuse rapidly to the gettering sites.These sites are extended crystallographic defects, interfaces, vacancies, or enhanced solubility regions. Figure 9.15 summarizes the gettering mechanism, which could be divided in three steps :

1°/ Extraction of impurities from substitutional sites, from complexes or from precipitates. The extracted impurities go to interstitial positions.
2°/ Fast diffusion through the crystal.
3°/ Capture by gettering sites.

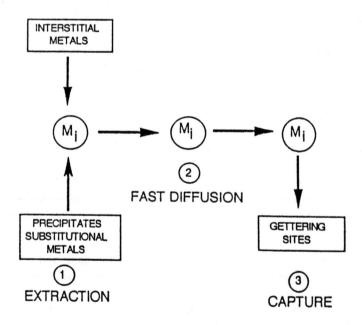

Figure 9.15 The three basic steps of the gettering mechanism of metallic impurities.

The extraction of impurities from substitutional sites, from precipitates and complexes could be facilitated by the injection in the bulk of self interstitials in excess, which participate to the well known kick-out mechanism of substitutional metal atoms and to the shrinkage of precipitates (Ourmazd 1985 , Kang 1989).

Phosphorus diffusion within the surface, from a $POCl_3$ source, is a well known technique of gettering which presents the advantage to be one of the preparation steps of cells made with p-type silicon wafers.

During such a diffusion a high phosphorus concentration is achieved near the surface, which exceeds the solubility limit. Precipitates of SiP are formed and at the highly stressed interfaces between the orthorombic SiP and the silicon matrix, dislocations are created which behave as gettering sites.

Due to the molar volume expansion, the formation of SiP precipitates induces also an excess of self interstitials which are injected in the bulk.

Additional gettering sites could also be created by the dislocation network which is produced by the stress introduced by the atomic radius difference between phosphorus and silicon atoms. Such a dislocation network is created when the surface concentration of phosphorus atoms reaches 1 to 2.10^{16} cm^{-2} (Tamura 1987 , Stojadinovic 1979), and this is the case when the phosphorus diffusion is carried out at 900°C for several hours.

Finally the highly N^+ doped region is a region of enhanced solubility because it contains a large density of vacancies which can trap interstitial metal atoms.
Indeed after phosphorus diffusion the effective electron diffusion lengths are always increased.

A clear example of the capability of the phosphorus gettering is given by the following results which concern a series of 200 μm thick samples, cut adjacently from the same Polix ingot. In these wafers the density and distribution of defects are very similar due to the columnar structure of the material. Arrays of mesa diodes, made at the same place in the different samples allow to measure L_{eff}, and to evaluate the gettering effect. Light Beam Induced Current (LBIC) scan maps at $\lambda = 940$ nm give informations on the evolution of the diffusion length within the grains, in the highly dislocated regions and at GBs.

Figure 9.16 shows the evolution of effective diffusion lengths in a series of matched samples gettered at 900°C for 60, 120 and 240 mn (Périchaud 1992a).
It appears that the effective electron diffusion lengths measured at the same place in the different samples increase neatly with diffusion time and that values of

L_{eff} are frequently comparable to the sample thickness (i.e. 200 μm). The improvement normally is as better as longer is the time, however some regions of the sample in which the initial values of L_{eff} were below 60 μm are less improved even after the longer treatment.

SIMS measurements indicate that in the N^+ region of the investigated samples heterogeneous of Fe, Ni and Cu is found and this segregation is strongly enhanced in samples which have been gettered for the longest time as shown by figure 9.17.

These impurities are extracted certainly from the grains, but also probably from the extended defects.

To obtain so high values of L_{eff} in the investigated material it is necessary, following the model of (El Ghitani 1989a), that the recombination strength of dislocations be reduced and that the homogeneous regions of the grains be improved. In fact the effective diffusion length, L_{eff} , is depending on the dislocation density N_{dis}, on their recombination strength S_d and on the value of the diffusion length in the homogeneous region of the grains L_n. In turn L_n is essentially limited by dissolved metallic atoms and S_d by impurity atoms segregated at the dislocations.

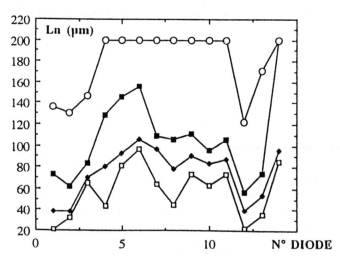

Figure 9.16 Evolutions of effective diffusion lengths in matched samples before (□) and after phosphorus diffusion at 900°C for 60 mn (◆) ; 120 mn (■) and 240 mn (○) The values greater than 200 μm have been arbitrary limited at 200 μm i.e. the thickness of the samples [from Périchaud (1992a)].
(Reproduced by permission of Ed. de Physique)

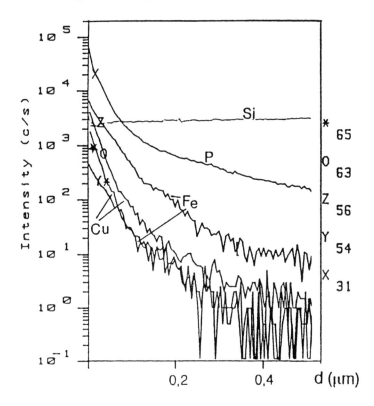

Figure 9.17 SIMS profile of iron (54 ; 56), copper (63 ; 65), and nickel (58 ; 60) in the N^+ region of a phosphorus gettered sample [from Périchaud (1992a)]. (Reproduced by permission of Ed. de Physique)

In addition grain boundary recombination centers must also be neutralized. These defects play a secondary role in a large grained material containing active dislocations but their influence cannot be neglected when the grains have been neatly improved.

The preceding assumptions are verified by means of LBIC measurements. A typical example is given for one of the diodes of the figure 9.16. Figure 9.18a is a microphotograph after chemical etching of this diode in the virgin sample which shows the presence of grain boundaries, subgrain boundaries and twins, while dislocation etch pits are revealed in the grain (figure 9.18b).

Figure 9.19 shows the LBIC scan maps corresponding to the virgin sample and to the gettered samples for 240 mn. It is clear that after the treatments the current in the grains are increased, and that the current attenuations at GBs are drastically

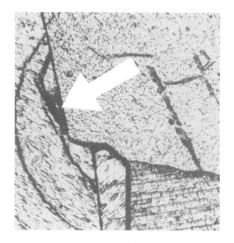 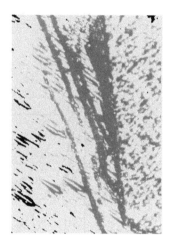

Figure 9.18a. Microphotograph of one diode of figure 16 after chemical etching. A highly dislocated region is indicated by an arrow.(⊢——⊣ 200 μm)

Figure 9.18b Magnification of the highly dislocated region of figure 9.18a. (⊢——⊣ 100 μm)

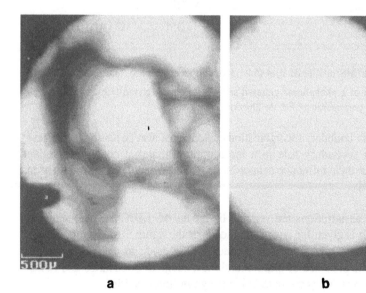

a   b

Figure 9.19. LBIC mapping at λ = 940 nm of the same diode before (a) and after phosphorus diffusion at 900°C for 240 mn (b) [from Périchaud (1992b)]. (Reproduced by permission of Ed. de Physique)

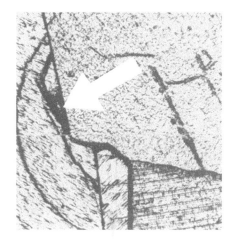

Figure 9.18a. Microphotograph of one diode of figure 16 after chemical etching. A highly dislocated region is indicated by an arrow.(⊥———⊥ 200 μm)

Figure 9.18b Magnification of the highly dislocated region of figure 9.18a. (⊥———⊥ 100 μm)

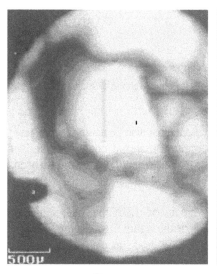

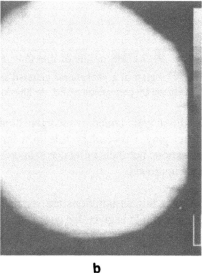

a   b

Figure 9.19. LBIC mapping at λ = 940 nm of the same diode before (a) and after phosphorus diffusion at 900°C for 240 mn (b) [from Périchaud (1992b)].
(Reproduced by permission of Ed. de Physique)

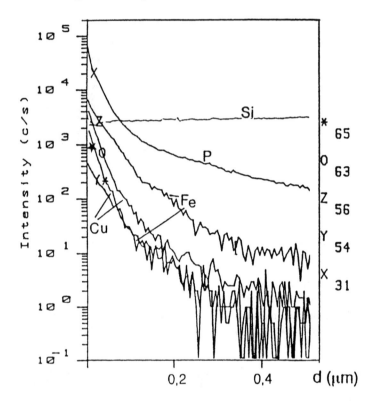

Figure 9.17 SIMS profile of iron (54 ; 56), copper (63 ; 65), and nickel (58 ; 60) in the N^+ region of a phosphorus gettered sample [from Périchaud (1992a)]. (Reproduced by permission of Ed. de Physique)

In addition grain boundary recombination centers must also be neutralized. These defects play a secondary role in a large grained material containing active dislocations but their influence cannot be neglected when the grains have been neatly improved.

The preceding assumptions are verified by means of LBIC measurements. A typical example is given for one of the diodes of the figure 9.16. Figure 9.18a is a microphotograph after chemical etching of this diode in the virgin sample which shows the presence of grain boundaries, subgrain boundaries and twins, while dislocation etch pits are revealed in the grain (figure 9.18b).

Figure 9.19 shows the LBIC scan maps corresponding to the virgin sample and to the gettered samples for 240 mn. It is clear that after the treatments the current in the grains are increased, and that the current attenuations at GBs are drastically reduced. In addition, the passivation of dislocations is neatly shown, when we compare the LBIC scan maps of the dislocated region of the diode corresponding to the figure 9.18b (Périchaud 1992b).

The gettering process we have described appear to be very efficient and could be applied to different multicrystalline silicon materials, like EFG ribbons (Micheels 1991).

As this treatment is currently used in the solar cell fabrication it does not increase significantly the cell cost, and the material becames as good as CZ crystals after 4 hours of phosphorus diffusion at 900°C.

Using the gettering process described in this paper Le Quang Nam (1992) have obtained a conversion efficiency of 15.6 % for 2 x 2 cm^2 cells with multicrystalline polix wafers. More recently (Rohatgi 1992) reported that they have achieved 17.7 % for cells made with OTC multicrystalline wafers, after they have been submitted to phosphorus and aluminium gettering treatment, in agreement with the results already reported by Green (1990).

Notice that the phosphorus gettering works fairly well with Polix and Eurosolare wafers but not so much with Silso material.
It appears also that the gettering efficiency is reduced and even cancelled when a too large concentration of dissolved oxygen is contained in the wafers (i.e. 10^{18} cm^{-3}).

9.5   CONCLUSION

In the present paper, we have shown that the properties of large grain polycrystalline materials are strongly influenced by the presence of extended defects, which, in turn, are a consequence of the rather high growth rate of the casting processes.

It has been however demonstrated that most of the electrical properties of these materials are depending more on the impurities segregated at extended defects than on the presence of dangling bonds. Among the different impurities, metallic impurities play a major role, but oxygen and carbon are also involved directly or indirectly
In fact, depending on the relative concentration of oxygen and carbon, both the density and the recombination activity of intragrain defects could be very different.

Moreover, as the oxygen and carbon content is strongly depending on the features of the growth process adopted, each type of material considered, i.e. Silso, Polix, Eurosolare, exhibits different electrical properties.

Considering that the generality of polycrystalline silicon produced now presents very large grains, dislocations more than grain boundaries are therefore the main sources of recombinations.

Among the different techniques which could be foreseen for upgrading the properties of polycrystalline silicon, hydrogenation and external gettering are demonstrated to work very efficiently. Hydrogenation is a true passivation technique which limits the recombination activity of recombination centres, while external gettering is a material cleaning procedure, capable to getter metallic impurities at the external surface after having extracted them from grains and from extended defects.

By the use of an hydrogen implantation step following phosphorus gettering at 900°C, as an example, diffusion length values in excess of 200 μm can be easily obtained, which compare with those of CZ wafers.

As the phosphorus diffusion is one of the processing steps of the solar cell fabrication and is currently used in the photovoltaic industry, this treatment appears to be particularly attractive. It could be applied to several materials (Polix, Eurosolare, EFG ribbon etc.) even if it works differently, depending on the nature and density of defects and impurities contained in the bulk. For example, the effect on Silso wafers are generally poor, probably due to the high purity of this material which is practically not contaminated by the crucible during the directional growth.

Unfortunatly, gettering as well as hydrogen passivation techniques are not so effective when the minority carrier diffusion lengths are too small i.e. below few tens of μm, due to a high density of defects and impurities. This will be particularly true for wafers made with metallurgical grade silicon.

In spite of the presence of extended defects and of a great number of recombination centers due to the interaction between defects and impurities, large grain polycrystalline silicon is neverthess the best material to prepare low cost solar cells, and thanks to the use of simple upgrading techniques the way to achieve high conversion efficiencies is opened.

9.6   REFERENCES

Agarwal A, Bana D and Singal C M 1985 *Int. J. Electron.* **5 8** pp.769-783.
Ammor L and Martinuzzi S 1986, *Solid State Electronics* **2 9**, 1-6.
Authier B A 1975 *German Patent* (DOS) nr 2508803.
Brenneman R K and Tomlison T A 1988 *Proc. 20th IEEE PVSC* pp.1395-99.
Card H C and Yang E S 1977 *IEEE Trans. Elec. Dev.* ED2 4 pp.397-402.
Chattak C P and Schmid F 1987 "Silicon ingots by HEM" : *in Silicon Processing for Photovoltaics* vol II, C P Chattak, K V Ravi (Elsevier Sci. Publ.) pp.153-66.
Dube C, Hanoka J I and Sandstrom D B 1984 *Appl. Phys. Lett* **4 4**, 77-81.
Doolittle W A, Rohatgi A and Brenneman R 1990 *Proc. of 21th IEEE Photovoltaic Spec. Conf.* Kissimee, pp.681-86.
Dorrity I A, Garrard B J and Hukin D A 1991 *Proceed. 10th EC Photovoltaic Solar Energy Conf.* Lisbonne (Kluwer-Academic Press) pp.317-19.
Dugas J and Oualid J 1987 *Rev. Phys. Appl.* **2 2** 677-85.
El Ghitani H and Martinuzzi S 1989a *J. of Appl. Physics* **6 6** pp.1717-1722.
El Ghitani H and Martinuzzi S 1989b *J. of Appl. Physics* **6 6** pp.1723-1726.
El Ghitani H. and Martinuzzi S. 1989c *Materials Science and Engineering*, **B 4**, pp.153-156.
Fally J and Guenel C 1980a *Proc. 3rd EC Photovoltaic Solar Energy Conf.*, Reidel Publ. pp.598-601.
Fally J, Fabre E and Chabot B 1987b *Rev. Phys.* **2 2** pp.529-35.
Goesele U 1986 Oxygen, carbon, hydrogen and nitrogen in silicon *ed. J C Mikkelsen et al* (Pittsburgh : MRS) pp.419-431.
Green M A 1990 *Proc. of 5th Int. Photovoltaic Science and Engineering Conf.* Kyoto pp.603-605.
Helmreich D 1985a Materials and new processing technologies for photovoltaics ed V K Kapur, J P Dismukes, S Pizzini (Pennington : *The Electrochem. Soc.*) pp.317-325.
Helmreich D 1987b *Silicon Processing for Photovoltaics* vol II, ed C P Chattak, K R Ravi (Amsterdam : Elsevier) ch2 pp. 97-151.
Herzer H 1991 *Proc. 10th EC Photovoltaic Solar Energy Conference*, Lisbon, April 1991 (Dordrecht : Kluwer-Academic) pp 501-506.
Hu S M 1991a Defects in Silicon II ed. W M Bullis et al (Pennington : *The Electrochm. Soc.*) pp.211-236.
Hu S M 1992b *J. Electrochem. Soc.* **1 3 2** pp.2066-2075.
Inoue N, Wilsen C W and Jones K A 1981 *Solar Cells* **3**, 35-42.
Kalejs J P 1987 *Silicon Processing for photovoltaics* vol II ed C P Chattak, K R Ravi (New-York : Elsevier) ch4 pp.185-254.
Kaneko K, Misalia T and Tabata K 1990 *Proc. of Intern. Photovoltaic Solar Energy Conf.* 5 Kyoto pp.201-204.
Kang J S and Schroder D K 1989 *J. Appl. Phys.* **6 5** , 2974-85.

Kazmerski L L and Russel P E 1982a *J. de Phys.* 43, 1 0, 171-77.

Kazmerski L L 1985b *Proc. of 18th IEEE Photovoltaic Spec. Conf.* Las Vegas pp.993-98.

Koch 1992 *Proc. of 11th European Photovoltaic Solar Energy Conf.* Montreux, p.518.

Kveder V V and Ossipyan Y A 1985 in *Dislocations in Solids Univ. of Tokyo Press* pp.395-98.

Lam A 1982 *Appl. Phys. Lett.* 4 0 pp.54-56.

Le Quang Nam, Rodot M, Ghannam M, Coppye J, De Schepper P, Nijs J, Sarti D, Périchaud I and Martinuzzi S 1992 *Int. J. Solar Energy* 11 , 273-279.

Lindmayer J 1976 *Proc. 12th IEEE PVSC* pp.82-87.

Lu C Y and Tsai N J 1986 *J. Electrochem. Soc.* 133 pp.847-853.

Margadonna D, Muleo A, Ferrari Degrada M F 1989 *Proceed. 9th EC Photovoltaic Solar Energy Conf.* Freiburg (Amsterdam : Kluwer Academic Press) pp.750-753.

Martinuzzi S 1987a *Rev. Phys. Appl.* 2 2 , 637-43.

Martinuzzi S, El Ghitani H, Ammor L, Pasquinelli M and Poitevin H 1987b *Proc. of 19th IEEE Photovoltaic Spec. Conf.* pp.1069-74.

Maurice J L 1991 in *Polycrystalline Semiconductors II* ed. by J H Werner and H P Strunk, Springer Verlag pp.166-177.

Micheels R H and Serres R 1991 *Proc. of 22nd Photovoltaic Spec. Conf.* Las Vegas pp.869-872.

Muller J C, Hussian E and Siffert P 1989, *Proc. of 9th European Community Photovoltaic Solar Energy Conf.* ed. by W Palz G T Wrixon and P Helm Kluwer Acad. Pub. pp.407-410.

Ourmazd A and Schröter W 1985 in *Impurity Diffusion and Gettering in Silicon* ed. by R B Fair C N Pearce J Washburn MRS Symposium Proc. 36 pp.25-30.

Panitz J K G, Sharp D J and Seager C H 1984 *Thin Solid Films* 111, 277-83.

Pearton S J 1990 in *"Hydrogen in semiconductors"* ed. by J I Pankove and N M Johnson Academic Press, New-York 24 pp.65-89.

Périchaud I and Martinuzzi S 1992a *J. de Phys. III* 2 , 313-324.

Périchaud I, Stemmer M and Martinuzzi S 1992b *J. de Phys. IV* C 6, 199-204.

Pichaud B and Mariani G 1992 *J. Phys. III* 2 pp.61-68.

Pizzini S Gasparini M and Rustioni M 1981a *Brevet Franc.* n° 23007A/21.

Pizzini S, Sandrinelli A, Beghi M, Narducci D and Allegretti F 1988b *J. Electrochem. Soc.* 135 pp.155-165.

Pizzini S, Binetti S., Acciarri M., Acerboni S.1993c, *Phys. Stat. Sol. (a)*, 138, pp.451-464..

Pizzini S 1989d Point and extended defects in semiconductors, ed. G Benedek et al (New-York : *Plenum*) pp.105-121.

Rohatgi A, Sana P and Salami J 1992 *Proc. of 11th European Photovoltaic Solar Energy Conf.* to be published.

Saito T, Shimura A and Ichikawa S 1981 *Proc. 15th IEEE PVSC* Orlando (New-York : IEEE) pp.576-580.

Schmid F and Viechnicki D 1972a *US Patent* n° 3653342.

Schmid F 1975b *US Patent* n° 3898051.

Seager C H 1981a *J. Appl. Phys* **52** pp.3960-66.

Seager C H, Sharp D J, Panitz J K G and Hanoka J I 1982b *J. de Phys.* **43** C1, 103-109.

Sopori B L 1988a *Proc. of 20th IEEE Photovoltaic Spec. Conf.* Las Vegas pp.591-95.

Sopori B L, Jones K M, Deng X, Matson R, Al Jassim M M, Tsuo S, Doolittle A and Rohatgi A 1991b *Proc. of 22nd Photovoltaic Spec. Conf.* Las Vegas, pp.833-841.

Stojadinovic S D 1979 *Phys. Stat. Sol.* (a) **54** , K5-10.

Tamura M 1977 *Phil. Mag.* 35, **3** , 663-691.

Taylor W J Tan T Y and Goesele U 1991 *Defects in Silicon II* ed. W M Bullis et al (Pennington : *The Electrochem. Soc.*) pp.255-262.

Wald F V 1991 *Solar Energy Mat.* **23** 175-81.

10

Properties, Analysis and Modelling of Polysilicon TFTs

P. Migliorato and M. Quinn

10.1 INTRODUCTION

Amorphous silicon thin-film transistors (a-Si TFTs) are currently employed in commercial Active Matrix Liquid Crystal Displays (AMLCDs). However, polysilicon offers the advantage of a higher field effect mobility that allows one to fabricate integrated logic circuits with adequate speed as well as individual switching elements. Moreover, polysilicon devices are less sensitive to light and more stable than a-Si because of their higher temperature process. Originally, polysilicon TFTs with good performances were only obtained by including high-temperature treatments (~1000°C) in their fabrication process. This required quartz substrates, rendering the process unsuitable for large-area low-cost products. In 1986, GEC reported a low-temperature process to obtain high performance polysilicon TFTs (Bryer 1986, Meakin 1987a). This led to the development of a new commercial polysilicon reactor (Meakin 1989). Other low-temperature (~600°C) techniques have since been investigated, and an increasing number of applications have been demonstrated. These include AMLCDs with fully integrated drive electronics, readout logic for image sensors, and logic circuits for advanced printer heads (both non-impact and thermal). These devices are expected to provide more versatile and lower cost human-machine interfaces. However, they require electronic circuits of a complexity comparable to some conventional integrated circuits, but on much larger areas and on cheap substrates. For this reason, this new technology is often referred to as Large Area Microelectronics. The role of polysilicon TFTs in Large Area Microelectronics is similar to that of bulk single-crystal silicon MOSFETs in traditional microelectronics.

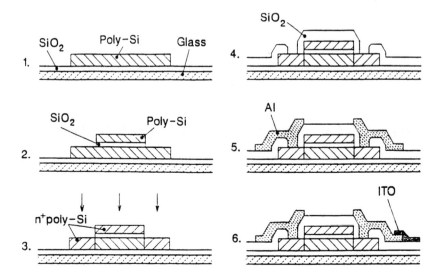

Figure 10.1 Self-aligned fabrication process for a polysilicon TFT used in an Active Matrix Display. [Courtesy of the Society for Information Display.]

The fabrication process most widely used for polysilicon TFTs closely resembles the one employed in SOI technology. For an active matrix display pixel transistor, a process with 5 levels of mask is used:

(1) Polysilicon film growth by low-pressure chemical vapour deposition (LPCVD) and patterning into islands (Mask 1). Films 1000 Å thick are normally used, although ultra-thin (300 Å) layers have also been employed (Noguchi 1986).

(2) Deposition of gate oxide (~1000 Å) and polysilicon gate electrode (3000 Å), followed by patterning of polysilicon gate and gate tracks (Mask 2).

(3) Source and drain polysilicon doping by ion implantation[*]. followed by thermal anneal. (This results in self-aligned structures.)

(4) Deposition of SiO_2 passivation layer, (~5000 Å) and contact vias etching (Mask 3).

(5) Aluminium contacts deposition and patterning (Mask 4).

(6) ITO deposition and patterning of pixel electrodes (Mask 5).

---

[*] Alternatives to ion implantation have been investigated, namely deposited contacts (Kobayashi 1989) and laser doping (Coxon 1986), the latter yielding self-aligned structures.

The analysis and theoretical modelling of polysilicon TFTs is much more complex than in the single crystal case, because of the presence of a continuum of states in the energy gap. Polysilicon films consist of crystalline regions (grains), typically between a hundred to a few thousand angstroms in size, separated by grain boundaries. A high density of dangling bonds is expected at grain boundaries, as revealed by ESR measurements (Hirose 1979). Weak or reconstructed bonds can form between dangling bonds. Both dangling and reconstructed bonds give rise to electronic energy levels in the energy gap, and the dangling bond states are believed to be located near midgap. In addition, other gap states can be introduced as a consequence of random fluctuations of the crystalline potential (Mott and Davis 1979). These states, known as disorder-induced localised states, are located in energy near the conduction and valence bands, and constitute the so-called tail states. In polysilicon, these states are probably associated with non-uniform strain fields, and may be distributed over the grain volume. As the Density-of-States (DOS) is spatially non uniform, the theoretical treatment of the field-effect in polysilicon is complex. It has however been shown that the properties of polysilicon TFTs can be adequately described by assuming an effective DOS, uniform over the grain volume (Fortunato and Migliorato 1986, Migliorato and Meakin 1987). That is, if N_{GB} is the grain boundary DOS ($cm^{-2}eV^{-1}$), the effective DOS is $a^{-1}N_{GB}$, where 'a' is the average grain size.

In the present paper we review the properties of polysilicon transistors, with emphasis on the effect of the DOS on the transistor characteristics. A theoretical analysis is presented, which leads to a generalised semi-analytical MOSFET model, and provides a new tool for device analysis. Structural properties of polysilicon are reviewed in Section 2. The effect of the DOS on the I-V characteristics is discussed in Section 3. The model is presented in Sections 4 and 5. The applications of this model to device analysis is described in Section 6. The basic equations of the theory and methods for the determination of the DOS are discussed in the Appendix, which also includes a list of symbols.

10.2 STRUCTURAL PROPERTIES AND DENSITY OF STATES

Films deposited in conventional polysilicon reactors (P_{SiH_4} =0.18 torr, T=625°C) have a <110> preferred orientation in the direction normal to the surface. The typical grain size is in the range 100-500 Å. It has been shown (Meakin 1987b) that individual grains consist of parallel twin laminae, a few angstroms thick (microtwins), the twin plane being a <111> plane. These microtwins have the same orientation within each grain (coherent twins), but are rotated by an arbitrary angle around the <110> axis in adjacent grains. The importance of these defects had gone unnoticed until it was discovered that microtwins are associated with a considerable amount (3%) of lattice parameter distortion

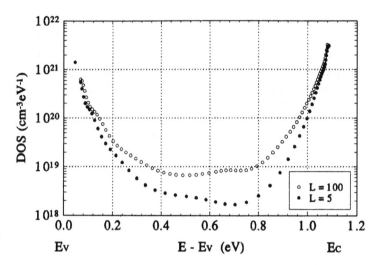

Figure 10.2 The Density Of States of two polysilicon TFTs with equal widths but different lengths. The graph highlights the effect of hydrogenation on the Density of States.

(Meakin 1987*b*). The presence of voids and dangling bonds at incoherent twin boundaries was first demonstrated in the case of the silicon ribbon (Cunningham 1982). In polysilicon films a large number of incoherent twins is present at grain boundaries and the strain associated with voids and dangling bonds can result in lattice parameter distortion. This in turn introduces strong fluctuations in the crystalline potential, leading to disorder-induced localised states in the band gap (Migliorato and Meakin 1987).

High performance polysilicon TFTs have been fabricated using a process known as Very Low Pressure Chemical Vapour Deposition (VLPCVD), which was originally developed at GEC. Both microtwinning and lattice parameter distortion were absent from VLPCVD films (Meakin 1987*b*). This process has recently been used to produce 6.2"-diagonal displays on glass substrates (Haws 1993). The effect of pressure on the film texture was investigated by Joubert *et al.* (1987) and, more recently, by Miyasaka *et al.* (1991). In the latter work it was shown that the preferred orientation changes from <110> to <111> when the pressure is reduced from 20.8 mtorr to 0.4 mtorr. Other methods for producing high quality films have been developed, namely thermal recrystallisation (see, for instance, Brotherton 1991 and references therein) and laser recrystallisation (Kuriyama 1991, Masumo 1990). It is worth noting that for the last two methods the best results appear to have been obtained for films with <111> or no preferential orientation (Mimura 1989, Kuriyama 1991). This confirms the importance of avoiding microtwinning, which is favoured by the <110> orientation.

Measurement of the DOS is essential for material characterisation and process control. The polysilicon DOS has been determined both by field-effect analysis of the I-V characteristics at a fixed temperature (Fortunato and Migliorato 1986) and by the temperature dependence of the field-effect (Fortunato 1988). The latter method is more accurate, as it allows simultaneous measurement of the bulk Fermi Energy. Both methods are described in the Appendix.

The DOS of polysilicon is shown in Figure 2 for hydrogenated TFTs with equal widths but different lengths. States in the conduction or valence band tails are, by analogy with the a-Si case, attributed to crystalline potential fluctuations associated with lattice distortion, while states located near midgap are believed to be associated with dangling bonds. The devices used for Figure 2 were plasma-hydrogenated. Diffusion of hydrogen into the channel is impeded by the gate electrode and occurs mainly from the regions near source and drain (Mimura 1989). This explains the differences between the two curves.

10.3   EFFECT OF THE DOS ON THE I-V CHARACTERISTICS

A high DOS results in a slow turn-on of the transfer characteristics. This makes it difficult to define a threshold voltage (V_T) and a field effect mobility (μ) in the usual way, since a linear I_D-V_{GS} regime may not be present at all. This is illustrated in Figures 3a-c. The apparent threshold voltage V_T and mobility μ are deduced here from the $I_D = 0$ intercept and slope, respectively, of the tangent to the G-V_{GS} curve. The figure shows that V_T varies from 4V to 7V for V_{GS} in the range 10V-25V. Nevertheless, it has been customary to quote V_T and μ as figures of merit as is normal for single crystal devices. This leads to difficulties in interpreting experimental results. We analyse below the validity of these parameters for polysilicon TFTs.

The effect of the DOS on the G-V_{GS} characteristics can be studied in detail by using the following set of equations, which are deduced in the Appendix:

$$\frac{dG}{d\psi_s} = q\mu_n \frac{n(\psi_s) - n_o}{F_s} \qquad (10.1)$$

$$F_s^2 = \frac{2q}{\varepsilon} \int_{E_v}^{E_c} N(E)H(\psi_s, E, E_F)dE + A \int_{E_c}^{\infty} (E - E_c)^{1/2} H(\psi_s, E, E_F)dE \qquad (10.2)$$

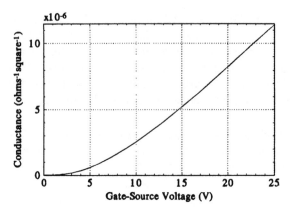

Figure 10.3(a)   Experimental conductance characteristic (G) for an n-type polysilicon TFT, taken from low-V_{DS} transfer curve measurements.

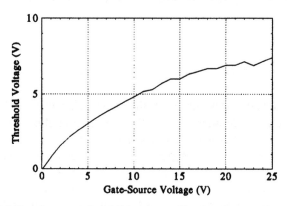

Figure 10.3(b)   Apparent threshold voltage (V_T) for the transistor of Figure 10.3(a), calculated from the intercept of the G-V_{GS} characteristic with the x-axis.

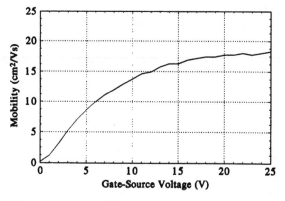

Figure 10.3(c)   Apparent mobility (μ) for the transistor of Figure 10.3(a), calculated from the slope of the conductance characteristic.

$$V_{GS} - V_{FB} = \psi_s + V_{ox} = \psi_s + \frac{t_{ox}}{\varepsilon_{ox}} (\varepsilon F_s - Q_{ss}) \qquad (10.3)$$

We assume for N(E), in the upper half of the forbidden gap, the expression:

$$N(E) = N_T \exp\left(\frac{E - E_c}{kT_T}\right) + N_D \exp\left(\frac{E - E_c}{kT_D}\right) \qquad (10.4)$$

The two terms on the right hand side of Equation 4 account for the tail states and the deep states respectively.

We have calculated the G-V_{GS} characteristics for fixed tail states ($N_T = 10^{21}$ cm^{-3} eV^{-1}, $T_T = 150$K), and different deep states distributions ($T_D = 2000$K, $N_D = 10^{18}, 10^{19}$ and 10^{20} cm^{-3}eV^{-1}). Plots of V_T and μ are shown in Figures 4a-b. Only in the case of the lower N_D, can the threshold voltage and mobility be regarded as reasonably constant in the range of operation. The effect of varying the width of the tail states (T_T=150K and 400K) is shown in Figures 5a-b, which indicates a strong dependence of V_T and μ upon V_{GS} for $T_T = 400$K. These results can be understood by looking at the generalised expression relating the conductance to the fixed charge (Q_{fixed}), which is deduced in the Appendix (Migliorato 1992):

$$G = \mu_n \frac{\varepsilon_{ox}}{t_{ox}} \left[V_{GS} - V_{FB} - \left(\psi_s + t_{ox} \frac{Q_{fixed}}{\varepsilon_{ox}}\right) \right] \qquad (10.5)$$

In the case of single crystal, ψ_s saturates to the value E_i-E_F for sufficiently high V_{GS}. The term within brackets, therefore, becomes a constant. The threshold voltage is then defined as :

$$V_T = \psi_s + t_{ox} \frac{Q_{fixed}}{\varepsilon_{ox}} + V_{FB} \qquad (10.6)$$

In the case of polysilicon however, this situation is never achieved. As undoped or lightly doped films are employed, E_i-$E_F \approx 0$. The maximum band bending is determined by the width of the tail states, but this value can only be achieved if the concentration of deep states is low. This is the case in Figures 4a-b and 5a-b for $N_T = 10^{21}$ cm^{-3}eV^{-1}, $T_T = 150$K and $N_D = 10^{18}$ cm^{-3}eV^{-1}, $T_D = 2000$K, which corresponds to the DOS observed in devices processed at high temperature. So, for low temperature devices, which are the most important for future applications, the assumptions of a constant threshold voltage and mobility are not applicable. We will show in the next section how this problem can be overcome.

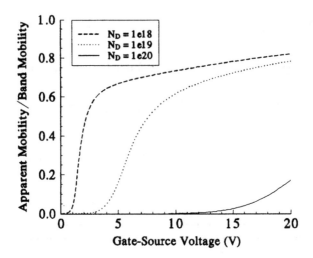

Figure 10.4(a) Apparent mobility (normalised to the band mobility) for a given distribution of tail-states ($N_T = 10^{21}$ cm^{-3}eV^{-1}, $T_T = 150$K), but different deep-states concentrations. ($T_D = 2000$K, $N_D = 10^{18}, 10^{19}$ and 10^{20} cm^{-3}eV^{-1}). Only for the case of the lower deep-states concentrations can the mobility be regarded as constant over the range of operation.

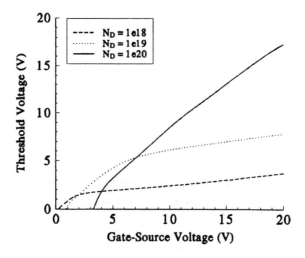

Figure 10.4(b) Apparent threshold voltage for a given distribution of tail-states ($N_T = 10^{21}$ cm^{-3}eV^{-1}, $T_T = 150$K), but different deep-states concentrations. ($T_D = 2000$K, $N_D = 10^{18}, 10^{19}$ and 10^{20} cm^{-3}eV^{-1}). Only in the case of the lower deep-states concentrations can the threshold voltage be regarded as constant over the range of operation.

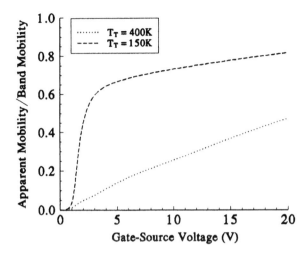

Figure 10.5(a) Apparent mobility (normalised to the band mobility) for a given distribution of deep-states ($N_D = 10^{18}$ cm^{-3}eV^{-1}, $T_D = 2000$K), but different tail-states distributions ($N_T = 10^{21}$ cm^{-3}eV^{-1}, $T_T = 150$K and 400K). Only for $T_T = 150$K can a mobility be defined.

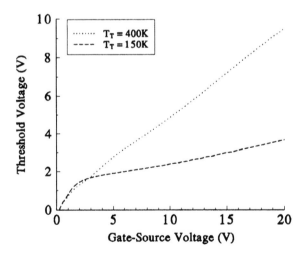

Figure 10.5(b) Apparent threshold voltage for a given distribution of deep-states ($N_D = 10^{18}$ cm^{-3}eV^{-1}, $T_D = 2000$K), but different tail-states distributions ($N_T = 10^{21}$ cm^{-3}eV^{-1}, $T_T = 150$K and 400K). Only for $T_T = 150$K can a threshold voltage be defined.

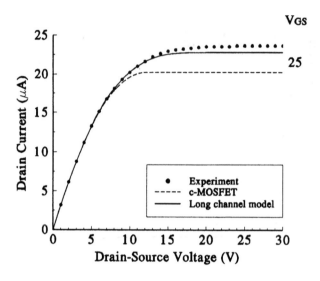

Figure 10.6(a) Comparison of Experimental output curves at $V_{GS} = 25V$ with a single crystal approximation (c-MOSFET) and the Long Channel approach (Equation 9 in the text) for an n-type TFT with $L = 100\mu m$.

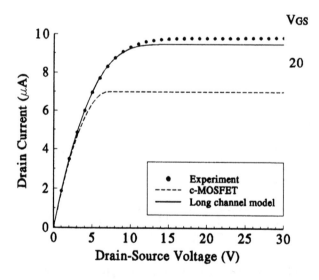

Figure 10.6(b) Comparison of Experimental output curves at $V_{GS} = 20V$ with a single crystal approximation (c-MOSFET) and the Long Channel approach (Equation 9 in the text) for an n-type TFT with $L = 100\mu m$.

10.4   CURRENT-VOLTAGE CHARACTERISTICS

The field in the direction parallel to the channel (F_y) is related to G through the relationship[*]:

$$F_y = \frac{dV(y)}{dy} = \frac{I_D}{WG[V_{GS} - V_{FB} - V(y)]} \qquad (10.7)$$

Equation 7, by integration, gives:

$$I_D y = W \int_0^{V(y)} G(V_{GS} - V_{FB} - V)dV \qquad (10.8)$$

and, for y = L: $$I_D = \frac{W}{L} \int_0^{V_{DS}} G(V_{GS} - V_{FB} - V)dV \qquad (10.9)$$

Equation 9 can be used to calculate the current-voltage characteristics at all drain voltages. For this it is sufficient to know $G(V_{GS})$, which in turn is obtained from the I_D-V_{GS} characteristics at low drain voltage. No analytical approximation is required, as one can use a polynomial interpolation for $G(V_{GS})$ without sacrificing calculation speed (Izzard 1990). The characteristics calculated by this approach are compared with the experiment and those calculated by the single crystal model in Figures 6a-b. Clearly only the present method (which is hereafter referred to as the long channel model) gives a good fitting of the experimental data. The data of Figures 6a-b are for a long channel device (100μm). However, for shorter channel lengths a new approach is required.

High electric fields present at the drain junction can affect the output characteristics, the leakage current and the device stability both in thin-film and single crystal MOSFETs. In polysilicon TFTs these effects manifest themselves for thicker gate oxides and longer channel lengths than in their single crystal counterparts, due to the higher fields resulting from a high concentration of gap states. Therefore, deviations from the gradual channel approximation, even at drain voltages well below saturation, are observed in polysilicon TFTs.

Our new approach involves using an 'effective' conductance, G_{eff}, obtained from Equation 1 in the form of a polynomial, to account for these discrepancies

---

[*] See Figures A1 and A2 in the Appendix for details of the co-ordinate system.

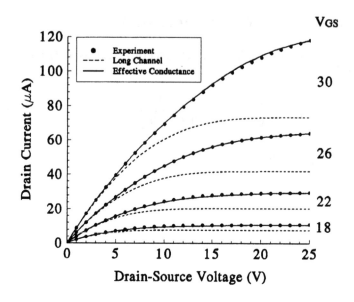

Figure 10.7(a)   Comparison of Experimental I_D-V_{DS} curves with the Long Channel approach and Effective Conductance method over a wide range of gate voltages for an n-type TFT with W/L = 10μm/20μm.

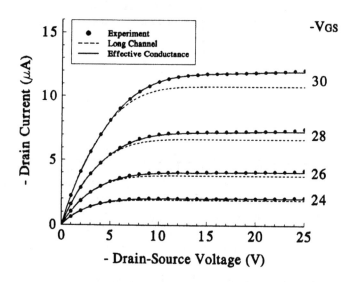

Figure 10.7(b)   Comparison of Experimental I_D-V_{DS} curves with the Long Channel approach and Effective Conductance method over a wide range of gate voltages for a p-type TFT with W/L = 10μm/20μm.

(Quinn 1993*a*). A very important feature of this model is the absence of discontinuity in the function of I_D and its derivative. This is essential to ensure convergence during the numerical solution of the non-linear equations in a circuit simulator. In Figures 7*a-b* we present a comparison between our new model and experimental output curves for n and p-channel TFTs. Also included in the figure is the long channel model (given by Equation 9) to highlight the improvement over our previous approach. An excellent agreement over a wide range of gate voltages is obtained for both types of device. Use of this method for the analysis of the channel potential and field profiles will be discussed in Section 6.

10.5   CAPACITANCE VOLTAGE CHARACTERISTICS

The nodal capacitances of a crystalline MOSFET are accounted for by the Meyer model (1971), which assumes that these capacitances are constant in each operating region of the device. The absence of very sharply defined operating regions in polysilicon TFTs (as discussed in Section 3) makes this approach unsuitable. We follow here the procedure developed by Izzard and co-workers (1990, 1991).

From Gauss's law we have:

$$Q_G = \iint_{\text{Gate area}} \varepsilon_{ox} F_{ox} dz dy = W \int_0^L \varepsilon_{ox} F_{ox} dy \qquad (10.10)$$

Changing the variable of integration gives:

$$Q_G = W \int_0^{V_{DS}} \varepsilon_{ox} F_{ox} \frac{dy}{dV} dV \qquad (10.11)$$

$$F_{ox} = V_{ox} / t_{ox} = (V_{GS} - V - \psi_s) / t_{ox} \qquad (10.12)$$

$$Q_G = W^2 \frac{\varepsilon_{ox}}{t_{ox}} \int_0^{V_{DS}} \frac{(V_{GS} - V)}{I_D} G(V_{GS} - V) dV, \qquad (10.13)$$

having assumed that:

$$\psi_s \ll V_{GS} - V$$

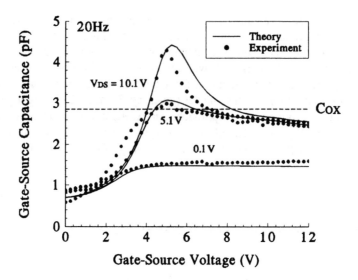

Figure 10.8 Experimental and Theoretical C_{GS} capacitance results at 20Hz for $V_{DS} = 0.1V$, 5.1V, and 10.1V. The oxide capacitance overshoot disappears with decreasing V_{DS}.

The terminal capacitances C_{GS} and C_{GD} can then be calculated from:

$$C_{GS} = \frac{\partial Q_G}{\partial V_S}\bigg|_{V_G, V_D \text{ const}} \qquad C_{DS} = \frac{\partial Q_G}{\partial V_D}\bigg|_{V_G, V_S \text{ const}} \qquad (10.14)$$

Since a TFT is a true three-terminal device (i.e. no body terminal), C_{GS} and C_{GD} are sufficient to define the transient behaviour.

We have measured the terminal capacitances of polysilicon TFTs, using a technique similar to that of Yao and co-workers (1988), and analysed the results using the present model. These experimental and theoretical results are shown in Figure 8. A considerable success of the model is its ability to predict the correct capacitance characteristics, and in particular, the presence of a peak (capacitance overshoot) in C_{GS} vs V_{GS} (Tam 1993). This peak, in excess of the oxide capacitance, is a consequence of the device AC gain. We consider here explicitly the dependence of the channel potential on the drain and gate voltages. Substituting Equation 11 into Equation 10 we have :

$$Q_G = C_{ox} W \int_0^L \left[V_{GS} - V_y (V_{GS}, V_{DS}) \right] dy \qquad (10.15)$$

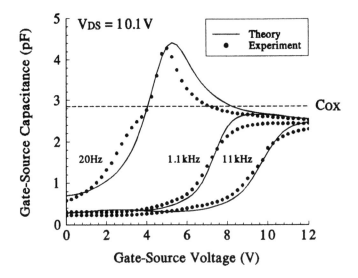

Figure 10.9 Experimental and Theoretical C_{GS} capacitance results for $V_{DS} = 10.1V$ at 20Hz, 1.1kHz, and 11kHz. The oxide capacitance overshoot disappears with increasing the frequency and there is a shift towards higher gate voltages.

$$C_{GS} = C_{ox} WL \left[\frac{\partial}{\partial V_{GS}} \overline{V_y} + \frac{\partial}{\partial V_{DS}} \overline{V_y} \right] \qquad (10.16)$$

where,

$$\overline{V_y} = \frac{1}{L} \int_0^L V_y \, dy$$

is the average channel potential. For small variations of the source voltage V_S the first term on the right hand side of Equation 16 can be bigger than unity. This is because a small variation of the gate to source voltage can strongly alter the channel potential profile, when the profile is strongly non-linear. This also explains why the peak in C_{GS} increases for increasing V_{DS}.

To explain the frequency dependence, the effect of the distributed channel resistance and capacitance needs to be considered. When the TFT is operating in the linear region, it can be modelled by a transmission line and the problem can be solved analytically (Tam 1993). However, at higher V_{DS}, the approximation of a linear channel potential fails. Therefore, we have developed a new approach, the 'subtransistor' method, where each TFT is considered as a combination of shorter length ones, whose capacitance and channel conductances are given by Equations 13 and 14 (Tam 1993, Quinn 1993*b*).

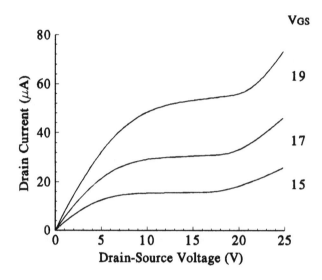

Figure 10.10   Output characteristics for a practical (10μm) channel length n-type TFT, highlighting the importance of the "kink" effect.

The solution obtained for 5 subtransistors over a range of frequencies is given in Figure 9. The results show that excellent agreement with the experiment can be obtained with a relatively small number of subtransistors. Therefore, this approach is ideally suited for implementation in a circuit simulator.

10.6   ELECTRIC FIELD AT THE DRAIN AND "KINK" EFFECT

Equations 7 and 8 can be used to calculate the channel potential and field profiles of polysilicon TFTs (Migliorato 1992). The channel potential profile V(y) is plotted in Figures 11a-b for 20 μm gate length n and p-type TFTs biased in the linear, saturation, and kink regions. Since the conductance is a strong function of the channel potential, deviation from linearity in the potential profile is exaggerated by the presence of localised states. Lack of saturation in the output characteristics, the 'kink' effect, is normally observed for gate lengths below 20 μm and has been attributed to impact ionisation in the high field region near the drain (Reita 1992, Hack and Lewis 1991). The experimental output curves of Figure 10 highlight the importance of this effect for circuit design.

The field profiles for the devices of Figure 10, biased in the kink region, are given in Figure 12. The kink region is characterised by a reduction in field near the drain. (corresponding to a change in slope of the potential profiles of

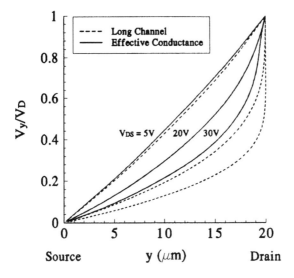

Figure 10.11(a) Normalised channel potential profiles for a 20 μm gate length n-type TFT, for $V_{GS} = 26V$, in the linear, saturation, and kink regions. The kink region is characterised by a change in slope of the potential profiles.

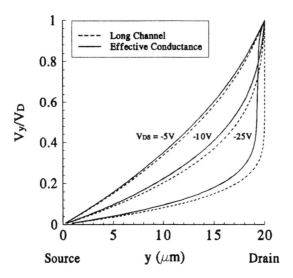

Figure 10.11(b) Normalised channel potential profiles for a 20 μm gate length p-type TFT, for $V_{GS} = -26V$, in the linear, saturation, and kink regions. The kink region is characterised by a change in slope of the potential profiles.

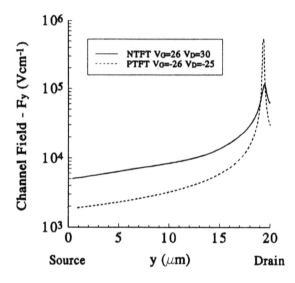

Figure 10.12 Channel field profiles for 20 μm gate length n and p-channel transistors operating in the kink region. The peak field ratio is consistent with impact ionisation being the mechanism responsible for the kink.

Figures 11*a-b*). Interestingly, the maximum field in a p-channel device is higher (by a factor of five) than for an n-channel device. This is consistent with the fact that for a given field the electron ionisation coefficient is higher than that for holes (Van Overstraeten and De Man 1970). These predictions are in agreement with recent measurements of the channel potential (Shirai and Serikawa 1992). The results of Shirai and Serikawa, although based on laser recrystallised material, can be compared with our low temperature glass substrate process in the case where the laser recrystallised material contains a relatively high concentration of defects.

10.7 CONCLUSIONS

Considerable progress has been made over the past decade in the analysis of the electrical properties of polysilicon TFTs. Since the characteristics of these devices are strongly affected by states in the bandgap, methods for measuring the density of these states in finished devices are very important tools for device characterisation and process control. The field effect method, reviewed in the present paper, is in our opinion, the best approach for this purpose. However, work in this area is by no means exhausted. Increased accuracy in the determination of the band tails is required, and the relative contributions of surface and bulk states should be determined. At present, no satisfactory methods exist to solve these problems.

The polysilicon TFT model developed in the present paper provides very accurate simulation results for both n and p-channel devices. When coupled with our 'subtransistor' approach, it also accounts for the frequency dependence of the nodal capacitances and this is, to our knowledge, the only model which has achieved such a result. We believe that our approach is not only restricted to the case of polysilicon TFTs. It has potential for use with single crystal Silicon-On-Insulator MOSFETs to overcome the difficulties of standard models, particularly in the simulation of the terminal capacitances. The parameter extraction routine required by the present method is a very straightforward one. We have, in fact, implemented the present model into a circuit simulator with excellent results. Finally, the usefulness of the present theoretical approach to gain a better insight into the physical mechanisms controlling the device operation has been demonstrated through its ability to predict channel potential and field profiles consistent with experimental observations.

10.8   APPENDIX

10.8.1   List of Symbols

C_{GD}	Gate-drain capacitance
C_{GS}	Gate-source capacitance
C_{ox}	Gate dielectric capacitance per unit area (ε_{ox}/t_{ox})
d	Semiconductor film thickness
ε	Semiconductor permittivity
ε_o	Permittivity in vacuum
ε_{ox}	Gate dielectric permittivity
E	Energy
E_c	Conduction band energy
E_{co}	Conduction band energy in the bulk
E_v	Valance band energy
E_F	Fermi energy
F	X-component of the electric field
F_s	Value of F at the surface (x = 0)
F_y	Y-component of the electric field
G	Conductance
G_{eff}	Effective conductance
k	Boltzman constant
I_D	Drain current
L	Channel length
μ	Apparent mobility

μ_n	Free carrier mobility
n	Free electron concentration
n_o	Free electron concentration in the bulk
N(E)	Density of states in the gap ($cm^{-3}\ eV^{-1}$)
ψ	Electrostatic potential
ψ_s	Electrostatic potential at the surface
q	Electronic charge = $+1.6 \times 10^{-19}$
Q_G	Channel charge
Q_{ss}	Interface charge density ($C\ cm^{-2}$)
ρ_{sc}	Bulk space charge density
t_{ox}	Gate dielectric thickness
T	Temperature (K)
V	Channel potential
V_{DS}	Drain-source voltage
V_{FB}	Flat-band voltage
V_{GS}	Gate-source voltage
V_{ox}	Potential drop across the gate dielectric
W	Channel width

10.8.2   Relationship between ψ_s, F_s, G, V_{GS}

We use throughout the "distributed trap approximation" (Depp 1982, Hirose 1979, Fortunato and Migliorato 1986), whereby the gap states associated to the grain boundary defects are considered as uniformly distributed over the grain volume. A sketch of the TFT and its band diagram are shown in Figures A1 and A2 respectively. The following treatment is based on Migliorato (1989, 1992). The band bending in the polysilicon is determined by Poisson's equation:

$$\frac{d^2\psi}{dx^2} = \frac{q}{\varepsilon}\ [N_{trap}(x) + n(x) - n_o] \qquad (10.A1)$$

where N_{trap} is the net concentration of negatively charged centres, and n(x) -n_0 is the change in electron concentration with respect to equilibrium. These quantities can be written as functions of ψ:

$$N_{trap}(\psi) = \int_{E_v}^{E_c} N(E)[f(E,E_F+q\psi)-f(E,E_F)]dE \qquad (10.A2)$$

$$n(\psi)-n_0 = A\int_{E_c}^{\infty}(E-E_c)^{1/2}(f-f_0)dE \qquad (10.A3)$$

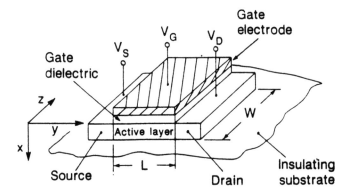

Figure 10.A1 Sketch of TFT showing the co-ordinate system employed in the present paper. [From Migliorato 1989, Courtesy of Academic Press Inc.]

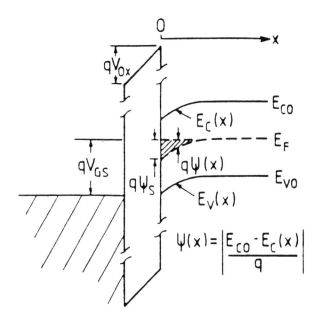

Figure 10.A2 Band diagram of a polysilicon TFT in the direction orthogonal to the channel for $V_{DS} = 0$. [From Migliorato 1989, Courtesy of Academic Press Inc.]

where:

$$A = \frac{2N_c}{\pi^{1/2}kT^{3/2}}$$

$$f(E, E_F + q\psi) = \left[1 + \exp\left(\frac{E - E_F - q\psi}{kT}\right)\right]^{-1}, \qquad (10.A4)$$

and

$$f_0 = f(\psi = 0)$$

Multiplying both sides of Equation A1 by $2d\psi/dx$ and integrating from $x = 0$ (oxide/semiconductor interface) to $x = d$ (the unmodulated neutral bulk where $d\psi/dx = 0$) one obtains:

$$F_s^2 = \frac{2q}{\varepsilon} \int_0^{\psi_s} d\psi \int_{E_v}^{E_c} N(E)[f(E, E_F + q\psi) - f(E, E_F)]dE +$$

$$\frac{2q}{\varepsilon} A \int_0^{\psi_s} d\psi \int_{E_c}^{\infty} (E - E_c)^{1/2}[f(E, E_F + q\psi) - f(E, E_F)]dE \qquad (10.A5)$$

where $F_s = -d\psi/dx$ at $x = 0$ is the surface electric field. By changing the order of integration, Equation A5 reduces to:

$$F_s^2 = \frac{2q}{\varepsilon} \int_{E_v}^{E_c} N(E)H(\psi_s, E, E_F)dE + \frac{2qA}{\varepsilon} \int_{E_c}^{\infty} (E - E_c)^{1/2}H(\psi_3, E, E_F)dE \quad (10.A6)$$

where $H(\psi_s, E, E_F) = \frac{kT}{q} \ln\left\{f_0\left[\exp\left(\frac{q\psi_s}{kT}\right) + \exp\left(\frac{E - E_F}{kT}\right)\right]\right\} - \psi_s f_0$

The conductance G is defined as:

$$G = q\mu_n \int_0^d n(x)dx \qquad (10.A7)$$

By changing the variable of integration from x to ψ and taking the derivative of both sides with respect to ψ_s one has:

$$\frac{dG}{d\psi_s} = q\mu_n \frac{n(\psi_s) - n_0}{F_s} \qquad (10.A8)$$

$$\frac{dG}{d\psi_s} = q\mu_n n_0 \frac{\exp(q\psi_s / kT) - 1}{F_s} \tag{10.A9}$$

Gauss's Law can be written as:

$$V_{GS} - V_{FB} = \psi_s + V_{ox} = \psi_s + \frac{t_{ox}}{\varepsilon_{ox}} (\varepsilon F_s + Q_{ss}) \tag{10.A10}$$

Here, Q_{ss} is the charge in the interface states at the oxide/semiconductor interface. Equations A6, A8 and A10 can be used to calculate the G-V_{GS} characteristic once N(E) is known.

Equation A10 can be re-written in terms of the fixed charge (Q_{fixed}), by using the relationships:

$$F_s = \frac{1}{\varepsilon} \int_0^d (qn(x) + \rho_{sc}) dx \tag{10.A11}$$

$$F_s = \frac{G}{\varepsilon \mu_n} + \frac{1}{\varepsilon} \int_0^d \rho_{sc}(x) dx = \frac{G}{\varepsilon \mu_n} + \frac{Q_{sc}}{\varepsilon}, \tag{10.A12}$$

having indicated the bulk space charge density with ρ_{sc}. Substitution of Equation A12 into Equation A10, and using the definition $Q_{fixed} = Q_{sc} + Q_{ss}$ gives Equation 5 in the text.

10.8.3 Determination of the DOS

We consider here the case n $\ll$ N$_{trap}$, so that the second term on the right hand side of Equation A5 is negligible. Taking the second derivative of both sides and using the identity (Migliorato and Meakin 1987):

$$\frac{\partial\ f(E, E_F + q\psi_s)}{\partial\ \psi_s} = -q\frac{\partial\ f(E, E_F + q\psi_s)}{\partial E} \tag{10.A13}$$

one has:

$$\frac{\partial\ ^2 F_s^2}{\partial\ \psi_s^2} = -\frac{2q^2}{\varepsilon} \int_{E_v}^{E_c} N(E)\frac{\partial\ f(E, E_F + q\psi_s)}{\partial E} dE \tag{10.A14}$$

Equation A14 indicates that once $F_s(\psi_s)$ is known, $N(E)$ can be determined by deconvolution. The problem simplifies when the Fermi function can be approximated by a step function (i.e. For $T \to 0$). Under these conditions:

$$\frac{\partial\ f(E, E_F + q\psi_s)}{\partial\ E} \to \delta\ (E_F + q\psi_s)$$

and
$$\frac{\partial\ ^2 F_s^2}{\partial\ \psi_s^2} \to \frac{2q^2}{\varepsilon} N(E_F + q\psi_s) \qquad (10.\text{A}15)$$

There are currently 3 methods of deducing $F_s(\psi_s)$ from the I_D-V_{GS} characteristic at low V_{DS}, all neglecting the interface states density Q_{ss}:

(1)   THE STEP-BY-STEP INTEGRATION METHOD (Hirae 1980)

This consists of approximating Equation A9 by finite differences. By using Equation A10, and neglecting Q_{ss}, one obtains the recurrent relationship:

$$\psi_{si+1} = \psi_{si} + \frac{d}{t_{ox}}\ \frac{\varepsilon_{ox}}{\varepsilon}\ \frac{G_{i+1} - G_i}{G_0}\ \frac{V_{Gi} - V_{FB} - \psi_{si}}{\exp(q\psi_{si}/kT) - 1} \qquad (10.\text{A}16)$$

where $G_0 = q\mu n_0 d$ and the flat band voltage V_{FB} can be approximated from the minimum of the G-V_{GS} characteristic for 'long' devices (i.e. when junction effects can be neglected). Equation A16 yields $\psi_s(V_{GS})$. $F_s(\psi_s)$ may then be deduced from Equation A10.

(2)   THE INTEGRAL METHOD (Grunewald 1980)

Integrating Equation A9 by separation of variables, one obtains:

$$\int_{V_{FB}}^{V_{GS}} F_s dG = q\mu_n n_0 \left[\frac{kT}{q}\ \exp\left(\frac{q\psi_s}{kT}\right) - \frac{kT}{q} - \psi_s \right] \qquad (10.\text{A}17)$$

Using Equation A10 to express F_s as a function of V_{GS}, one has, for $\psi_s \ll V_{GS}$:

$$\frac{\varepsilon_{ox}}{\varepsilon t_{ox}} \int_{V_{FB}}^{V_{GS}} (V_{GS} - V_{FB}) \frac{dG}{dV_{GS}}\ dV_{GS} = q\mu_n n_0 \left[\frac{kT}{q}\exp\left(\frac{q\psi_s}{kT}\right) - \frac{kT}{q} - \psi_s \right] \qquad (10.\text{A}18)$$

Equations A18 and A10 define $F_s(\psi_s)$.

(3)   THE TEMPERATURE METHOD (Fortunato 1988)

Taking Equation A9, for $q\psi_s/kT \gg 1$, and multiplying both sides by $d\psi_s/dV_{GS}$, one obtains:

$$\frac{dG}{dV_G} \cong \frac{G_0}{d} \frac{\exp(q\psi_s/kT)}{F_s(dV_G/d\psi_s)} \qquad (10.A19)$$

This indicates that $\psi_s(V_{GS})$ can be deduced from a plot of $\log(dG/dV_{GS})$ vs $1/T$.

10.9   REFERENCES

Brotherton S D, 1991, *Microelectronic Engineering*, **15**, 333.

Bryer N, Coxon P, Fortunato G, Kean R, Meakin D, Migliorato P, Rundle P, and Urwin M, 1986, *Proceedings of Japan Display*, Tokyo, 80.

Coxon P, Lloyd M, and Migliorato P, 1986, *Appl. Phys. Lett.*, **49**, 1785.

Cunningham B, Strunk H, and Ast D G, 1982, *Appl. Phys. Lett.*, **40**, 237.

Depp S W, Huth B G, Juliana A, and Koepcke R W, 1982, *Grain boundaries in semiconductors*, Pike/Seager/Leamy, Editors, Elsevier Science Publishing, New York, 297.

Fortunato G, and Migliorato P, 1986, *Appl. Phys. Lett.*, **49**, 1025.

Fortunato G, Meakin D B, Migliorato P, and LeComber P, 1988, *Phil. Mag. B*, **5**, 573.

Grunewald M, Thomas P, and Wurtz D, 1980, *Phys. Stat. Sol. B*, **100**, K139.

Hack M, and Lewis A G, 1991, *IEEE Trans. Electron Dev. Lett.*, **12**, 203.

Haws S, Fluxman S, and Rundle P, 1993, *Society for Information Display Int. Symposium Digest*, Seattle, **24**, 895.

Hirae S, Hirose M, and Osaka Y, 1980, *J. of Appl. Phys.*, **51**, 1043.

Hirose M, Taniguchi M, and Osaka Y, 1979, *J. Appl. Phys.*, **50**, 377.

Izzard M J, Migliorato P, and Milne W I, 1991, *Jpn. J .Appl. Phys.*, **30**, L170.

Izzard M J, 1990, *PhD Thesis*, Cambridge University.

Joubert P, Loisel B, Chouan Y, and Haji L, 1987, *J. Electroch. Soc.*, **134**, 2543.

Kobayashi K, Nijs J, and Mertens R, 1989, *J. Appl. Phys.*, **65**, 2541.

Kuriyama H, Kiyama S, Noguchi S, Kuwahara T, Hishida S, Nohda T, Sano K, Iwada H, Tsuda S, and Nakano S, 1991, *Int. Electron Dev. Meet. Digest*, 563.

Masumo K, Kunigita M, Takafuji S, Nakamura N, Iwasaki A, and Yuki M, 1990, *Extended Abstracts of the 22nd Int. Conf. on Solid State Dev. and Materials*, Sendai, 975.

Meakin D B, Coxon P A, Migliorato P, Stoemenos J, and Economou N A, 1987a, *Appl. Phys. Lett.*, **50**, 1984.

Meakin D B, Stoemenos J, Migliorato P, and Economou N A, 1987*b*, *J. Appl. Phys.*, **61**, 5031.

Meakin D B, Ferry A, Palmer D, Saunders M, 1989, *Society for Information Display Int. Symposium Digest*, 159.

Meyer J E, 1971, *RCA Review*, **32**, 42.

Migliorato P, and Meakin D B, 1987, *Appl. Surf. Sci.*, **30**, 353.

Migliorato P, 1989, *Encyclopaedia of Physical Science and Technology*, Academic Press, 598.

Migliorato P, 1992, *Microelectronic Engineering*, **19**, 89.

Mimura A, Konishi N, Ono K, Ohwada J, Hosokawa Y, Ono Y A, Suzuki T, Miyata K, and Kawakami H, 1989, *IEEE Trans. Electron Dev.*, **36**, 351.

Miyasaka M, Kazawa T, and Ohshima H, 1991, *Int. Electron Dev. Meet. Digest*, 559.

Mott N F, and Davis E A, 1979, *Electronic Processes in Non-crystalline Materials*, 2nd ed., Clarendon Press, Oxford.

Noguchi T, Hayashi H, and Ohshima T, 1986, *Jpn. J. Appl. Phys.*, **25**, L121.

Van Overstraeten R, and De Man H, 1970, *Solid State Electronics*, **13**, 583.

Quinn M, Migliorato P, Tam S and Reita C, 1993*a*, to be published, *Proceedings of the 3rd Int. Conf. on VLSI and CAD*, Taejon, Korea.

Quinn M, Migliorato P, Reita C, Pecora A, Tallarida G, and Fortunato G, 1993*b*, To be published, *IEE Proceedings - Part G*.

Reita C, Migliorato P, Pecora A, Fortunato G, and Mariucci L, 1992, *Microelectronic Engineering*, **19**, 183.

Shirai S, and Serikawa T, 1992, *IEEE Trans. Electron Dev.*, **39**, 450.

Tam S W-B, Migliorato P, Izzard M J, Reita C, 1993, *Proceedings of ESSDERC*, Edition Frontieres, Gif-sur-Yvette Cedex, France, 535.

Yao C T, Peckerar M, Friedman D, and Hughes H, 1988, *IEEE Trans. Electron Dev.*, **35**, 384.

11

Application and Technology of Polysilicon Thin Film Transistors for Liquid Crystal Displays.

Chris Baert

11.1   INTRODUCTION

The term 'Thin Film Transistor' or 'TFT' covers a broad range of electronic devices whose working principle is identical to the MOSFET, widely used in integrated circuits (Grove 67). The basic device structure of a Si TFT differs from that of a conventional Si MOSFET device by the fact that a Si thin film channel layer situated on top of a non-Si material (usually SiO_2) is used instead of bulk crystalline Si. Examples of this technology are c-Si TFT's in Silicon-On-Insulator technology (Colinge 91) and polysilicon TFT's used as a load transistor in SRAM's (Yamanaka et al 88). In this chapter, we will restrict the definition of TFT to those devices deposited on a thick SiO_2 substrate such as glass or quartz. This technology, which allows the fabrication of millions of TFT's on a large, transparent glass or quartz substrate, has resulted in some entirely novel products. The most eye-catching are today's high-performance colour liquid crystal displays (LCD), which rival the quality of cathode ray tubes at only a fraction of their weight.

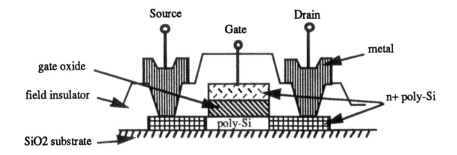

Figure 11.1 Crossectional view of a poly-Si Thin Film Transistor

Due to the non-crystalline nature of the glass or quartz substrate, the channel material is fabricated in the amorphous or polycrystalline state (in Figure 11.1: polycrystalline silicon). Despite their crystallographic imperfection, electronic properties of both amorphous Si (a-Si) and polycrystalline Si (poly-Si) can be made sufficiently good for the use as TFT's. There are however major differences in technology and device performance. TFT's made in a-Si:H (a-Si with a large H-content) can be processed at very low temperatures (< 400 °C), but have a limited electron mobility (typically less than 1 $cm^2/V.s$). Poly-Si TFT's need higher fabrication temperatures (at least 450 °C), but offer a mobility exceeding 10 or even 100 $cm^2/V.s$. The difference in mobility reflects the higher degree of lattice order found in poly-Si. At present, the a-Si:H TFT technology, which is more mature, is widely being used for TFT applications. It is expected however that the poly-Si TFT, with its inherently larger mobility, may become the TFT of choice in many applications.

In this review, we will first detail the field of application of poly-Si TFT's, and the requirements in different display applications (section 2). In section 3, the technology of the channel thin film polysilicon fabrication and its relation with polysilicon properties is discussed. Section 4 deals with the technology and performance of standard coplanar as well as advanced poly-Si TFT processes.

11.2   APPLICATIONS OF POLYSILICON THIN FILM TRANSISTORS

11.2.1   Principles of TFT-LCD displays

The majority of TFT use today is for colour-matrix Liquid Crystal Displays (LCD). For a detailed review of the operating principles of LCD's, we refer to literature (Kaneko 87). In Figure 11.2, we show a cross-sectional view of one picture element of a colour matrix LCD. The LCD consists of a sandwich of two glass substrates with a Liquid Crystal (LC) layer of the Twisted Nematic type in between. The colour function is achieved by using three pixels and a RGB (red-green-blue) colour filter assembly for each picture element. Two polarisers, which are turned 90 ⁰ with respect to one another, are put on the top and bottom of the display. Pixel electrodes on the lower plate and a common electrode on the upper plate control the voltage over the liquid crystal in between. The state of the liquid crystal, which is dependant on the voltage over it, determines to what extent the pixel is transparent to light, in the following way. At the innermost surfaces, 2 alignment layers, also turned 90⁰ with respect to one another, orient the TN Liquid Crystal such that it is twisted over 90⁰. Hence, when no voltage is applied over the LC, light passing the lower polariser 'follows' the 90⁰ twist of the LC and can pass through the upper polariser: the pixel is transparent for light

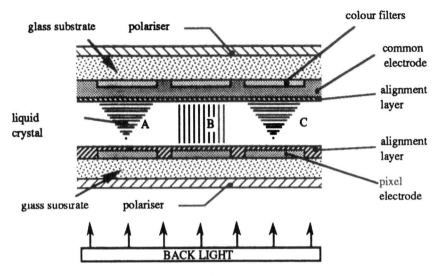

Figure 11.2 Crossectional view of one picture element of a color-matrix LCD. Pixels A and C are shown in the transparant state, pixel B in the non-transparent state.

and hence white (pixel A and C in Figure 11.2). When a sufficiently high voltage is applied over the LC however, the LC unwinds and aligns itself perpendicular to the plates. Therefore, light passing the lower polariser passes through the LC unchanged and cannot pass the upper polariser: the pixel is non-transparent and hence dark (pixel B in Figure 11.2). Applying an intermediate voltage will result in a grey pixel. In conclusion, the state of transparency of any individual pixel can be controlled by the voltage at the pixel electrode. By controlling the voltage at all pixel electrodes of a large matrix of picture elements as that in Figure 11.2 individually, 2-dimensional colour pictures can be displayed.

In order to display video pictures on the LCD, we have to control the voltage on each pixel at a certain repetition rate (e.g. 60 fields/sec). During one such period (a frame time), the lines of the matrix are addressed one after the other, by means of a scanning bus driver (Figure 11.3). During the time that line I is addressed by the scanning circuit (the line time), the TFT's at line I are in the conducting state, and charge is transferred from the data bus into the capacitors formed over the LC by the pixel electrodes and common electrode. The rest of the frame time, the TFT's of line I are non-conducting, and ideally, the charge on the pixels of line I is kept. In practice, some part of the stored charge will leak away through the TFT and/or the liquid crystal. Many displays therefore incorporate a storage capacitor in parallel with the liquid crystal capacitor, in order to reduce the voltage drop over the LC due to leakage, and to reduce crosstalk caused by parasitic capacitance.

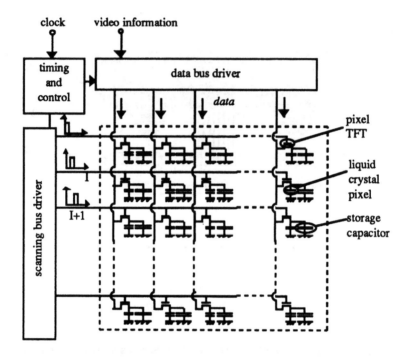

Figure 11.3 Schematic overview of a transistor-adressed LCD consisting of a matrix of pixels, scanning bus drivers and data bus drivers. The voltage-time waveform is shown for each scanning line.

11.2.2   Requirements for driver and pixel thin film transistors

From the above, the requirements for the TFT present at each pixel become clear. In the on-state, its impedance should be sufficiently low to allow charging of the liquid crystal pixel capacitor within the line time. In the off-state, the voltage drop over the LC due to leakage through the TFT should be small. This results in a certain on/off current ratio requirement. This requirement becomes more stringent with increasing number of lines and grey scale levels in the display. The absolute value of the current is less stringent since the designer can (to some extent) shift the levels of required On and Off-current upward by increasing the storage capacitor value, while the On/Off ratio requirement remains essentially unchanged (Firester 90).

As it is impractical to make long interconnects to the several hundreds of line and column electrodes on a display, the driving circuit (scanning bus and data bus drivers) must be located on the display or in its vicinity. This can be implemented in two ways. A common approach is to use LSI chips which are mounted by Chip On Glass technique or Chip On Board technique. With increasing display density however, this approach becomes less attractive due to the cost of the LSI circuits. Also, for small size, high density displays in particular, where the electrode pitch gets very small (<100 µm), the interconnection becomes difficult from a technological point of view. An obvious alternative is to implement the driving circuit (or at least the most slow part of it, namely the scanning bus circuit), using the TFT technology already used for the pixel TFT's. Whether this is possible or not depends on the max. operating frequency of circuit made in TFT's. Due to their low mobility, a-Si TFT's are generally too slow for this purpose. In poly-Si, the maximum operating frequency is of course still lower than that of single-crystal Si transistors, but 25 MHz operation has been reported (Emoto 90). The speed of poly-Si TFT's can be sufficient for the implementation of the scanning bus drivers of a display for High Definition Television (Yokozawa et al 91). For displays of a lower density, a fully integrated driving circuitry (scanning bus and data bus drivers) has been achieved (Mimura et al 1987, Faughnan et al 1988).

11.2.3 Direct view and projection displays.

From the users point of view, one can distinguish two groups of LCD displays: direct view displays and projection type displays. In a direct view display, the picture size equals the display size, and the user is looking directly to the display as presented in Figure 11.2. Colour LCD displays, typically 10" diag. in size, as used in laptop computers, are examples of direct view displays. In a projection-type display however, the picture formed on the LCD display is magnified by an optical system and projected on a screen (Figure 11.4). The picture the user sees is many times larger than the LCD display, which is kept as small as technologically possible due to economical considerations. The projection-type display in Figure 11.4 is for use in a High-Definition Television (HDTV) set, where a 55" picture is made from a 4.55" LCD panel.

While there are no fundamental differences in the mode of operation of both types of displays, the difference in display size affects the choice of the TFT technology (Takeda et al 1990). In the case of a 10" direct view display for example, the relatively large substrate dictates the use of cheap substrates. This rules out the use of quartz as a substrate. Since the cheapest glass substrates are soda-lime glasses which should not be processed above 400 °C, it is preferable to use an a-Si:H TFT technology. In a projection display however, the substrate cost

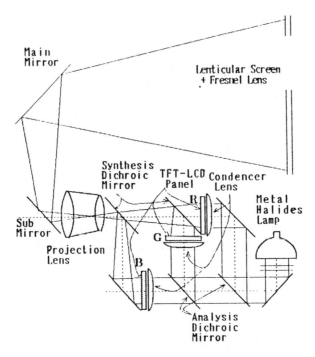

Figure 11.4 Set-up of a projection-type LCD display for HDTV application. One LCD is used for meach color. (Yokozawa et al 91, permission for reprint, courtesy Society for Information Display)

is relatively less important, and borosilicate glass substrates, which can sustain about 650 °C, or even quartz substrates which can be processed up to 1000 °C, may be acceptable. For projection displays, poly-Si TFT technology is affordable and is also preferred, in particular for displays for HDTV. Firstly, as already mentioned, the narrow electrode pitch and the large number of lines make the use of LSI driving circuit difficult and expensive. Incorporation of poly-Si TFT driving circuits on the display is very attractive in this case (Takeda and Hotta 1990). Secondly, the optical transmission of a LCD display is a critical design issue in projection displays, where a high light output is required. Therefore, novel liquid crystals such as polymer dispersed LC, which can operate without polarisers and hence result in a higher light throughput, are being investigated. At present however, polymer dispersed LC's have a higher leakage than TN LC's, necessitating the use of a large storage capacitor. This increases the On-current requirement of the TFT to such an extent that the low mobility a-Si:H TFT's cannot be applied (Kunigata et al 90).

11.3   TECHNOLOGY OF POLYSILICON THIN FILMS

11.3.1   Requirements for polysilicon thin films

The properties of the poly-Si TFT are determined by the density of states which exist in the forbidden gap due to the presence of grain boundaries and intragrain defects. These states lower the field effect mobility of carriers (Levinson et al 82) and result in anomalously high leakage currents (Greve et al 85, Fossum et al 85). For use as a pixel TFT, a high on/off current ratio is required, and for use as a driver TFT, a high on-current is necessary. Therefore, one has to minimise the density of states in poly-Si.

States are introduced in the bandgap by the disruption of the lattice periodicity at grain-boundaries. At a grain-boundary, midgap states are introduced in the bandgap due to Si dangling bonds, and tail states due to strained bonds (Khan and Pandya 90). Therefore, one will try to reduce the relative area of the grain boundaries, by using large grains. Further, the number of defects per unit area at a grain-boundary can be decreased by high temperature processing and hydrogen passivation.

In the following, we will first review the quality of poly-Si thin films made by 3 different fabrication technologies: direct deposition, solid phase crystallisation and laser crystallisation. In order to compare the defect density of different poly-Si materials, we will consider the field effect mobility of n-channel TFT's made in the poly-Si materials. Although this parameter has the drawback that it is also influenced by factors different than the poly-Si defect density (e.g. gate oxide quality), it is sufficient to observe some relations between fabrication technology and defect density. Next, we will discuss how high temperature processing and hydrogen passivation improve the poly-Si defect density.

11.3.2   Direct deposition of polysilicon thin films

Poly-Si thin films can be made by Low Pressure Chemical Vapour Deposition (LPCVD) from SiH_4 at a temp. between 575 and 650 °C. This technology is widely used for the deposition of gate polysilicon in VLSI processing (Sze 88). A cross-sectional Transmission Electron Microscopy (XTEM) photograph of a polysilicon thin film deposited by LPCVD at a pressure of 160 mTorr is shown in Figure 11.5. The poly-Si material consists of laminar crystals and has a considerable surface roughness (up to 0.1 μm). The grain-size of poly-Si deposited in this pressure range is small (30 nm at 200 mTorr) (Table 11.1). Due

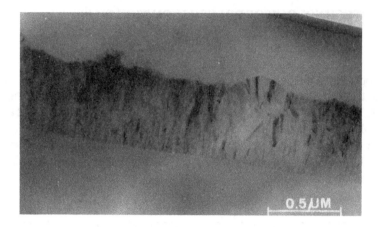

Figure 11.5 XTEM of a poly-Si thin film deposited directly by LPCVD from SiH$_4$ at 610 °C, 160 mTorr (Pattyn 92).

to the presence of these small grains as well as microtwins (Dimitriadis et al 92), TFT's made in this poly-Si material exhibit a very low mobility (less than 1 cm^2/V.s). Also, the surface roughness results in a low breakdown voltage of oxides thermally grown on the poly-Si (Mitra et al 91c). By decreasing the deposition pressure down to 40 mTorr, the grain size can be increased up to 100 nm and the twin density can be reduced, resulting in an increased mobility of about 10 cm^2/V.s (Table 11.1). Lowering the pressure even further is however not effective any more, since the TFT mobility is degraded by the increase in surface roughness with decreasing deposition pressure (Dimitriadis et al 92).

Table 11.1 also lists some results for poly-Si grown by e-beam evaporation or Plasma-enhanced CVD (PECVD). Since these approaches do not rely on the thermal decomposition of SiH$_4$ for the generation of Si precursors, they can be applied at lower deposition temperatures, which decreases the thermal budget of the TFT process. These techniques result in a mobility comparable to those by LPCVD.

We can conclude that TFT's made in directly deposited poly-Si thin films have a limited mobility. It seems that in order to improve the TFT mobility further, it is necessary to use high temperature processing and thermal oxidation to increase the grain size (see 11.3.3) . Even then, this leads to a mobility of only 28 cm^2/V.s (ref. Ohshima '89 in Table 11.1).

Table 11.1 Grain-size and electron field-effect mobility of poly-Si TFT's using different poly-Si thin film materials. LT and HT stand for glass-compatible low temperature (< 650 °C) and quartz-compatible high temperature (>900 °C) processes.

Deposition technology	Reference	Grain-size	Mobility (cm^2/V.s) LT	HT
DIRECT DEPOSITION:				
LPCVD	Meakin et al 87a 87b	30 nm	< 1	
(SiH$_4$, 630 °C, 200 mTorr)	& Dimitriadis et al 92			
LPCVD		100 nm	10	
(SiH$_4$, 630 °C, 40 mTorr))				
LPCVD (SiH$_4$, 600 °C)	Ohshima 89	150 nm		28
E-beam (520 °C)	Oana 84		15	
E-beam (550 °C)	Schmolla et al 89		10	
PECVD (500 °C)	Hirai et al 83		3	
SOLID PHASE CRYST.:				
Amorphisation of poly-si	Noguchi et al 86,	1 µm	60	
by Si ion implantation	Nakamura et al 88	up to 6 µm		166
LPCVD (SiH$_4$, 550 °C)	Hatalis and Greve 88,	400 nm	44	
	Hseih 88			
	Pattyn et al 90	550 nm	56	
LPCVD (Si$_2$H$_6$, 460 °C)	Nakazawa 91	up to 5 µm	120	
PECVD (180 °C)	Liu and Fonash 90	1 µm	60	
	Takenaka et al 90	up to 2 µm		158
E-beam (<100 °C)	Katoh 88	3.5 µm		165
LASER CRYST.:				
Pulsed Laser (ArF) of a-Si	Pattyn 92	100	141	
Scanning Laser	Serikawa et al 89	40-200	350	
(CW Ar of a-Si)				
Scanning Laser	Nishimura et al 82	several µm		320
(CW Ar of poly-Si)				

11.3.3 Solid phase crystallisation of polysilicon thin films

A solution to the problems of grain-size and surface roughness encountered in the direct deposition technique is obtained when the process of Si deposition and grain growth are separated. In this approach, a film firstly is fabricated, containing some Si nuclei but being mainly amorphous-like. When this material is annealed, the amorphous material will crystallise, and a poly-Si film is

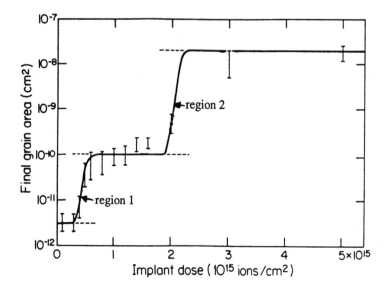

Figure 11.6a Grain-size of poly-Si made by implantation and annealing of poly-Si vs. implant dose. The solid line is a simulation based on channeling theory (Iverson and Reif 85, reproduced by permission of the Am. Inst. Phys.)

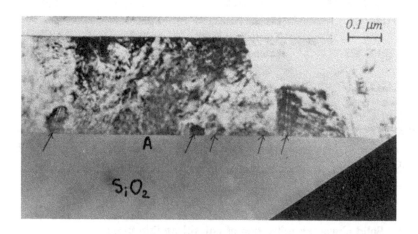

film deposited by LPCVD from SiH$_4$ at 550 °C The arrows indicate initiated grains at the substrate interface. A grain grown through the film is indicated by A. (Pattyn 92).

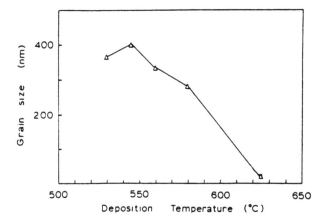

Figure 11.6c Grain size of a poly-Si thin film made by annealing of an a-Si film deposited by LPCVD from SiH$_4$, as a function of deposition temperature (Hatalis and Greve 88, reproduced by permission of the American Institute of Physics).

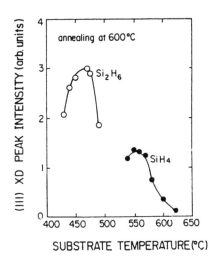

Figure 11.6d Crystallinity, as observed by X-ray diffraction, of poly-Si thin film made by annealing of an a-Si film deposited by LPCVD from SiH$_4$ and Si$_2$H$_6$ (Nakazawa 91, reproduced by permission of the American Institute of Physics)

obtained by Solid Phase Crystallisation (SPC). The grain-size of the crystallised material is determined by the initially present number of nuclei, and can be up to several microns.

One way to obtain the starting material is to subject a poly-Si film to a high dose Si ion implant (Kwizera and Reif 82, Iverson and Reif 85). The ion implant will amorphise most of the poly-Si film, except for certain crystals. The number of crystals surviving the ion self-implant, and hence the final grain-size is controlled by the ion dose (Figure 11.6a). In a first transition region (region I), grains which are not aligned along the implant direction are amorphised. In a second transition region (region II), also aligned grains are amorphised. Under optimum conditions of ion dose and crystallisation time, films with a grain size up to 5 μm have been reported (Hayashi 86).

Due to this large grain-size, a significantly higher mobility (60 cm^2/V.s for low temperature processing) has been reported by this technique as compared to the direct deposition technique (Table 11.1). A drawback of very large grains however can be that the TFT mobility properties become inhomogeneous when the grain size gets close to the device dimensions (Yamauchi et al 89). This problem is overcome by the 'sentaxy' technique (Kumoma and Yonehara 90): if the ion implant is done locally, one can put the nuclei at a defined position (e.g. the sites for the pixel TFT of a LCD), resulting in a narrow distribution of the electrical device properties.

In practice however, self-implantation of poly-Si is not a very economical way of obtaining a-Si films. A better approach is to use conventional LPCVD from SiH_4, but reduce the deposition temperature such that an amorphous Si layer instead of poly-Si is deposited (Harbeke et al 83). This technique was developed by Mimura et al 87 and by Hatalis and Greve 88. In this approach, grains originate from the substrate interface upon annealing the a-Si layer, and some dominant orientations grow wider towards the film surface, as can be seen by XTEM (Figure 11.6b). An optimum deposition temperature of about 545 °C exists, where the nucleation rate is minimum and the resulting grain-size maximum (about 400 nm) after crystallisation (Figure 11.6c). The final grain-size increases with increasing film thickness (Pattyn et al 90). Apart from the larger grain size, another important advantage of the SPC films as compared to the directly deposited ones is their smooth surface (Harbeke 83). Using poly-Si made by SPC of a-Si deposited from SiH_4, TFT's with a mobility of around 50 cm^2/V.s have been realised without high temperature processing (Table 11.1).

Even larger grains, comparable in size to those made by the amorphisation technique, are obtained when the LPCVD is done by Si_2H_6 gas. The optimum temperature for minimum nucleation rate is then about 460 °C (Figure 11.6d), and a grain-size of up to 5 μm has been reported, resulting in a mobility as high as 120 cm^2/V.s without high temperature processing (Table 11.1). Films with a grain-size over 1 μm have also been achieved by other techniques, such as e-beam deposition and PECVD (Table 11.1). However, the latter methods are not so commonly used as compared to LPCVD, because they have a lower throughput.

We can conclude that films with a smooth surface and large grain size can be obtained by the SPC technique. TFT's processed at low temperatures have demonstrated a mobility in excess of 100 cm^2/V.s. The SPC technology is a very attractive candidate for the fabrication of poly-Si TFT's for LCD displays.

11.3.4 Laser crystallisation of polysilicon thin films

An alternative to solid phase crystallisation is the use of a laser beam which heats an a-Si (or poly-Si) layer until the melting point. The molten silicon then recrystallises into poly-Si. The crystallography of that film is dependant on the starting material (a-Si or poly-Si) and the laser annealing method (pulsed versus scanning). We will review the formation and properties of the types of poly-Si material most important for display applications

<u>Laser crystallisation of a-Si and a-Si:H.</u>

The crystallographic structure of poly-Si made by pulsed laser crystallisation of LPCVD a-Si has recently been investigated in detail (Bachrach et al 90, Mathe et al 92). Above a certain threshold energy density (typically in the range 0.1 - 0.2 J/cm^2), the upper part of the film is molten and crystallised. The irradiated film has a two layer structure: a crystallised poly-Si layer on top, and non-crystallised a-Si layer below (Figure 11.7a). With increasing laser energy, the depth over which the upper part of the film is melted gradually increases, and an increasing part of the film (eventually down to the substrate interface) is crystallised into poly-Si. This is observed as an increase in grain size with laser energy density, since the poly-Si grain-size increases with the poly-Si film thickness (Figure 11.7b). Fully crystallised films have a grain size in the order of 100 nm (Table 11.1).

For LCD applications, it is attractive to use a-Si:H as a starting material instead of a-Si. In this approach, only at the edges of the display, where the TFT's for the driver circuits are to be constructed, the a-Si:H is crystallised into poly-Si, whereas in the display area the a-Si:H TFT's are used as such for the pixel TFT's (Sera 89). A more complex crystallographic structure than the above two-layer

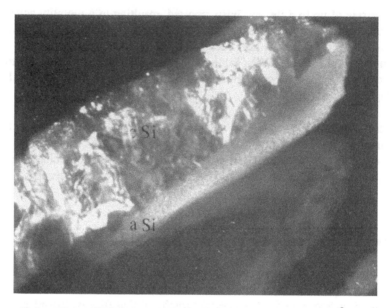

Figure 11.7a XTEM of a poly-Si thin film made by 0.135 J/cm^2 ArF pulsed laser annealing of an a-Si film deposited by LPCVD (Mathe et al 92)

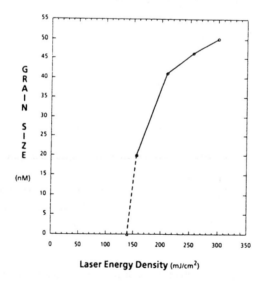

Figure 11.7b Grain-size of a poly-Si thin film made by excimer pulsed laser annealing of an a-Si film deposited by LPCVD (Bachrach et al 90, reprinted with permission from Journal of Electronic Materials, Vol. 19, No 3, p. 244, a publication of the Minerals, Metals & Materials Society, Warrendale, Pennsylvania 15086).

structure is obtained when starting from a-Si:H. In general, a three-layer structure is observed: a top layer of poly-Si (similar as in the previous case, but thinner) followed by a micro crystalline Si layer between the poly-Si layer and the remaining a-Si:H layer (Sera et al 89, Bachrach et al 90, Mathe et al 92). The origin of this structure is related to the mechanism of crystallisation from the liquid phase in a-Si and a-Si:H. The crystallisation involves a phenomenon known as explosive crystallisation. Indeed, the amorphous to crystalline phase transition is accompanied by a change in the Gibbs free energy (Geiler et al 84). Thus, latent heat is released upon crystallisation of the first amorphous layers, initially melted by the laser pulse. This latent heat results in the explosive crystallisation of the underlying amorphous layers (Thompson et al 84). It is known that the presence of H in a-Si results in an increased nucleation rate (Wagner et al 90). Therefore, the material crystallised by the explosive crystallisation will have a fine grain structure (micro crystalline) in the case of a-Si:H, whereas in the case of a-Si it will preserve the grain-structure of the melt-crystallised poly-Si layer on top.

A difficulty in the pulsed laser crystallisation of a-Si:H is the sudden release of H from the Si network upon laser heating, resulting in the presence of H_2 bubbles in the poly-Si grains and a high surface roughness (Mathe et al 92). This problem can be avoided by using multistep laser anneal sequences, which release the hydrogen in a more gradual manner (Sameshima et al 89).

A-Si(:H) can also be crystallised by scanning laser instead of pulsed laser crystallisation. Yuki crystallised a-Si:H by CW Ar-laser and obtained a grain-size of about 10-100 nm (Yuki et al 89). Serikawa reported a poly-Si material with a grain size of 40-200 nm (Serikawa et al 89). Although the crystallisation mechanism has not been investigated in detail, the grain-size suggests a mechanism similar to that in the case of pulsed laser crystallisation of a-Si(H).

Laser crystallisation of poly-Si.

A different grain structure is obtained by melting poly-Si starting material using a scanning laser (Johnson et al 82, Seki et al 87). The energy used for crystallisation (> 100 J/cm^2) is higher than that in the case of a-Si, due to the higher melting point of poly-Si. Unlike a-Si, the molten poly-Si phase is not undercooled and hence will nucleate less quickly. Also, explosive crystallisation does not occur. Therefore, the melting of the film over its complete thickness is preceding the nucleation and grain growth. The growth proceeds laterally in the direction of the scanning resulting in a poly-si material with a grain-size in the micron-range, with the grains being elongated along the scanning direction (Nishimura et al 82).

From Table 11.1, we observe that poly-Si obtained by laser crystallisation results in TFT's with a mobility in excess of 100 cm^2/V.s even with low temperature processing. Comparing this to SPC-made TFT's, we conclude that there is no unique correlation between grain-size and mobility: laser-crystallised poly-Si material with a small grain size can have a much higher mobility than SPC-made films with comparable small grain size. This can be attributed to the lower defect densities at a grain boundary obtained by laser crystallisation (Seki 87) or a lower intragrain defect density.

We can conclude that a polysilicon film of high quality can be obtained by laser crystallisation. Scanning laser crystallisation of poly-Si, while resulting in very large grains, has to be done on a quartz substrate, which is disadvantageous from the cost point of view. Pulsed or scanning laser crystallisation of a-Si or a-Si:H however is compatible with cheap glass substrates. If a buffer layer is used, the temperature of the glass does not exceed 170 °C during pulsed laser crystallisation (Sameshima 89). Additionally, it has been demonstrated that this technique allows the combination of poly-Si TFT's and a-Si:H pixel TFT's on the same glass substrate. This makes the laser crystallisation of a-Si(:H) technology very attractive for fabrication of TFT's for peripheral driver circuits. For fabrication of entire displays however, the capability of laser crystallisation technology to process large areas, and the uniformity are at present poor as compared to SPC techniques (Little 91).

11.3.5   Decrease of polysilicon thin film defect density

We explained above how the defect density in a poly-Si thin film is determined to a large extent by the deposition technique for the poly-Si film. Still, the material quality can be improved by subsequent processing, namely high-temperature processing and hydrogen passivation.

Inspection of Table 11.1 showed already that TFT devices which are made by high temperature processing generally have a higher mobility than devices whose process temperature did not exceed 700 °C. This is due to two reasons. Firstly, thermal treatment improves the defect density of poly-Si films. Although a high temperature anneal at 900 - 1100 °C in an inert ambient does not affect the grain-size, local reorganisations around the boundary allow for a reduction in the dangling bond density. A temperature treatment of 1000 °C for example reduces the Electron Spin Resonance signal (which relates to the dangling bond density) by about 3 times (Aoyama et al 90). Secondly, if the gate oxide is grown thermally, the oxidation will decrease the poly-Si defect density by several mechanisms. One effect of oxidation is an increase in grain-size (Noguchi et al

85). Another one is the passivation of defects at grain boundaries by oxygen (Mitra et al 91): oxygen can diffuse quickly along grain boundaries and form inter granular oxide (Irene et al 80). Therefore, a thermal oxidation of poly-Si thin films will result in an oxidation and hence passivation of the grain boundaries. Additionally, grain-boundary oxidation facilitates the transport of hydrogen along the grain boundaries, leading to more efficient hydrogen passivation (Moore and Ast 90).

We can conclude that high temperature processing is beneficial to device performance. The gain in TFT performance has however to be balanced with the cost increase associated with the use of quartz substrates.

The electrical activity of some defects can also be passivated by atomic hydrogen. Atomic hydrogen has a high diffusion coefficient in poly-Si, and passivates Si at relatively low temperatures, typically below 400 ºC. At least two reactions play a role. Si dangling bonds are neutralised by bonding with atomic hydrogen. This reduces the trap state density around midgap by a factor of typically 3 (Jousse et al 88, Khan et Pandya 90). It is however observed that the concentration of hydrogen found in poly-Si passivated until saturation, which is in the order of 10^{20} cm^{-3} (Johnson et al 82, Rodder et al 87) is several decades higher than the density of dangling bonds, which is typically in the order 10^{17} - 10^{18} cm^{-3}. Additional hydrogen is accommodated in the relaxation of strained bonds. It is known that a strained Si network can be structurally relaxed by hydrogen, by breaking strained Si-Si bonds and then passivating them with hydrogen (Kampas 82). This reduces the band tail state density in poly-Si (Wu et al 91, Pandya and Khan 87).

H-passivation has a strong influence on the on and off current (Mitra et al 91a). It has therefore become a common process step in both high and low temperature poly-Si TFT processes. Since temperatures in excess of 400 ºC result in exodiffusion of the bonded hydrogen and hence degradation of the TFT (Mimura et al 89), the introduction of H into the TFT is usually done at the back end of the TFT process. The atomic hydrogen can be generated by several methods.
1) The sintering of Al contacts at 400 - 450 ºC, which is standardly done for obtaining a low contact resistivity of the Al/Si contact, results in some H-passivation. In particular when the sintering is done before the patterning of the Al layer, some increase in on/off ratio has been reported (Pattyn et al 90). However, the amount of hydrogen that can be generated in this method is rather limited: about 10^{19} cm^{-3} (Pattyn 92).
2) When a PECVD Si_3N_4 layer is deposited over the device and then annealed at 400-500 ºC, H is freed from the Si_3N_4 film and diffuses into the poly-Si TFT (Pollack et al 84). Devices with good on/off ratio have been reported. However, the H-content of the passivated poly-Si is below 2.10^{19} cm^{-3} (Miyauchi et al 87).

3) Devices are inserted in a H-plasma at 300 - 350 ºC. Atomic hydrogen generated in the plasma diffuses into the poly-Si. The plasma system is usually a conventional radio-frequency (rf) plasma. Using rf-plasma's, a hydrogen concentration of over 10^{20} cm^{-3} can be reached in poly-Si thin films (Nakazawa et al 87). Rf H-plasma passivation is the technique most widely used for H-passivation today. Excellent on/off ratio's have been reported for TFT's passivated until saturation, which may however take several hours (Mitra et al 91b). Recent research has suggested that high-density plasma's such as ECR-plasma, are able to perform H-passivation in a shorter time span than the conventional rf-plasma (Baert et al 93, Hseih et al 90).

In all 3 methods, the generated atomic H has to diffuse through the TFT-structure before it reaches the poly-Si channel which has to be passivated. A recent study demonstrated that in a poly-Si gate coplanar TFT, the preferential diffusion path is a lateral diffusion through the (quartz) substrate (path C in Figure 11.8). This preferential diffusion path is due to the much higher diffusion constant of H in SiO_2 as compared to that in Si. In practice, this means that long channel devices take more time to be fully passivated than short channel ones (Mitra et al 91b, Wu et al 89). In devices with an Al gate however, the preferred diffusion path is path A, because of the high diffusion coefficient of H through Al (Hashimoto and Kino 83). Hence, the passivation time is much shorter, and independent of channel length (Moore et al 90, Hawkins 85).

11.4.   POLYSILICON TFT DEVICE STRUCTURE

In the previous section, we have reviewed how a poly-Si thin film layer with a low defect density can be made by solid phase or laser crystallisation, and how the defect density is improved by subsequent processing. The On and Off properties of the TFT are determined by the final defect density of the passivated poly-Si thin film. Using high quality poly-Si, standard self-aligned coplanar TFT's with a high on/off current ratio have been reported. However, when very low off-currents are mandatory and/or lower quality poly-Si material is used, other device structures such as offset-drain TFT and dual gate TFT are also widely being used.

11.4.1   Standard coplanar TFT process

Fabrication of the coplanar TFT shown in Figure 11.1 can be done by a self-aligned process asking 4 mask steps. The process starts with the fabrication of a

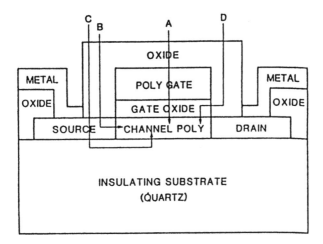

Figure 11.8 Paths for hydrogen transport to channel polysilicon (Mitra 91, reproduced by permission of the Electrochemical Society)

poly-Si thin film. It is advantageous to use relatively thin films, since thinner films result in larger on-current as well as smaller off-current (Ohshima 89). On the other hand, film thicknesses much below 1000 Å make the processing more critical. After patterning poly-Si into islands (mask 1) a gate oxide is fabricated. Thermally grown oxides are widely used for this purpose in c-Si MOSFET processing because of the excellent properties of the Si/SiO_2 interface. This is also true in the case of poly-Si, and additionally, thermal oxidation also improves the bulk poly-Si layer properties (see 11.3.5). However, the drawback of thermal oxidation is that the temperature of the process is too high for glass substrates, unless special oxidation techniques such as high pressure steam oxidation are used (Mitra et al 91c). In order to be glass compatible, low temperature deposited oxides are also being used. The most commonly used deposited oxide, Atmospheric Pressure CVD oxide, results only in mediocre device performance because of its poor interface with Si. Better results are obtained by using higher quality oxides such as TOMCATS (Pattyn et al 90), sputtered oxide or ECR-CVD oxide (Izawa et al 90). On top of the gate oxide, a gate electrode is deposited in poly-Si or metal. After patterning and etching the gate (mask 2), ion implantation is performed for Source and Drain, using the gate as a mask. During the subsequent dopant activation step, lateral diffusion of the dopants can occur (Madan and Antoniadis 85), if a high temperature process is used. This diffusion is beneficial since it decreases the peak electric field in the reversed drain junction during the off-state, and thereby the off-current (Mitra 91a). Finally, the device is sealed by deposition of a SiO_2 field insulator. Contact holes are made in the insulator (Mask 3) and metal contacts are deposited and patterned (Mask 4). H-passivation is performed on the finished device.

Table 11.2 Comparison of the device parameters of a device (a) made by high temperature process using SPC of amorphised poly-Si (W/L =32/8 μm, Vds =4V) (Stupp et al 90); (b) a laser crystallization process and (c) a low temperature SPC process (Pattyn 92).

Device	Mobility (cm^2/V.s)	Threshold Voltage (V)	Subthreshold Slope (V/dec)	Ioff (A/μm)
a	> 100	1	0.25	5E-14
b	141	1.5	0.5	1E-13
c	56	5	1.05	5E-12

Table 11.2 compares device characteristics of poly-Si TFT's made by the standard coplanar TFT process as outlined above, for three types of process sequences:
(a) SPC poly-Si from a-Si by amorphisation of poly-Si (max. processing temp. 1000 °C)
(b) pulsed laser crystallised poly-Si (max. processing temp. 630 °C)
(c) SPC poly-Si from a-Si by LPCVD from SiH$_4$ (max. processing temp. 630 °C)

The lower defect density in case a and b as compared to c, results in a higher mobility, a lower threshold voltage, a lower subthreshold slope and lower leakage. The on/off ratio of a device processed according to sequence a is more than 9 decades.

11.4..2 Low leakage TFT structures.

Reaching the above mentioned on/off current ratios in standard coplanar poly-Si TFT's, asks for a very low poly-Si defect density, which is particularly difficult to achieve when using low temperature process sequences. There has been significant progress in methods to achieve very low off-currents by other means. This is possible because the off-current in poly-Si TFT's is not only dependent on the defect density, but also on the electric field existing in the reversed n-p junction formed by the drain and channel under reverse bias conditions (Fossum et al 85, Greve et al 85). The electric field in the reversed junction results from the drain-to-source and the gate-to-drain applied voltages. The designer can therefore reduce the off-current by choosing a design to lower this electric field.

If the drain is laterally separated from the channel and the gate, a reduced off-current can be expected. This is achieved by inserting a lowly doped (or undoped) 'offset drain' region between the channel and the N$^+$ drain (Figure 11.10a). Of

(a)

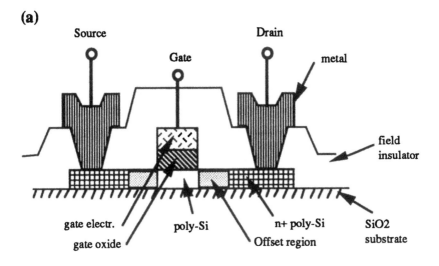

(b)

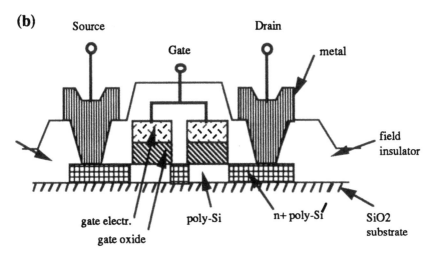

Figure 11.10 Crossectional view of TFT device structures designed for low leakage: (a) offset region TFT, (b) dual -gate TFT.

course, the associated series resistance also reduces the on-current. However, by carefully optimising the resistance of the offset region, the on/off current ratio can be significantly improved [Nakazawa et al 90]. The drawback of the technique is that the fabrication of the offset region usually requires an extra implantation step and hence an extra mask step.

An alternative method to the use of an ion-implanted offset region is the 'field induced offset region', in which a n- layer is electrostatically induced in an undoped offset region by means of an extra 'field' electrode (Huang et al 90, Tanaka et al 90). Low off-currents have been demonstrated almost without any reduction of the On-current, but the technique has several disadvantages which make it unsuited for display applications: non-self-aligned structure, extra process and mask steps, large gate-drain overlap, extra area and an extra voltage supply line.

Another way to reduce the electrical field in the drain is to make use of the dual (triple...)-gate TFT (Figure 11.10b). The device is fabricated by splitting the gate pattern of a standard coplanar device into two parts. The dual gate TFT operates essentially as a series connection of two standard coplanar TFT's whose gates are driven by the same voltage. It is clear that under this configuration, the electrical fields in both of the drain junctions will be reduced as compared to the standard coplanar TFT (Proano 89). By consequence, the off-current is reduced, again at some lesser expense of the on-current. From a fabrication point of view, the dual gate TFT has all the advantages of the standard coplanar process. The only drawback is an increased area of the transistor, but this does not need to be a critical issue if one can implement the transistor around the gate bus line (Matsueda et al 90).

11.5   SUMMARY

Poly-Si TFT's, which can be fabricated on glass or quartz substrates, are an important candidateto be used in today's TFT-LCD's. Poly-Si TFT technology is particularly attractive for small pitch, high density displays, such as those required for HDTV projection television. This is because the poly-Si TFT's mobility can be sufficiently high to allow fabrication of the peripheral driving circuits using poly-Si TFT technology already present for the pixels, thereby allowing a reduction in display cost.

Due to the many defects present in the fine grain poly-Si deposited directly by LPCVD, poly-Si TFT's made in this material exhibit a lower mobility and higher leakage current than their single-crystalline counterparts. Since a high on-current and a low off-current are mandatory for TFT's to be used in high density displays, the fabrication of lower defect density poly-Si thin film is a key issue in the fabrication of poly-Si TFT's. Poly-Si thin films with a larger grain-size and lower defect density can be grown by solid phase crystallisation of a-Si deposited by LPCVD from Si_2H_6. Alternatively, laser crystallisation techniques also result in low defect density poly-Si thin films, but the technique is more difficult. The defect density of the poly-Si thin film is further reduced by hydrogen passivation. Using solid phase crystallisation or laser crystallisation, poly-Si TFT's with a high mobility (> 100 cm^2/V.s) have been fabricated at temperatures compatible with borosilicate glass. TFT's processed at higher temperatures generally offer even better performance, but requires the use of expensive quartz substrates.

Using very low defect density poly-Si thin film material, standard coplanar TFT's can reach on/off ratios up to 10 decades. High on/off ratios can also be obtained by using specific device configurations such as the dual-gate TFT and the offset-drain TFT, which do not rely so much on the poly-Si thin film quality.

11.6 REFERENCES

Aoyama T, Mochizuki Y, Kawachi G, Oikawa S and Miyata K 1990 *Jap. J. of Appl. Phys.* **30** 84

Bachrach R, Winer K, Boyce J, Ready S, Johnson R and Anderson G 1990 *J. of Electron. Mat.* **19** 241

Baert K, Murai H, Kobayashi K, Namizaki H and Nunoshita M 1993 *Jpn. J. Appl. Phys. Part 1* **32** 2601

Colinge J 1991 *Silicon on Insulator Technology: Materials to VLSI* (Boston, Kluwer)

Dimitriadis C, Coxon P, Dozsa L, Papadimitriou L and Economou N 1992 *IEEE Trans. Electr. Dev.* **39** 598

Emoto F., Senda K., Fujii E., Nakamura A., Yamamoto A., Uemoto Y. and Kano G. 1990 *IEEE Trans. Electr. Dev.* **37** 1462

Faughnan B, Stewart R, Ipri A, Stewart W, Jose D, Plus D, Kaganowicz G, Furst D, Valochovic J, Pancholy D, Medwin M, Fischer J, Glock T, Hasili J, Cuomo F and Weisbrod S 1988 *Proc. of the Soc. Information Display Vol 29/4* p 279

Firester A 1990 *Soc. Information Display 1990 Intern. Symp.* (Playa del Rey: Soc. Information Display) p. M-8.7

Fossum J, Ortiz-Conde A, Shichijo H and Banerjee S 1985 *IEEE Trans. Electr. Dev.* **32** 1878

Geiler H, Glaser E, Gotz G and Wagner M 1986 *J. Appl. Phys.* **59** 3091

Greve D, Potyraj P and Guzman A 1985 *Solid-State Electronics* **28** 1255

Grove A 1967 *Physics and Technology of Semiconductor Devices* (New York: J.Wiley) pp. 317-333

Harbeke G, Krausbauser L, Steigmeier E, Widmer A, Kappert H and Neugebauer G 1983 *Appl. Phys. Lett.* **43** 249

Hashimoto E and Kino T, *J. Phys. F: Met. Phys.* **13** 1157

Hatalis M and Greve D 1988 J. *Appl. Phys.* **63** 2260

Hawkins W 1985 *Mat. Res. Soc. Symp. Proc.* **49**, pp 443-448

Hayashi H, Noguchi T, Ohshima T, Negishi M and Hayashi Y 1986 *Ext. Abstr. of the 18th Conf. on Solid State Devices and Materials, Tokyo, Aug. 20-22*, pp 549-552

Hirai Y, Osada Y, Komatsu T, Omata S, Aihara K and Nakagiri T 1983 *Appl. Phys. Lett.* **42** 701

Hseih B, Hatalis M and Greve D 1988 *IEEE Trans. Electr. Dev.* **35** 1842

Hseih B, Hawkins G and Ashok S 1990 *Mat. Res. Soc. Symp. Proc.* **182** 369

Huang T, Wu I, Lewis A, Chiang A and Bruce R 1990 *IEEE Electr. Dev. Lett.* **11** 244

Irene E, Tierney E and Dong W 1980 *J. Electrochem. Soc.* 705

Iverson R and Reif R 1985 *J. Appl. Phys.* **57** 5169

Izawa H, Nishi Y, Okamoto M, Morimoto H and Ishii M 1990 *Ext. Abstr. of the 22nd Conf. on Solid State Devices and Materials, Sendai, Aug. 22-24*, pp 183-186

Johnson N, Biegelsen D and Moyer M 1982 *Appl. Phys. Lett.* **40** 882

Johnson N, Biegelsen D, Tuan H, Moyer M and Fennell L 1982 *IEEE Electr. Dev. Lett.* **3** 369-372

Jousse D, Delage S, Iyer S and Crowder M 1988 *Mat. Res. Soc. Symp. Proc.* **106** pp 359-364

Kaneko E 1987 *Liquid Crystal TV Displays: Principles and Applications of Liquid Crystal Displays* (Tokyo: KTK Scientific Publishers) pp. 211 - 278

Khan B and Pandya R 1990 *IEEE Trans. Electr. Dev.* **37** 1727

Kumoma H and Yonehara T 1990 *Ext Abstr. of the 22nd Conf. on Solid State Devices and Materials, Sendai, Aug. 22-24*, pp 1159-1160

Kunigata M, Hirai Y, Ooi Y, Niiyama S, Asakawa T, Masumo K, Kumai H, Yuki M and Gunjima T 1990 *Soc. Inf. Display Digest* pp 227-230

Kwizera P and Reif R 1982 *Appl. Phys. Lett.* **41** 379

Levinson J, Sheperd F, Scanlon P, Westwood W, Este G and Rider M 1982 *J. Appl. Phys.* **53** 1193

Little T, Koike H, Takahara K, Nakazawa T and Ohshima H 1991 *Conf. Rec. of the 1991 Int. Displ. Res. Conf., San Diego, Oct 15-17* (Playa del Rey: Soc. Information Display) pp 219-222

Liu G and Fonash S 1990 *Ext. Abstr. of the 22nd Conf. on Solid State Devices and Materials, Sendai, Aug. 22-24* pp963-965

Madan S and Antoniadis D 1985 *IEEE Trans. Electr. Dev.* **33** 1519-1528

Mathe E, Naudon A, Elliq M, Fogarassy E and de Unamuno S. 1992 *Appl. Surf. Sc.* **54** 392

Matsueda Y, Ashizawa M, Aruga S, Ohshima H and Morozumi S 1990 *Soc. Inf. Display Digest* pp 315-318

Meakin D, Coxon P, Migliorato P, Stoemenos J and Economou 1987b *Appl. Phys. Lett.* **50** 1894

Meakin D, Stoemenos J and Economou N 1987a *J. Appl. Phys.* **61** 5031

Migliorato P and Meakin D 1987 *Appl. Surf. Sc.* **30** 353

Mimura A, Konishi N, Ono K, Ohwada J, Hosokawa Y, Ono Y A, Suzuki T, Miyata K, and Kawakami H 1987 *Proc. of the Int. Electr. Dev. Meeting, Washington D.C. December 6-9* 436

Mimura A, Konishi N, Ono K, Ohwada J, Hosokawa Y, Ono Y, Suzuki T, Miyata K and Kawakami H 1987 *Proc. of the Int. Electr. Dev. Meeting, Washington D.C. Dec. 6-9* pp 436-439

Mimura A, Konishi N, Ono K, Ohwada J, Hosokawa Y, Ono Y, Suzuki T, Miyata K and Kawakami H 1989 *IEEE Trans. Electr. Dev.* 36 351

Mitra U, Khan B, Venkatesan M and Stupp E 1991a *Conf. Rec. of the 1991 Int. Display Research Conf, San Diego, October 15-17* pp 207-210

Mitra U, Rossi B and Khan B 1991b *J. Electrochem. Soc.* 138 3420

Mitra U, Chen J, Khan B and Stupp E 1991c *IEEE Electr. Dev. Lett.* 12 390

Miyauchi M, Setsune K, Hirao T, Kagawa K and Yasui J 1987 *Ooyoo Butsuri* 5 1371 (in Japanese)

Moore C and Ast D 1990 *Mat. Res. Soc. Symp. Proc.* 182 pp 341-346

Morin F, Coissard P, Morel M, Ligeon E and Bontemps A 1982 *J. Appl. Phys.* 53 2879

Morozumi S 1989 *Soc. Information Display 1990 Intern. Symp.* (Playa del Rey: Soc. Information Display) pp 148-151

Nakamura A, Emoto F,, Fujii E, Uemoto Y, Yamamoto A, Senda K and Kano G 1988 *Proc. Int. Conf. on Solid State Devices and Materials, Tokyo, Aug. 24-26,* pp 189-192

Nakazawa K 1991 *J. Appl. Phys.* 69 1703

Nakazawa K, Arai H and Kohda S 1987 *Appl. Phys. Lett.* 51 1623

Nakazawa K, Tanaka K, Suyama S, Kato K and Kohda S 1990 *Soc. Inf. Displ.Digest* 311

Nishimura T, Akasaka Y, Nakata H, Ishizu A and Matsumoto T 1982 *Proc. of the Soc. for Information Display* 23 pp 209-214

Noguchi T, Hayashi H and Ohshima T *1985 Jap. J. of Appl. Phys.* 24 434

Oana Y 1984 *Digest Soc. Information Display Conf.* pp.312-315

Ohshima H and Morozumi S 1989a *Int. Electr. Dev. Meeting, Washington D.C., Dec 3-6* pp 157-160

Ohshima H and Morozumi S 1989b *Techn of the Int. Electr. Dev. Meeting, Washington D.C., Dec 3-6* pp 157-160

Pandya R and Khan B 1987 *J. Appl. Phys.* 62 3244

Pattyn H 1992a *Ph.D. Thesis: Polysilicon Thin Film Transistors for use in Active Matrix LCD's* (Univ. of Leuven, Belgium) ch 3

Pattyn H, Baert K, Debenest P, Heyns M, Schaekers M, Nijs J and Mertens R 1990 *Ext. Abstr. of the 22nd SSDM Conf., Sendai, Aug. 22-24,* pp 959-962

Pollack G, Richardson W, Malhi S, Bonifield T, Shichijo H, Banerjee S, Elahy M, Shah A, Womack R and Chatterjee P 1984 *IEEE Electr. Dev. Lett.* 5 468

Proano R, Misage R and Ast D 1989 *IEEE Trans. Electr. Dev.* 36 1915

Rodder M, Antoniadis D, Scholz F and Kalnitsky A 1987 *IEEE Electr. Dev. Lett.* **8** 27

Sameshima T, Hara M and Usui S 1989, *Jap. J. of Appl. Phys.* **28** 2131

Schmolla W, Diefenbach J, Blang G, Ocker B and Senske W 1989 *Solid-State Electronics 32* 391

Seki S, Kogure O and Tsuiyama B 1987 *IEEE Electr. Dev. Lett.* **8** 368

Serikawa T, Shirai S, Okamoto A and Suyama S 1989 *IEEE Trans. Electr. Dev.* **36** 1929

Stupp E, Mitra U, Carlson A, Sorkin H, Venkatesan M, Khan B, Janssen P and Stroomer M 1990 *Soc. Inf. Displ. Eurodisplay Conf,* pp 52

Sze S 1988 *VLSI Technology* (New York: McGraw-Hill) ch 6

Takeda M and Hotta S 1990 *Ext. Abstracts of the 22nd Conf. on Solid State Devices and Materials, Sendai, Aug. 22-24,* pp 1043-1046

Takenaka S, Masafumi K, Oka H and Hajime K 1990 *Ext. Abstr. of the 22nd Int. Conf. on Solid State Devices and Materials, Sendai, Aug. 22-24,* pp 955-958

Tanaka K, Nakazawa K, Suyama S and Kato K 1990 *Ext. Abstr. of the 22nd Solid State Devices and Materials Conf, Sendai, Aug. 22-24,* pp 1011-1014

Thomson M, Galvin G, Mayer J, Peercy P, Poate J, Jacobson D, Cullis A and Chew N 1984 *Phys. Rev. Lett.* **52** 2360

Wagner S, Wolf S and Gibson J 1990, *Mat. Res. Soc. Symp. Proc.* **164** 161

Wu I, Huang T, Jackson W, Lewis A, Chiang A, *IEEE Trans. on Electr. Dev.* **12** 181

Wu I, Lewis A, Huang T and Chiang A 1989 *IEEE Electr. Dev. Lett.* **10** 123

Yamanaka T, Hashimoto T, Hashimoto N, Nishida T, Shimizu A, Ishibashi K, Sakai Y, Shimogashi K and Takeda E 1988 *Techn. Dig. Int. Electr. Dev. Meeting, San Fransisco, Dec 11-14,* pp 48-51

Yamauchi N, Hajjar J and Reif R 1989 *Tech. Dig. Int. Electr. Dev. Meeting, Washington D.C., Dec. 3-6,* pp 353-356

Yokozawa M, Okamoto N, Matsumoto T, Fujimura R and Hirashima T *Conf. Rec. of the 1991 Int. Display Research Conf., San Diego, October 15-17* (Playa Del Rey: Soc. Information Display) 4

Yuki M, Masumo K and Kunigata M 1989 *IEEE Trans. Electr. Dev.* 1934

12

The use of polycrystalline silicon and its alloys in VLSI applications
M. Y. Ghannam

12.1 INTRODUCTION

Polycrystalline silicon (polysilicon) is widely used in several very large scale integrated (VLSI) circuit elements, for example as the gate material of metal-oxide-semiconductor (MOS) capacitances and transistors, as interconnect lines, in fuse and anti-fuse elements, and for filling deep isolation trenches and via plugs. Polysilicon is also used for the emitter and base contacts of bipolar transistors, for high value load resistors and diodes in static memory circuits, and in many integrated circuit processing steps. Sensors and mechanical micro-elements compatible with IC technology are also made from polysilicon. Finally, as detailed in a separate chapter of this book, thin film MOS transistors (TFT's) are now commonly fabricated in polysilicon thin film substrates deposited on an oxidized silicon surface.

Kamins (1988) treated in an elegant way different aspects of polysilicon including some of its applications. In the present chapter the structural, electrical and technological properties of polysilicon deposited by low pressure chemical vapor deposition (LPCVD) are briefly reviewed. Different applications of polysilicon in VLSI technology are discussed and updated. Finally two sections are devoted to semi-insulating polysilicon (SIPOS) and to polycrystalline silicon/germanium alloys.

12.2 DEPOSITION AND STRUCTURAL PROPERTIES OF POLYSILICON

Polysilicon is composed of small crystallites (grains) separated by grain boundaries as depicted in Fig.1. Inside the grains the silicon follows a periodic crystalline orientation while the grain boudaries are highly damaged regions composed of disordered and partially bonded silicon atoms.

Polysilicon can be deposited by sputtering, by vacuum evaporation, by Plasma Enhanced Chemical Vapor Deposition (PECVD), by laser-assisted CVD, by molecular beam deposition in an ultrahigh vacuum (UHV). The cost of polysilicon deposition using these techniques is, however, too high and the throughput is too small. Atmospheric pressure chemical vapor deposition (APCVD) of polysilicon films at high temperatures (up to 1200°C) offers a relatively high throughput and a conformal step coverage but with a modest film uniformity.

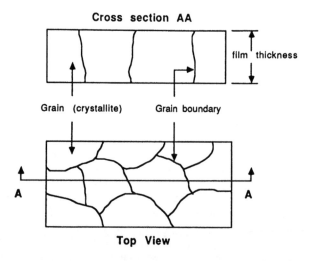

Figure 12.1 A Schematic illustration of the structure of polycrystalline silicon showing the crystallites and the grain boundaries.

Low pressure chemical vapor deposition (LPCVD) is most commonly used in planar microelectronics technology to deposit uniform polycrystalline silicon films on stand-up closely-packed wafers. Polysilicon is deposited by LPCVD as a result of the thermal decomposition (pyrolysis) of silane

$$SiH_4 \longrightarrow Si + 2H_2$$

at a relatively low pressure (0.1 to 1 Torr) in the temperature range 580-630°C. Besides its high throughput, LPCVD offers film thickness uniformity of 1 to 3% within the wafer compared to 4 to 8% for APCVD and 2 to 5% from wafer to wafer compared to 6 to 12% for APCVD (Rosler 1977). Also a very good reproducibility is insured with the LPCVD technique. The deposited layer is columnar and has a rough surface. The grain size is usually limited by the film thickness typically in the order of 0.1-0.5 μm. The grain growth is governed by silicon self diffusion, is affected by the deposition conditions, and is strongly enhanced by heavy P doping and annealing at high temperatures (Wada and Nishimatsu 1978). Although the grains are usually considered crystalline, they might demonstrate significant lattice distortion (+/- 3%) and a high density of microtwinning.

Large grain polysilicon can be obtained by solid phase recrystallization of amorphous silicon deposited at a temperature below 580°C by LPCVD or of polysilicon amorphized by means heavy Si^+ ion implantation. The final average grain size of the recrystallized film is influenced by the annealing (recrystallization) temperature, deposition temperature, deposition rate and film thickness (Hatalis and Greve 1988). Large grain polysilicon is commonly obtained today by annealing the amorphous silicon films for 24 hours at a relatively low temperature, typically 600°C.

Large polysilicon grains can also be achieved by recrystallization in the liquid phase by means of laser-induced melting and laser-assisted zone melting. The laser-assisted zone melting method (Mertens *et al* 1988) is commonly used for the recrystallization of relatively thick films (10 μm). Selective laser annealing (Colinge *et al* 1982), in which the grain boundaries in the polysilicon film are controlled by a conventional lithographic step, has been successfully used to grow very large nearly rectangular (20x400 μm) grains on a 500 nm polysilicon film deposited on SiO_2. Such films are very useful for TFT and SOI applications.

Finally, selective deposition of polysilicon has been achieved based on the technology of selective epitaxial growth by adjusting the deposition conditions and choosing the proper gas system (Furumura 1986). Using this method P doped polysilicon films have been deposited in a well controlled way (Oshita 1991), whereas As doped polysilicon films have been deposited selectively using the rapid thermal processing chemical vapor deposition which combines RTP and CVD techniques Hsieh (1989). Selective deposition of polysilicon is of great interest for many applications involving self-aligned structures.

12.3 TECHNOLOGICAL PROPERTIES OF LPCVD POLY-SILICON FILMS

12.3.1 Polysilicon doping

LPCVD polysilicon films used in VLSI are doped by diffusion or by ion implantation. For the latter, conventional furnace or rapid thermal annealing should follow the implantation in order to activate the implanted dopant atoms. It has been proven that most dopants diffuse and redistribute much more rapidly in polysilicon than in crystalline silicon (Kamins *et al* 1972, Horiuchi and Blanchard 1975, Tsukamoto et al 1977, Ryssel *et al* 1981, Swaminathan *et al* 1982, Ghannam *et al* 1987a, and others). The dopant diffusivity in the grain boundaries is enhanced by the high defect density occuring in these regions. The effective diffusivity in polysilicon is thermally activated and combines intra-grain diffusivity (bulk silicon) and grain boundary diffusivity. The activation energy associated with the diffusivity is lower for most dopant atoms in polysilicon than in monocrystalline silicon. The enhancement of the dopant diffusivity due to heavy doping is less pronounced in polysilicon than monocrystalline silicon due to the domination of the grain boundary diffusion as well as to the dopant segregation and clustering at the grain boundaries. It is worth mentioning that the dopant diffusivity may be influenced by the condition of the silicon surface prior to the polysilicon deposition. It has been reported that the diffusivity of As is higher in LPCVD polysilicon films deposited on natural silicon dioxide than in films deposited on freshly etched silicon surfaces (Ryssel et al 1981).

In-situ doping is sometimes used to reduce the number of processing steps and to achieve very high doping concentrations. In this case, however, the polysilicon deposition rate and the homogeneity of the film are strongly affected by the highly reactive chemical species included in the gas doping source, mainly phosphine PH_3, arsine AsH_3 and diborane B_2H_6. *In-situ* phosphorus doping of polysilicon results in reduced deposition rates, film growth suppression and strong inhomogeneities while *in-situ* boron doping leads to an increased deposition rate. In addition, a severe pile-up of the dopant species at the interface between *in-situ* doped polysilicon films and the underlying oxide usually takes place during deposition.

12.3.2 Polysilicon as a diffusion source

It has been found that dopant atoms redistribute uniformly in LPCVD polysilicon films after a short annealing of a few seconds at a high temperature. Such doped films can be used during subsequent high temperature cycles as continuous source for dopant diffusion into an underlying substrate. By this way, extremely shallow and defect-free pn junctions can be realized and applied to the source and drain of MOS transistors, to the emitter and to the intrinsic and extrinsic base regions of npn bipolar transistors.

Epitaxial alignment of the polysilicon to the underlying substrate, however, might occur during the high temperature diffusion step and must be taken into consideration when using doped polysilicon as a source for diffusion. Using Rutherford-Back-Scattering channeling experiments, Tsaur and Hung (1980) showed that undoped polysilicon deposited by LPCVD on a monocrystalline <100> silicon substrate and subjected to a high temperature treatment regrows in the solid phase and epitaxially aligns to the underlying substrate, as depicted in the TEM photograph of Fig.2 (photo a). The alignment is initiated when the thin interfacial oxide layer breaks up forming oxide islands at the interface as shown in the High Resolution TEM microphotograph displayed in Fig.2 (photo b). The epitaxial alignment rate is found to be enhanced by heavy doping (Ghannam and Dutton 1987b, Hoyt *et al* 1987). The epitaxial alignment of boron doped polysilicon is found to be associated with an activation energy of 4.82 eV (Ghannam and Dutton 1986) which is very close to that of silicon self-diffusion. The rate of epitaxial alignment of boron doped polysilicon can be expressed as (Ghannam and Dutton 1987b)

$$R = [9.73 \times 10^{19} + 3.57 \times 10^{17} \, (p/n_i)^2] \, \exp \, (-4.82/kT) \, \text{nm/min}$$

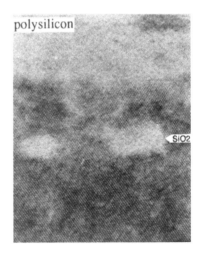

a   b

Figure 12.2   a) TEM photograph illustrating a partially recrystallized (epitaxially regrown) polysilicon film on a monocrystalline silicon substrate, and b) High Resolution TEM microphotograph illustrating the disrupted interface between the polysilicon layer and the monocrystalline substrate and the balling-up of the interfacial oxide.

where p is the hole density and n_i is the intrinsic carrier concentration at the corresponding annealing temperature. Note that this expression reflects the fact that, like for silicon self-diffusion, epitaxial alignment predominantly occurs via doubly charged point defects.

It has been reported that the interfacial oxide breaking-up and the starting of the epitaxial alignment procedure are accelerated if the annealing is carried out in an oxidizing ambient (Ghannam and Dutton 1987b) especially if the annealing temperature is below 950°C.

12.3.3 Polysilicon   oxidation

Thermal oxide grown on polysilicon (poly-oxide) is widely used in VLSI MOS and bipolar technologies as an interlevel dielectric between overlapping layers and electrodes. In some applications the function of the interlevel dielectric is to provide electrical isolation between the two conducting levels, e.g. in double poly self-aligned processes. In other applications, the function of the poly-oxide is to allow a well controlled conduction current to flow between the electrodes, e.g. for charging and discharging floating gate non-volatile memory cells (EPROM and EEPROM).

Because the surface of the polysilicon is an aggregation of small surfaces with different crystallographic orientations, the thickness of the thermal oxide grown on the polysilicon surface is non-uniform. Other factors, such as dopant segregation and higher diffusivity at the grain boundaries lead to a large difference between the oxidation rates of the grains and of the grain boundaries (Saraswat and Singh 1982). Irene et al (1980) demonstrated the presence of undulations in the poly-oxide thickness replicating the grain boundary structure and a thinner oxide over the grain boundaries. Macroscopically, under the same oxidation conditions, the thickness of the oxide grown on polysilicon lies between that of oxide grown on <100> and that grown on <111> monocrystalline silicon substrates. This is true especially for very thin oxides for which the oxidation rate is mainly limited by the reaction at the surface.

The enhancement of the oxidation rate caused by heavy P and As doping is less pronounced in polysilicon than in monocrystalline silicon. A similar enhancement is observed with heavy B doping. In addition to classical doping-enhanced oxidation, dopant segregation enhances the oxidation rate at the grain boundaries and results in a doped oxide with electrical properties markedly inferior to those of pure SiO_2. During oxidation of doped polysilicon films, a thermal equilibrium segregation of impurities is established at the poly-oxide/polysilicon interface. Atomic segregation coefficients of 0.41 (depletion) and 35 (pile-up) have been reported for boron and arsenic, respectively (Suzuki and Kataoka 1991).

Compared to oxides grown on monocrystalline silicon, poly-oxides have a higher conductivity and a lower breakdown voltage (DiMaria and Kerr 1975). This deficiency in the insulating properties of poly-oxides is not inherent to the oxide layer itself but is attributed to an asperity-induced field enhancement factor (Anderson and Kerr 1977) and to non-uniformities in the poly-oxide thickness (Irene *et al* 1980) which are both related to the roughness of the polysilicon surface at the poly-oxide/polysilicon interface. Many models have been elaborated in order to determine the field enhancement factor taking into account the variation of the asperity factor with polysilicon doping, with the ambient gas, with the oxidation temperature, as well as with the temperature of the LPCVD polysilicon deposition. Due to oxidation-induced surface roughness, more degradation in the insulating properties is observed in thick thermally grown poly-oxides (Faraone 1986). Better insulating properties are observed in poly-oxides thermally grown on polysilicon films initially deposited in the amorphous phase.

Conductive thin poly-oxide layers are used mainly in non-volatile memories to charge and discharge the floating gate. The conduction in these thin poly-oxide layers is governed by Fowler-Nordheim tunneling mechanism. The conduction current is very sensitive to the field enhancement and electron trapping phenomena, and is strongly non-uniform accross the oxide surface area (Groeseneken and Maes 1986).

12.3.4 Polysilicon etching

Dry directional etching such as plasma reactive ion etching (RIE) is usually used to etch polysilicon. Several gas systems mainly fluorine-based, chlorine-based, and bromine-based have been investigated and adopted. The choice of a specific gas system depends on the requirements imposed by the technology. For example, in a sub-half-micron VLSI technology a high etching rate, very sharp vertical edge profiles, a highly selective etching of polysilicon with respect to the underlayer especially silicon dioxide, a good photoresist integrity and a low surface damage are necessary. Very narrow lines with a width as low as 70 nm have been demonstrated (Tang and Wilkinson 1991) using reactive ion etching of polysilicon in a silicon tetrachloride ($SiCl_4$) plasma system. Several modified versions of RIE techniques have been proposed aiming at the improvement of the selectivity and directionality of polysilicon etching, such as the reactive ion beam etching (RIBE) in which the reactive ionized gases are accelerated in an ion gun, the chemically assisted ion beam etching (CAIBE) in which an inert gas plasma is used inside the ion gun. The surface bombardment by highly energetic particles is supressed in the magnetically enhanced reactive ion etching (MERIE) successfully implemented in the fabrication of sub-half-micron polysilicon gate electrodes (Nguyen *et al* 1989). Finally, the chlorine-based Electron Cyclotron Resonance (ECR) is proven to be highly anisotropic and to have a high etching selectivity for undoped and n^+ doped polysilicon with respect to silicon dioxide.

12.4 ELECTRICAL PROPERTIES OF POLYSILICON

12.4.1 Resistivity of doped polysilicon

In the low doping range, the resistivity of polysilicon is orders of magnitude larger than that of monocrystalline silicon. As the doping concentration increases, the resistivity of polysilicon decreases very fast and approaches that of mono-crystalline silicon, as illustrated in Fig.3. The larger resistivity of polysilicon in the low doping range is due to a reduced free carrier concentration inside the grains as a result of carrier trapping at the grain boundaries (Kamins *et al* 1971, Mandurah *et al* 1981). Furthermore, in order to maintain charge neutrality, a potential barrier is created at the grain boundaries, as depicted in Fig.4, leading to a decrease in the inter-grain carrier mobility and to an additional increase in the resistivity. Under these conditions thermionic conduction, a thermally activated process, is dominant. The activation energy is equal to the grain boundary barrier height and therefore is a strong function of the doping concentration and of the grain size (Kamins 1988). When increasing the doping level saturation of the traps at the grain boundaries takes place which leads to a significant increase in the free carriers inside the grains and to a decrease in the barrier height. These effects combined result in the sharp decrease in the resistivity of polysilicon at moderate doping levels depicted in Fig.3.

Other models for the conduction in polysilicon based on the carrier trapping model have been proposed and developed such as the thermoemission conduction model (Seto 1975) and the thermionic field emission conduction model (Seager and Pike 1979). Dopant segregation at the grain boundaries also leads to an increase in the resistivity of polysilicon since the segregated atoms are electrically inactive and do not provide free carriers for conduction. This effect occurs mainly in phosphorus and arsenic doped films and to a much less extent in boron doped films. Other conduction models assume that the conduction theory in amorphous semiconductors applies to the grain boundary regions.

At high doping levels the free carrier concentration is much larger than the trapped carrier density and inter-grain conduction is governed by tunneling across the grain boundary which leads to a quasi one to one correlation between the resistivities of polysilicon and of monocrystalline silicon. On the other hand, the potential fluctuation model has been proposed (Hirose at al 1979) to explain the conduction in heavily doped polysilicon. In this model besides the conduction in the extended

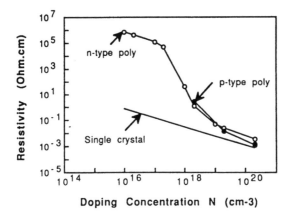

Figure 12.3 Resistivity of p-type (B doped, Ghannam 1988) and n-type (P doped, Kamins 1988) polysilicon as a function of the doping concentration.

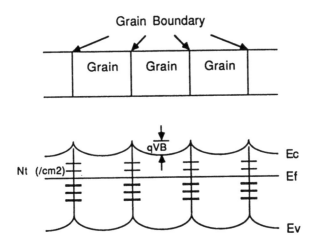

Figure 12.4 Band diagram of the polysilicon material indicating the grain boundary potential barrier V_B controlled by the charge at the interface grain boundary states.

states, conduction by carrier hopping via the localized states created as a result of the potential fluctuations at the grain boundaries takes place. According to this model the carrier conduction in polysilicon is characterized by two different activation energies as observed by Cressler *et al* (1989) for polysilicon heavily doped with arsenic.

It has been reported that the shift in the Fermi level towards the band edge upon doping the polysilicon material depends not only on the doping concentration but also on the doping species (Lifshitz 1985). This effect might have a serious impact in some applications such as on the threshold voltage of polysilicon gate MOSFET's directly affected by the metal-semiconductor workfunction difference ϕ_{ms}.

12.4.2 Minority carrier lifetime and recombination properties

The combined carrier recombination in the grains and at the grain boundaries determine the effective minority carrier lifetime in polysilicon. Within the grains, the lifetime is expected to be smaller than in monocrystalline silicon due to the presence of characteristic defects such as lattice distortion and microtwinning. Grain boundary recombination is determined by the barrier height and by the energetic distribution of the trap density, $N_t(E)$. The distribution of the grain boundary trap density determined from the activation energy of the resistivity (De Graaff *et al* 1982) indicates a minimum density at midgap, while optical absorption reveals a peak grain boundary trap density at midgap (Jackson *et al* 1983). Since a grain boundary represents an interface between two different grains, grain boundary recombination is usually treated using the concept of interface recombination velocity (Seager 1982). The grain boundary recombination rate increases with carrier depletion at the grain boundary and decreases with increasing carrier population and under illumination (Depauw *et al* 1984).

The density of traps in the grain boundary can be reduced by incorporating hydrogen atoms which saturate the silicon dangling bonds. Atomic hydrogen can be introduced in the grain boundaries by exposing the polysilicon film to a hydrogen containing plasma or by hydrogen ion implantation. Hydrogenation, however, causes a modification in the carrier trapping and in the grain boundary barrier height which may affect the resistivity of the material.

12.5 VLSI APPLICATIONS OF POLYSILICON

12.5.1 Polysilicon as a material for MOS gate electrodes

The substitution of metal by heavily doped polysilicon for the gates and for the interconnects of MOS devices allowed the development of self aligned and multilayered structures which contributed significantly in reducing the size of the devices and in increasing the integration level of silicon integrated circuits. In self-aligned MOS transistors the polysilicon gate is deposited and patterned on the thin gate oxide. The source and drain regions are formed by ion implantation with the polysilicon gate masking the implanted atoms from reaching the channel region. In this way, heavy uniform doping of the polysilicon gate and the formation of shallow source and drain junctions are carried out simultaneously. Due to the smaller workfunction difference between the polysilicon gate and the silicon surface, polysilicon gates result in smaller threshold voltages and hence offer the possibility of further down-scaling. Moreover, the workfunction of polysilicon gates can be adjusted by proper doping and consequently p-MOSFET's and n-MOSFET's with complemetary threshold voltages can be realized.

12.5.1.1 Special polysilicon gate technology

a. Polysilicon gates with enhanced hot carrier reliability

As mentioned earlier, one implantation step is necessary to dope the polysilicon gate of MOSFET's and to form shallow source and drain junctions self-aligned to the gate. During the implantation step, however, the gate oxide properties especially at the edges might be seriously degraded which causes serious reliability problems (Osburn *et al* 1982). Sidewall oxide spacers are commonly used to minimize the effect of such degradation since the implantation damage is shifted away from the gate edge which maintains the integrity of the very thin gate oxide. Moreover, sidewall oxide spacers help in reducing the gate/drain and gate/source overlap capacitances, and in reducing the electric field in the gate edge

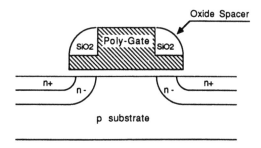

Figure 12.5 Inverted T polysilicon gate applied to an LDD MOSFET structure (Huang 1986)

spacer regions. Sidewall oxide spacer technology led to the development of the lightly doped drain (LDD) MOS structures (Tsang *et al* 1982).

The hot carrier reliability is significantly improved using the inverted T polysilicon gate displayed in Fig.5. Such a gate can be realized by partially etching polysilicon (Huang *et al* 1986) or by selective polysilicon deposition (Pfiester *et al* 1990). Finally, lower edge degradation can be obtained by doping the polysilicon film before patterning the gate (Taylor *et al* 1987).

b. p-type polysilicon gates

Standard CMOS technology uses heavily n^+ doped polysilicon gates and buried channel PMOS transistors. A surface threshold adjustment implant is necessary in order to account for the workfunction difference between the n^+ polysilicon gate and the silicon surface. Buried channel p-MOSFET's, however, seriously suffer from short channel effects, such as drain induced barrier lowering and punchthrough leakage currents because the doping concentration in the channel is low. These short channel problems are strongly alleviated in surface channel p-MOSFETs in which a p^+ polysilicon gate is used instead of the conventional n^+ polysilicon gate (Parrillo *et al* 1984). CMOS circuits with dual gate doping can be realized by combining the source/drain doping and the polysilicon gate doping in one implant of each type.

The penetration of boron from the p^+ doped polysilicon gate through the thin gate oxide during high temperature processing steps is a major problem encountered in p^+ polysilicon gate technology. Boron penetration through the oxide changes the surface active doping concentration and modifies the workfunction difference which lead to instabilities in the threshold voltage control. Moreover, boron diffusion causes generation of negative charges and of carrier traps, which also gives rise to threshold voltage shifts. Thin nitrided oxide is more appropriate than thin oxides as a gate dielectric for p^+ polysilicon gates since it acts as a barrier for boron diffusion. Nitrided oxide is, however, unable to block the boron penetration through the thin oxide in the presence of fluorine which limits their use when the p^+ polysilicon gate is doped by BF_2 implantation. In such a case thin reoxidized nitrided oxide (ONO) is a more appropriate gate dielectric to be used in combination with p^+ polysilicon gates (Cable *et al* 1991).

It is worth mentioning that a large amount of mobile sodium ions Na^+ is encountered in p^+ polysilicon gate technology and is the cause of serious reliability problems (Tanaka et al 1990).

12.5.2 Polysilicon interconnects

Heavily doped polysilicon is used for interconnecting the gates of neighbouring MOSFET's as well as other electrodes. The sheet resistance of the heavily doped polysilicon films is, however, too high (15-30 Ω/sq) to be useful in very high speed applications. Forming a polycide, a metal silicide on top of the heavily doped polysilicon line, is an efficient way to reduce the sheet resistance of polysilicon films (Murarka 1983, etc.). Refractory metal (Ti-Ta-Mo-W) silicides are compatible with standard polysilicon gate technology and result in a stable and reliable polysilicon/gate oxide interface. Several methods have been proposed to obtain an optimum polycide surface morphology. A smooth silicide/polysilicon interface is obtained by simultaneously depositing the refractory metal and silicon using e-beam evaporation or sputtering (Crowder and Zirinski 1979). High quality polycides have also been obtained by an LPCVD deposition of the silicide at 350°C, followed by a furnace or rapid anneal to allow interdiffusion between the silicide and silicon (Saraswat *et al* 1983). The LPCVD silicide deposition offers many advantages such as higher purity films, better conformal step coverage, higher throughput, and selective deposition of some silicides (such as WSi). Metal polycide formation by direct ion implantation of a high dose of transition metals followed by annealing has also been demonstrated (Kozicki and Robertson 1989).

Silicidation of the diffused source/drain regions (Scott et al 1981) is becoming a standard step in present CMOS processes. Considerable scaling-down of the devices is achieved by Self-Aligned Silicidation (Salicidation) of the gate and source/drain contacts (Lau et al 1982).

Implementing double and triple level of conducting polysilicon lines separated by dielectric layers leads to a significant reduction in die size. Heavily doped polysilicon is also used as a buried contact to overlaying interconnect structures and as via plugs to silicon in multilevel IC's (Vaidya 1986). The via plugs can be realized by depositing and etching-back a thick polysilicon films or by selective deposition of doped polysilicon.

12.5.3 Polysilicon in Memories

12.5.3.1 Dynamic Random Access Memory (DRAM)

Polysilicon is used in dynamic random access memory (DRAM) cells as the gate electrode of the access (switching) MOSFET and of the storage MOS capacitor. The major trade-off in very high density DRAM cells is to reduce the area while maintaining a sufficient storage capacity. The value of 1 Mbit has been suggested as the highest limit that can be reached in the storage capacity by reducing the thickness of the oxide of conventional gate capacitance down to minimum reliable values. Higher storage capacitances can be obtained using dielectric systems with higher dielectric constants or using three dimensional structures such as the stacked capacitor cell in which the storage capacitor is built on top of the access transistor, or the deep trench capacitor cell (Sunami *et al* 1982) represented in Fig.6. In the latter, the deep trench adjacent to the access MOSFET is separated from the bulk by a thin dielectric and is filled with heavily doped polysilicon which acts as the capacitor gate electrode. In the more advanced buried trench cell (Lu *et al* 1988), the access transistor is built in an epitaxial layer grown on top of the trench capacitor.

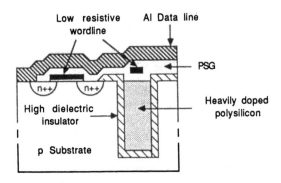

Figure 12.6 Polysilicon filled deep trench capacitor DRAM cell (Sunami 1982).

The storage capacity can be further increased by extending the areas of the capacitance electrode keeping the lateral extension constant. This can be done by intentionally creating microscopic uneven surfaces or surface asperities using various techniques. Texturing the surface of the polysilicon electrode using wet oxidation followed by wet oxide etching (Fazan and Lee 1990) results in more than 30% increase in the stored charge. Tailoring the roughness of the LPCVD polysilicon surface by changing the deposition temperature in the range 550-575°C is also used to deposit polysilicon films with a rugged surface (Yoshimaru *et al* 1990) as displayed in Fig.7. Such surfaces have an effective area up to 2.5 times larger than that of films deposited at 620°C. More than 2-fold increase in the storage area has been achieved using polysilicon electrodes with hemispherical grains deposited by LPCVD in a He-diluted silane ambient at a temperature of 550°C (Sakao *et al* 1990). Finally, surface asperities have also been intentionally created at the polysilicon gate surface by reactive ion etching (Mine *et al* 1989) or by exposing the surface to high power ultrashort excimer laser pulses (Yu and Mathews 1992).

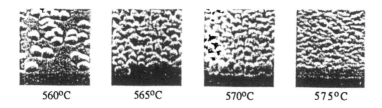

560°C   565°C   570°C   575°C

Figure 12.7 SEM photographs illustrating the difference in the surface texture of LPCVD polysilicon films deposited at different temepratures in the range 560-575°C (Yoshimaru 1990).

12.5.3.2 Static Random Access Memory (SRAM)

Lightly and moderately doped polysilicon films are used to fabricate high value resistors to be used as load elements in poly-R load static random access memory (SRAM) cells. Power dissipation in poly-R load SRAM cells can be significantly lower than in NMOS SRAM cells using enhancement load transistors and the packing density of ploy-R load cells can be considerably larger than full CMOS SRAM cells. Scaling of poly-R load cells is, however, largely bound by the trade-off existing between low standby current and high cell stability (noise margin and soft error immunity) requiring the flow of a relatively high current.

The high stability of CMOS SRAM and the high packing density of poly-R load SRAM cells are achieved in the three-dimensional stacked SRAM cells illustrated in Fig.8 (Hite *et al* 1985). In this cell a thin film p-MOSFET load is built in a polysilicon film on top of a bulk-Si n-MOSFET. With a relatively simple technology, a 4 Mb SRAM cell featuring a very low off current has been realized following this approach (Ikeda *et al* 1990). It is expected that 16 Mb or even higher density SRAM cells could be achieved when combining the vertical p-MOSFET load design with a sophisticated sub-half micron (0.25 μm) technology (Yamanaka *et al* 1990). Hydrogen passivation of the thin polysilicon film of such SRAM cells is essential since it leads to a four orders of magnitude increase in the ON/OFF current ratio of the p-MOSFET (Rodder and Aur 1991).

Bipolar cells are the fastest of all SRAM cells but are associated with a high power dissipation. Compact high speed non-saturating bipolar (ECL) SRAM cells have been developed with trench isolation and separate polysilicon levels for the interconnections and for the load resistors.

12.5.3.3 Non-volatile Electrically Erasable Programmable Read Only Memory (EEPROM)

Polysilicon has also been used as one of the key elements in non-volatile memory technologies, and more specifically the so-called floating gate memories.

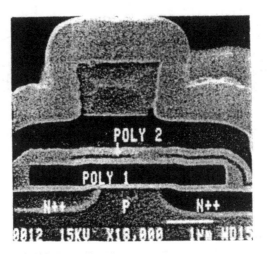

Figure 12.8 SEM photograph illustrating the stacked SRAM in which the load consists of a p-MOSFET fabricated in a polysilicon film on top of the bulk silicon n-MOSFET (Hite 1985).

As illustrated in Fig. 9, these memories are fabricated in a double polysilicon process. The first polysilicon layer is embedded completely inside dielectric layers and therefore acts as an electrically floating gate. The second polysilicon layer is stacked on top, and is acting as the external gate of the memory cell, usually referred to as the control gate.

Charge can be transported to the floating gate by Fowler-Nordheim tunneling through a thin tunneling oxide, or by channel hot electron injection. This charge induces a shift of the threshold voltage of the cell, thus providing a permanent

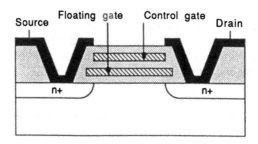

Figure 12.9 Floating gate non-volatile memory cell.

storage of information. Erasure, i.e. removal of the charge, can be performed either by UV illumination (like in UV-EPROM's), or electrically, by Fowler-Nordheim tunneling.

In the textured polysilicon floating gate cells (Klein et al 1979), one makes use of the enhanced local electric field caused by the fine surface texture of the polysilicon film and the enhanced conductivity of the overlaying poly-oxide in order to perform both program and erase operations. This is done by using the triple poly structure illustrated in Fig. 10, in which one polysilicon layer serves as the floating gate and is separated from the other polysilicon layers by a relatively thick poly-oxide. The first layer serves as the injecting layer during programming. The oxide grown on top of the second (floating) polysilicon layer is used to remove the charge from the polysilicon gate towards the upper (third) polysilicon layer in order to erase the memory cell.

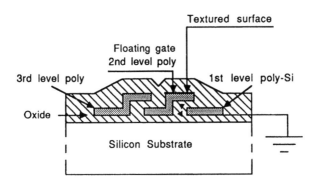

Figure 12.10 Textured polysilicon floating gate non-volatile memory cell

As mentioned earlier, poly-oxides with enhanced conductivity are used in EPROM's as an alternative to thin tunneling oxides usually associated with reliability problems. The main difficulty with this approach, however, is to grow poly-oxides with specific interface features (shape, size of the asperities) needed to insure required injection current characteristics. Furthermore, the strong non-uniformity in the distribution of the injection current leads to severe trapping problems and consequently to a limited number of write/erase cycles (Groeseneken and Maes 1986).

12.5.4 Polysilicon diodes

P/N diodes fabricated in heavily doped polysilicon show acceptable characteristics with, however, a breakdown voltage limited to the range 5-20 V. Reverse biased polysilicon diodes are used as anti-fuses in redundancy applications (Mano *et al* 1980). The resistance of these diodes changes from 10^9 Ω to 3000 Ω after being electrically destroyed. P/N junction polysilicon diodes show a forward bias current at least two orders of magnitude larger than similar monocrystalline p/n junction diodes and an ideality factor around two throughout the entire voltage range.

Polysilicon diodes are used as load elements in SRAM's. A 1 kbit ECL SRAM with polysilicon diode loads featuring an access time of 1.5 ns (Hwang *et al* 1989) has been demonstrated. In such a cell, the hold voltage is independent of the cell current and hence the operation can be maintained at very low standby current levels allowing a very small diode size, a very low power consumption and very low parasitics.

12.5.5 Polysilicon thin film resistors

Beside their use in poly-R load SRAM cells, polysilicon resistors are frequently used in analog applications. The proper design of a polysilicon resistor should take into account the sheet resistance of the polysilicon film and its temperature coefficient, the voltage nonlinearity, the uniformity and matching of nominally identical resistors with respect to position. It is worth mentioning that the resistance of the polysilicon film might undergo non-negligible changes during processing (Mandurah *et al* 1980, Ghannam and Dutton 1988) and as a result of post-fabrication treatments such as plasma hydrogenation (Ginley *et al* 1987).

During operation, polysilicon resistors may exhibit negative resistance characteristics due to thermal effects (Malhotra *et al* 1985). In addition, polysilicon resistors suffer from a large amount of low frequency 1/f noise (De Graaff and Huybers 1983) due to the electron mobility fluctuations occuring in the depletion regions near the grain boundaries. This noise is particularly significant at low doping levels and hence is a major drawback for high value resistors.

12.5.6 Polysilicon in bipolar technology

12.5.6.1 NPN bipolar transistor with polysilicon emitter contact

a. Curent gain enhancement

Polycrystalline silicon has originally been used in npn bipolar junction transistors as a diffusion source for the formation of shallow emitters. After the formation of the base region at the surface of a monocrystalline silicon wafer, a polysilicon film is deposited either *in-situ* doped (with phosphorus) as in the DOPOS process (Takagi *et al* 1972) or undoped followed by doping by ion implantation (of arsenic) as in the POLYSIL process (Graul *et al* 1976). During a subsequent high temperature step n-type dopant atoms out-diffuse from the polysilicon film into the base region to form the emitter-base junction. The emitter contact of these devices consists of a metal layer deposited on top of the n^+ polysilicon film as shown in Fig.11.

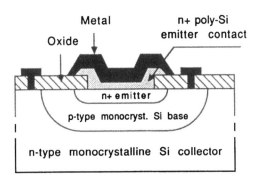

Figure 12.11 Npn bipolar transistor with polysilicon emitter contact.

The current gain of these transistors is found to be larger than that of similar transistors with a simple metal emitter contact. Three mechanisms have been proposed to explain this behavior.

First, the presence of a thin oxide layer at the polysilicon-monocrystalline silicon interface creates a potential barrier to the hole injection from the base into the emitter (De Graaff and De Groot 1979) as illustrated in the band diagram of Fig.12. The carrier transport between the emitter and the base is controlled by tunneling through the interfacial oxide layer. RCA transistors with a chemically grown interfacial oxide show a non-ideal base current (Wolstenholme *et al* 1988)

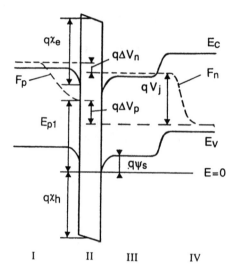

Figure 12.12   Band diagram of the polysilicon emitter contact npn bipolar transistor with an interfacial oxide (De Graaff 1979); I: n^+ polysilicon emitter, II: interfacial oxide, III: n^+ monocrystalline emitter, IV: p-type monocrystalline base.

and a much larger current gain compared to transistors in which the interfacial oxide has been removed in HF prior to the polysilicon deposition (HF transistors) (Soerowirdjo *et al* 1982), which tends to confirm the interfacial oxide theory. The current gain can be extremely large in transistors with thick interfacial oxide layers and in which the n^+ dopant is confined to the polysilicon region (Keyes and Tarr 1987). The base current of these devices, however, is severely non-ideal and may even display a negative resistance slope as depicted in Fig.13.

Second, Ning and Isaac (1980) have attributed the current gain enhancement in poly-emitter transistors to a lower hole mobility in n^+ polysilicon than in n^+ monocrystalline silicon. In this case, the slope of the hole profile at the monocrystalline emitter-base junction is reduced as illustrated in Fig.14. Consequently, the hole base current is decreased leading to an increase in the emitter efficiency and in the current gain of the transistor. This model predicts a direct relationship between the base current and the polysilicon layer thickness, a behavior that has not been systematically observed (Patton *et al* 1986). Furthermore, this model cannot explain the non-ideal base current characteristics frequently observed in polysilicon emitter transistors.

Finally, it has been proven that electrically active dopant atoms, especially As, segregate at the polysilicon-monocrystalline silicon interface (Patton *et al* 1986)

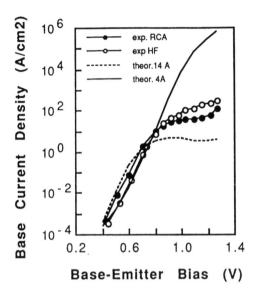

Figure 12.13 Experimental and theoretical plots of the base current in polysilicon emitter transistors as a function of the base-emitter bias (Wolstenholme 1988). Devices with a thicker interfacial oxides show a significant non-ideal behavior.

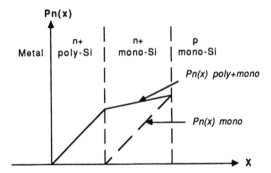

Figure 12.14 Hole distribution in the emitter of a forward biased emitter-base junction. (Ning 1980). The dashed line represents the hole profile in a metal emitter contact. The solid line represents the hole profile in a polysilicon emitter contact.

and create a potential barrier which impedes the flow of holes and consequently enhances the emitter efficiency and the current gain. According to this model, the hole injection from the base into the emitter of poly-emitter transistors is mainly controlled by thermionic emission across the interface (Crabbe *et al* 1986).

It is believed that all of the above mentioned mechanisms participate in the reduction of hole injection from the base into the emitter. For device engineering purposes, unified theoretical models which group several features specific to the polysilicon emitter into a lumped effective surface recombination velocity at the polysilicon-monocrystalline silicon interface have been proposed by Eltoukhy and Roulston (1982) and by Yu *et al* (1984).

b. Correlation between electrical parameters and interface properties

The High-Resolution TEM microphotograph displayed in Fig.2 (b) shows that the interfacial oxide, native or intentionally grown, is disrupted during annealing at high temperatures, giving rise to small oxide inclusions. This phenomenon is accelerated at high doping levels and during high temperature treatments. As discussed in section 3.2, the break-up of the interfacial oxide initiates the epitaxial alignment of the polysilicon film to the underlying monocrystalline silicon substrate. When the entire polysilicon film is epitaxially aligned, the emitter structure turns into a wide metal contacted monocrystalline emitter which results in a significant increase in the base current (Wolstenholme *et al* 1987).

Accurate predictions of the performance of polysilicon emitter transistors subjected to high temperature treatments during processing would be very valuable but are not straightforward. A theoretical model based on the solution of the transport equation including the kinetics of oxide break-up correlated to the processing conditions has been recently proposed (Ajuria *et al* 1992). The values of the base current and of the emitter resistance estimated using this model are in good agreement with the experimental results.

c. Trade-off between emitter resistance and current gain

Although RCA polysilicon emitter transistors feature a very low base current, they are associated with a considerably high emitter resistance (Stork *et al* 1985) which rules out their application in scaled-down VLSI high speed circuits. The HF transistors exhibit a lower specific emitter resistance at the expense of a slightly higher base current. Since the emitter resistance in polysilicon emitter transistors is essentially determined by the interface, breaking-up the native or intentionally grown interfacial oxide layer is necessary in order to achieve a reasonably low emitter resistance. Devices in which the interfacial oxide has been deliberately broken up during a pre-anneal prior to arsenic implantation showed an emitter resistance in the 10^{-7} $\Omega.cm^2$ range without resorting to very high implant doses or to very high emitter drive-in temperatures (Wolstenholme *et al* 1987). Such a process, however, is uncontrollable and leads to serious yield and reproducibility problems. Due to these practical limitations, polysilicon contacts are essentially used in today's VLSI bipolar technology as a tool for achieving advanced self-aligned bipolar processes rather than for boosting the current gain of the transistors.

12.5.6.2 Self-aligned bipolar process using polysilicon emitter and polysilicon base contacts

In self-aligned bipolar structures, polysilicon is used not only to contact the emitter but also to contact the base. The basic self-aligned double-poly process is displayed in Fig. 15 (Ning *et al* 1981). Boron doped polysilicon is used for the base contacts and is separated from the n⁺ polysilicon emitter contact by vertical oxide sidewall spacers. The spacing between the base and emitter contacts is determined by the thickness of the oxide spacer (usually less than 0.5 μm) which leads to a highly reduced base resistance. ECL gates with reduced RC time constants and propagation delays as short as 20 ps have been realized using state-of-the-art double-poly self-aligned bipolar transistors with deep sub-micron emitter

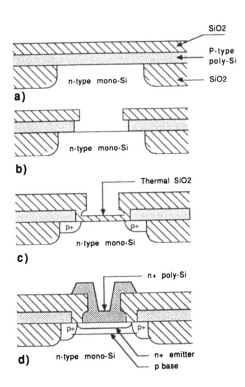

Figure 12.15 Self-aligned npn bipolar process using a double-poly technology (Ning 1981).

openings and with a spacer width as small as 0.1 μm. Single-poly processes which rely on the difference between the oxidation rates of heavily doped polysilicon and of the monocrystalline silicon substrate have also been successfully implemented (Cuthberson and Ashburn 1985).

Due to the oblique edges of the polysilicon emitter film and to the anisotropic dopant diffusivity in polysilicon, the perimeter of the monocrystalline emitter region might be depleted from dopant atoms with respect to the central emitter region (Kamins *et al* 1989). Implantation shadowing also leads to a non-uniform doping profile across the emitter region. On the other hand, the polysilicon sidewalls of deep-submicron emitters merge together as shown in Fig.16 to form a thick polysilicon emitter plug (Burghartz *et al* 1990). This effect causes the monocrystalline emitter to be shallower and more lightly doped than predicted and results in a reduced collector current and in a non ideal base current. These problems can be eliminated by *in-situ* doping the polysilicon emitter during deposition as developed and applied successfully by Burghartz *et al* (1991) and Cressler *et al* (1992).

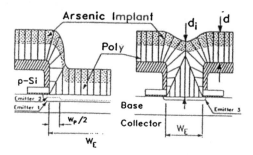

Figure 12.16 A Schematic illustration of the emitter plug effect in narrow emitter transistors (Burghartz 1990).

Polysilicon emitters have also been applied to heterojunction bipolar transistors with SiGe epitaxial base. Self-aligned phosphorus doped polysilicon emitter HBTs with a cut-off frequency of 73 GHz have been reported recently (Crabbe *et al* 1992).

12.5.6.3 PNP bipolar transistors with polysilicon contacts

High-speed low-power digital bipolar circuits require a complementary bipolar technology with high-performance PNP and NPN bipolar transistors. The challenge facing this technology is to realize a high-performance PNP transistor without degrading the performance of the NPN transistor. Vertical PNP polysilicon emitter bipolar transistors with a reasonably high current gain and a fairly low emitter resistance have been realized using *in-situ* boron doped polysilicon deposited on a clean silicon surface with no intentional interfacial oxide (Maritan and Tarr 1989). High speed PNP transistors with 0.5 µm emitters featuring a 27 GHz cut-off frequency have been fabricated in a double-poly self-aligned process with an As implanted intrinsic base, a boron doped polysilicon emitter and phosphorus-doped polysilicon base contacts (Warnock *et al* 1989). In these devices, the minimum base width is determined by the As implantation channeling tails. These tails can be eliminated by amorphizing the base by means of a Ge implant prior to the As implantation which allows thinner base regions to be achieved. Using this technology, PNP ECL gates with a propagation gate delays as short as 35 ps (Warnock *et al* 1990) have been successfully realized which confirms the feasibility of a high performance complementary bipolar technology.

12.5.7 Polysilicon in BiCMOS processes

As described above, polysilicon is used in bipolar and MOS technologies for a wide variety of purposes. In advanced BiCMOS technology, polysilicon is used for filling-in deep isolation trenches, for contacting the emitter and the base, for the gates of the CMOS devices as well as for realizing high value resistors. Describing all the BiCMOS processes in which polysilicon is used is beyond the scope of this chapter. It is worth mentioning, however, that combining polysilicon technology with advanced epitaxial SiGe technology resulted recently in a record performance for BiCMOS circuits with typically an ECL gate delay of 18.9 ps and a 0.25 µm-channel CMOS (Harame *et al* 1992).

12.5.8 Polysilicon fuses and anti-fuses

Polysilicon link elements are used in memory switching applications to insure the permanent transition to a lower resistance state as "anti-fuse" elements or to insure the permanent transition to a higher resistance (open circuit) state as fuse elements as in ROM and PROM cells. The programming of these elements determines the final current path and can be done electrically or by high power laser pulses.

Polysilicon gate electrodes are used in Field Programmable Gate Arrays (FPGA) utilizing capacitance antifuse elements. In these cells, a link is formed between the two metal lines after applying a high voltage, typically 14-16 V, accross the capacitance via a programming transistor. In the Polysilicon/Dielectric/Diffusion (PDD) antifuse technology proposed by Actel (Hamdy *et al* 1988), n^+ polysilicon is deposited on a gate dielectric and used as the top capacitance electrode while the bottom electrode is an n^+ diffused layer. The advantage of this technology lies in its compatibility with fine line (1.2 μm or less) CMOS processes. CMOS compatible antifuse cells with two polysilicon electrodes have also been presented (Liu *et al* 1991). Finally, a scaled 0.2 μm antifuse cell featuring poly-poly vertical structures in which the cell opening is determined by the thickness of the polysilicon layer has been proposed recently (Chen *et al* 1992).

12.5.9 Polysilicon as a useful material in IC technology

Aside from being an essential tool for the success of self-aligned processes, polysilicon is also used as a technology tool in many processes, e.g as a stress relief between the nitride layer and the pad-oxide in the poly-buffered local oxidation isolation technique (LOCOS) in CMOS processes. In this process, the nitride layer can be thicker which helps in blocking effectively the lateral diffusion of oxygen without creating silicon dislocations during the field oxide growth. Consequently, the bird's beak length is reduced down to typically 0.1-0.2 μm per side as demonstrated in a 0.8 μm CMOS technology (Chapman *et al* 1987). More bird's beak shrinkage is obtained in the Polysilicon Encapsulated Local Oxidation technique PELOX (Roth *et al* 1992), a modified version of the LOCOS isolation technique offering a good gate oxide integrity and a low diode leakage with low oxide encroachment. Moreover, polysilicon is sometimes used to mask the gate oxide from being etched during process steps utilizing HF. Last but not least, polysilicon is used as a hard-mask/spacer capable of achieving sub-half-micron contact definition using optical stepper conventional photoresist lithography (Sun *et al* 1991).

12.6 OTHER APPLICATIONS OF POLYSILICON

12.6.1 Polysilicon in power IC's

Polysilicon is widely implemented as a gate material for power MOS transistors, insulated gate bipolar transistors IGBT, and MOS-controlled thyristors.
Polysilicon thin-film, high-voltage transistors exhibiting a breakdown voltage larger than 100 V and capable of driving electroluminescent (EL) displays at low signal voltages have been successfully fabricated in LPCVD polysilicon deposited at 625°C on high grade transparent quartz substrates and recrystallized using a CW-Ar laser (Unagami and Kogure 1988).

12.6.2 Polysilicon sensors, microstructures and vacuum applications

Polyilicon is used in silicon sensors and strain gauges based on piezoresistive effects (Petersen 1982). Polysilicon membranes are used in highly sensitive surface micromachined mechanical sensors (Guckel *et al* 1990). Scaled ultrasensitive silicon pressure sensors are directly applicable to gas flowmeters (Cho *et al* 1992). Polysilicon is also used in polysilicon-bridge flow sensor, thermally excited actuators and chemical sensors such as the ion sensitive field effect transistor (ISFET). The latter technology is entirely compatible with CMOS technology.

Movable micro-elements, namely rotating and translating joints, brushing and sliding elements, gears, springs, cranks allowing some motion mechanisms to take place, have been successfully implemented on polysilicon using microfabrication technology (Fan *et al* 1988). This has given rise to the fabrication of polysilicon micromotors and cleared the way to more applications such as monolithic microrobotic systems, mechanical logic, micro-valves and pumps, etc...

The use of polysilicon has also been demonstrated in vacuum microelectronics especially in miniaturized micromachined light sources (microlamps) made using silicon IC technology (Sokolich *et al* 1990, Mastrangelo *et al* 1992).

12.7 SEMI-INSULATING POLYCRYSTALLINE SILICON (SIPOS)

Semi-Insulating polysilicon or SIPOS consists of an oxygen-rich polysilicon material or alternatively of a silicon-rich silicon dioxide. Undoped SIPOS is semi-insulating and can be deposited by APCVD or LPCVD in an N_2-SiH_4-N_2O gas mixture (Matsushita *et al* 1975). The semi-insulating property of SIPOS is changed to semiconducting by doping followed by a heat treatment at temperatures above 900°C. Doping the material is achieved by adding PH_3 or B_2H_6 to the gas mixture, or by implantation after the deposition. The deposition rate depends on the N_2O flow rate especially for phosphorus doped films. The resistivity of the P-doped SIPOS decreases with increasing Si content (decreasing the N_2O flow rate) and with increasing the post deposition annealing temperature. SIPOS is easily etched in H_2O_2:HF:NH_4F mixed solution and in a CF_4-O_2 plasma.

The SIPOS structure can be described as a mosaic of tiny crystalline silicon and SiO_x grains surrounded by disordered silicon (Hamasaki *et al* 1978). Amorphous phases of Si and SiO_2 have been detected in the SIPOS film (Dong *et al* 1978). For example, a 1 μm thick SIPOS film with 20% oxygen content annealed for 30 min at 900°C is found to contain 50% crystalline silicon, 26% SiO_x and 24% amorphous Si (Greenberg and Marshall 1988).

A combined amorphous/polycrystalline conduction seems to be the most plausible description for the conduction mechanism in SIPOS. Hopping through localized states and field-aided thermionic emission conduction through the grain boundary potential barrier are practically temperature independent and dominate the low temperature conduction. Extended state conduction and thermionic emission over the potential barrier prevail at higher temperatures. A strong increase in the SIPOS conductivity is observed at high fields (Comizzoli and Opila 1987) and is attributed to an enhanced carrier release from localized states (Frenkel effect).

It has been proven that the sheet resistance of ion implanted SIPOS films can be accurately controlled. Ohmic conduction in such doped SIPOS films implanted with B, P and As has been reported (Ozguz *et al* 1988a). Such films are very promising for high value resistors compatible with VLSI integrated circuit technology.

Undoped SIPOS layers have been used mainly for surface passivation of high voltage devices and as a resistive voltage grading layer. The current flowing in a SIPOS semi-resistive film deposited on top of an oxide layer generates a voltage drop between its edges and leads to the linearization of the surface potential. Such a linearizing field plate is beneficial in reducing the maximum electric field in high voltage devices. Using such a technique combined with a highly resistive top SIPOS passivating layer, 1000 and 1500 Volts planar power devices have been realized (Charitat *et al* 1990). Undoped SIPOS is also used as the insulator layer in high-speed power MISS Metal-Insulator-Silicon switches (Ang 1988).

In-situ doped or implanted n-type SIPOS layers have been used for the emitter contact of npn bipolar transistors (Oh-Uchi *et al* 1979). The emitter-base junction of the SIPOS emitter transistors is not a true heterojunction since P or As n-type dopant atoms diffuse from the SIPOS film into the underlying p-type monocrystalline silicon during annealing. Nevertheless the current gain of NPN SIPOS emitter transistors is found to be 50 times larger than that of a metal contacted emitter homojunction bipolar transistor having a similar base. This gain enhancement is mostly due to the presence of a thin interfacial tunneling oxide layer between the SIPOS and the monocrystalline silicon. The interfacial oxide could be existing on the silicon surface prior to the SIPOS deposition or could be grown during the post-deposition annealing step (Kwark *et al* 1984). Due to their high content of oxygen SIPOS films on monocrystalline substrates are less prone to epitaxial alignment than polysilicon. Epitaxially aligned hillocks, however, have been reported upon high temperature annealing (at 1100°C) of implanted SIPOS (Ozguz *et al* 1988b). The high specific resistance of the doped SIPOS emitter contact (10^{-3} $\Omega.cm^2$) limits its application to large area devices such as power transistors.

12.8 POLYCRYSTALLINE SILICON/GERMANIUM ALLOYS

Silicon/germanium ($Si_{1-x}Ge_x$) alloys are mainly used in the single crystalline form as a narrow bandgap base material for high performance silicon heterojunction bipolar transistors. On the other hand, polyscrystalline SiGe alloys have a workfunction which is smaller than that of polysilicon and decreases with increasing Ge content which renders p^+ doped polycrystalline $Si_{1-x}Ge_x$ very suitable to be used for the gates of surface channel p- and n-MOSFETs in advanced submicron CMOS technology (King *et al* 1990).

LPCVD techniques are used to deposit polycrystalline $Si_{1-x}Ge_x$ films in a conventional hot wall reactor with GeH_4 added to the SiH_4 gas source. The Ge mole fraction in the film increases monotonically with the percentage of GeH_4 present in the deposition source gas. The deposition temperature and pressure are very close to those used for the deposition of polysilicon. X-ray diffraction and TEM analyses confirm the deposition of SiGe alloys rather than clusters of Ge in a matrix of silicon. The grain structure of the deposited layer is essentially the same as that of polysilicon, with a tendency to larger grain size for films with higher Ge content. Up to 60% Ge mole fraction, the material behaves very similarly to polysilicon with respect to usual wet and dry processes encountered in VLSI technology, which makes it completely compatible with VLSI fabrication processes.

The resistivity of heavily boron doped poly-SiGe alloys is significantly lower than that of similarly doped polysilicon and decreases with increasing Ge content. This has been attributed to increased dopant activation and hole mobility (King *et al* 1990). The annealing temperature necessary to activate the implanted boron atoms decreases significantly with increasing Ge content. These properties allow less boron and a lower activation temperature to be used in the p^+ poly-SiGe gate technology and relax partially the problem of boron diffusion through the very thin gate oxide discussed in section 12.5.1.1.b

Finally thin film p-MOS transistors have been fabricated in poly-SiGe alloys (King *et al* 1991) and demonstrated well-behaved transistor characteristics after hydrogenation. The lower processing thermal budget is the main advantage of these devices compared to their polysilicon counterpart.

REFERENCES

Anderson R and Kerr D 1977 J. Appl. Phys. 48 4834-4836

Ajuria S, Gan C, Noel J and Reif R 1992 IEEE Trans. on Electron Dev. 39 1420-1427

Ang S 1988 IEEE Electron Device Lett. 9 1378-1381

Burghartz J, Sun J, Mader S, Stanis C and Ginsberg B 1990 VLSI Symposium on VLSI Technology Extended Abstract 5C-5 55-56
Burghartz J, Megdanis A, Cressler J, Sun J, Stanis C, Comfort J, Jenkins K and Cardone F 1991 IEEE Electron Device Lett. 12 679-681
Cable J, Mann R and Woo J 1991 IEEE Electron Device Lett. 12 128-130
Chapman R, Haken R, Bell D, Wei C, Havemann R, Tang T, Holloway T and Gale R 1987 IEDM Tech. Digest 362-365
Charitat G, Jaume D, Peyre-Lavigne A and Rossel P 1990 IEDM Tech. Digest 803-806
Chen K-L, Liu D, Misium G, Gosney W, Wang S, Camp J and Tigelaar H 1992 IEEE Electron Device Lett. 13 53-55
Cho S, Najafi K, Lowman C and Wise K 1992 IEEE Trans. on Electron Dev. 39 825-835
Colinge J-P, Demoulin E, Bensahel D and Auvert G 1982 Appl. Phys. Lett. 41 346-347
Comizzoli R and Opila R 1987 J. Appl. Phys. 61 261-270
Crabbe E, Swirhun S, Del Alamo J, Pease R and Swanson R 1986 IEDM Tech. Digest 28-31
Crabbe E, Comfort J, Lee W, Cressler J, Meyerson B, Megdanis A, Sun J and Stork J 1992 IEEE Electron Device Lett. 13 259-261
Cressler J, Hwang W and Chen T-C 1989 J. Electrochem. Soc. 136 794-804

Cressler J *et al* 1992 IEEE Electron Device Lett. 13 262-264
Crowder B and Zirinski S 1979 IEEE Trans. on Electron Dev. 26 369-371
Cuthberson A and Ashburn P 1985 IEEE Trans. on Electron Dev. 32 242-247
De Graaff H and De Groot J 1979 IEEE Trans. on Electron Dev. 26 1771-1776
De Graaff H, Huybers M and De Groot J 1982 Solid-State Electr. 25 67-72
De Graaff H and Huybers M 1983 J. Appl. Phys. 54 2504-2507
Depauw P, Mertens R and Van Overstraeten 1984 IEEE Electron Device Lett. 5 234-237
DiMaria D and Kerr D 1975 Appl. Phys. Lett. 27 505-507
Dong D, Irene E and Young D 1978 J. Electrochem. Soc. 125 819-823
Eltoukhy A and Roulston D 1982 IEEE Trans. on Electron Dev. 29 961-964
Fan L-S, Tai Y-C and Muller R 1988 IEEE Trans. on Electron Dev. 35 724-730
Faraone L 1986 IEEE Trans. on Electron Dev. 33 1785-1794
Fazan P and Lee R 1990 IEEE Electron Device Lett. 11 279-281
Furumura Y, Mieno F, Nishizawa T and Maeda M 1986 J. Electrochem. Soc. 133 379-383
Ghannam M, Dutton R and Novak S 1987a Mat. Res. Soc. Symp. 76 283-288
Ghannam M and Dutton R 1986 IEEE Bipolar Circuits and Technology Meeting BCTM 5-6
Ghannam M and Dutton R 1987b Appl. Phys. Lett. 51 611-613
Ghannam M and Dutton R 1988 Appl. Phys. Lett. 52 1222-1224
Ginley D, Hellmer R and Lum W 1987 J. Electrochem. Soc. 134 2078-2080
Graul J, Glasl A and Murrmann H 1976 IEEE J. Solid-State Circuits 11 491-495

Greenberg B and Marshall T 1988 J. Electrochem. Soc. 135 2295- 2299

Groeseneken G and Maes H 1986 IEEE Trans. on Electron Dev. 33 1028-1042

Guckel H, Christenson T, Skrobis K,Sniegowski J, Kang J, Choi B and Lovell E 1990 IEDM Tech. Digest 613-616

Hamasaki M, Adachi T, Wakayama S and Kikuchi M 1978 J. Appl. Phys. 49 3987-3992

Hamdy E, McCollum J, Chen S-O, Chiang S, Eltoukhy A.S, Chang J, Speers T and Mohsen A 1988 IEDM Tech. Digest 786-789

Harame D *et al* 1992 IEDM Tech. Digest 19-22

Hatalis M and Greve D 1988 J. Appl. Phys. 63 2260-2266

Hirose M, Taniguchi M and Osaka Y 1979 J. Appl. Phys. 50 377-382

Hite L, Sundaresan R, Malhi S, Lam H, Shah A, Hester R and Chatterjee P 1985 IEEE Electron Device Lett. 6 548-550

Horiuchi S and Blanchard R 1975 Solid St. Electron. 18 529-532

Hoyt J, Crabbe E, Gibbons J and Pease R 1987 Appl. Phys. Lett. 50 751-753

Hsieh T, Chun H and Kwong D 1989 Appl. Phys. Lett. 55 2408-2410

Huang T, Yao W, Martin R, Lewis A, Koyanagi M and Chen J 1986 IEDM Tech. Digest 742-745

Hwang B-Y, Bushey T, Kirchgessner J, Foertsch S, Stipanuk J, Marshbanks L, Hernandez J and Herald E 1989 IEEE J. Solid-State Circuits 24 504-510

Ikeda S, Hashiba S, Kuramoto I, Katoh H, Ariga S, Yamanaka T, Hashimoto T, Hashimoto N and Meguro S 1990 IEDM Tech. Digest 469-472

Irene E, Tierney E and Dong D 1980 J. Electrochem. Soc. 127 705-713

Jackson W, Johnson N and Biegelsen D 1983 Appl. Phys. Lett. 43 195-197

Kamins T, Manoliu J and Tucker T 1972 J. Appl. Phys. 43 83-91

Kamins T 1971 J. Appl. Phys. 42 4357-4365

Kamins T 1988 Polycrystalline Silicon for Integrated Circuit Applications (Boston, Kluwer)

Kamins T 1989 IEEE Electron Device Lett. 9 401-404

Keyes E and Tarr N 1987 IEEE Electron Device Lett. 8 312-314

King T-J, Pfiester J, Shott J, McVittie J and Saraswat K 1990 IEDM Tech. Digest 253-256

King T-J, Saraswat K and Pfiester J 1991 IEEE Electron Device Lett. 12 584-586

Klein R, Owen W, Simko R and Tchon W 1979 Electronics Oct. 11 111-116

Kozicki M and Robertson J 1989 J. Electrochem. Soc. 136 878-881

Kwark Y, Sinton R and Swanson R 1984 IEDM Tech. Digest 742-745

Lau C, See Y, Scott D, Bridges J, Perna S and Davies R 1982 IEDM Tech. Digest 714-717

Lifshitz N 1985 IEEE Trans. on Electron Dev. 32 617-621

Liu D, Chen K, Tigelaar H, Paterson J and Chen S 1991 IEEE Electron Device Lett. 12 151-153

Lu N, Rajeevakumar T, Bronner G, Ginsberg B, Machesney B and Sprogis E 1988 IEDM Tech. Digest 588-591

Malhotra V, Mahan J and Ellsworth D 1985 IEEE Trans. on Electron Dev. 32 2441-2449

Mandurah M, Saraswat K and Kamins T 1981 IEEE Trans. on Electron Dev. 28 1163-1171

Mandurah M, Saraswat K, Helms C and Kamins K 1980 J. Appl. Phys. 51 5755-5763

Mano T, Takeya K, Watanabe T, Ieda N, Kiuchi K, Arai E, Ogawa T and Hirata K 1980 IEEE J. Solid State Circuits 15 865-872

Maritan C and Tarr N 1989 IEEE Trans. on Electron Dev. 36 1139-1143

Mastrangelo C, Yeh J and Muller R 1992 IEEE Trans. on Electron Dev. 39 1363-1375

Matsushita T, Teruaki A, Takaji O, Hisayoshi Y, Hisao H, Masanori O and Yoshiyuki K 1975 IEEE Trans. on Electron Dev. 23 826-830

Mertens P, Wouters D, Maes H, Veirman A, and Van Landuyt J 1988 J. Appl. Phys. 63 2660-2668

Mine T, Iijima S, Yugami J, Ohga K and Morimoto T 1989 21st SSDM Extended abstract 137-

Murarka S 1983 Silicides for VLSI Applications (Academic, New York)

Ning T and Isaac R 1980 IEEE Trans. on Electron Dev. 27 2051-2055

Ning T, Isaac R, Solomon P, Tang D, Yu H, Feth G and Wiedman S 1981 IEEE Trans. on Electron Dev. 28 1010-1013

Nguyen S, Dobuzinsky D and Harmon D 1989 The Electrochem. Soc. Meeting Extended Abstract 89-1 (abstract 172) 248-249

Ohshita Y and Kitajima H 1991 J. Appl. Phys. 70 1871-1873

Oh-Uchi N, Hayashi H, Yamoto H and Matsushita T 1979 IEDM Tech. Digest 522-525

Osburn C, Cramer A, Schweigart A and Wordeman W 1982 Proc. Electrochem. Soc. Symp. VLSI Sci. Technol., C. Dell'Oca & W. Billis Eds., 354-361

Ozguz V, Wortman J, Hauser J, Littlejohn M, Rozgonyi G and Curran P 1988a J. Electrochem. Soc. 135 665- 670

Ozguz V, Posthill J, Wortman J, Littlejohn M 1988b J. Appl. Phys. 63 2831-2838

Parrillo L, Hillenius S, Field R, Hu E, Fichtner W and Chen M-L 1984 IEDM Tech. Digest 418-421

Patton G, Bravman J and Plummer J 1986 IEEE Trans. on Electron Dev. 33 1754-1768

Petersen K 1982 Proceedings of the IEEE 70 420

Pfiester J, Baker F, Sivan R, Crain N, Lin J, Liaw M, Seelbach C, Gunderson C and Denning D 1990 IEEE Trans. on Electron Dev. 11 253-255

Rodder M and Aur S 1991 IEEE Electron Device Lett. 12 233-235

Rosler R 1977 Solid State Technology 20-4 63-70

Roth S, Ray W, Mazure C, Kirsch H, Gunderson C and Ko J 1992 IEEE Trans. on Electron Dev. 39 1085-1089

Ryssel H, Iberl H, Bleier M, Prinke G, Haberger K and Kranz H 1981 Appl. Phys. 24 197-200

Sakao M, Kasai N, Ishijima T, Ikawa E, Watanabe H, Terada K and Kikkawa T 1990 IEDM Tech. Digest 655-658

Mandurah M, Saraswat K and Kamins T 1981 IEEE Trans. on Electron Dev. $\underline{28}$ 1163-1171

Mandurah M, Saraswat K, Helms C and Kamins K 1980 J. Appl. Phys. $\underline{51}$ 5755-5763

Mano T, Takeya K, Watanabe T, Ieda N, Kiuchi K, Arai E, Ogawa T and Hirata K 1980 IEEE J. Solid State Circuits $\underline{15}$ 865-872

Maritan C and Tarr N 1989 IEEE Trans. on Electron Dev. $\underline{36}$ 1139-1143

Mastrangelo C, Yeh J and Muller R 1992 IEEE Trans. on Electron Dev. $\underline{39}$ 1363-1375

Matsushita T, Teruaki A, Takaji O, Hisayoshi Y, Hisao H, Masanori O and Yoshiyuki K 1975 IEEE Trans. on Electron Dev. $\underline{23}$ 826-830

Mertens P, Wouters D, Maes H, Veirman A, and Van Landuyt J 1988 J. Appl. Phys. $\underline{63}$ 2660-2668

Mine T, Iijima S, Yugami J, Ohga K and Morimoto T 1989 21st SSDM Extended abstract **137-**

Murarka S 1983 Silicides for VLSI Applications (Academic, New York)

Ning T and Isaac R 1980 IEEE Trans. on Electron Dev. $\underline{27}$ 2051-2055

Ning T, Isaac R, Solomon P, Tang D, Yu H, Feth G and Wiedman S 1981 IEEE Trans. on Electron Dev. $\underline{28}$ 1010-1013

Nguyen S, Dobuzinsky D and Harmon D 1989 The Electrochem. Soc. Meeting Extended Abstract $\underline{89\text{-}1}$ (abstract 172) 248-249

Ohshita Y and Kitajima H 1991 J. Appl. Phys. $\underline{70}$ 1871-1873

Oh-Uchi N, Hayashi H, Yamoto H and Matsushita T 1979 IEDM Tech. Digest 522-525

Osburn C, Cramer A, Schweigart A and Wordeman W 1982 Proc. Electrochem. Soc. Symp. VLSI Sci. Technol., C. Dell'Oca & W. Billis Eds., 354-361

Ozguz V, Wortman J, Hauser J, Littlejohn M, Rozgonyi G and Curran P 1988a J. Electrochem. Soc. $\underline{135}$ 665- 670

Ozguz V, Posthill J, Wortman J, Littlejohn M 1988b J. Appl. Phys. $\underline{63}$ 2831-2838

Parrillo L, Hillenius S, Field R, Hu E, Fichtner W and Chen M-L 1984 IEDM Tech. Digest 418-421

Patton G, Bravman J and Plummer J 1986 IEEE Trans. on Electron Dev. $\underline{33}$ 1754-1768

Petersen K 1982 Proceedings of the IEEE $\underline{70}$ 420

Pfiester J, Baker F, Sivan R, Crain N, Lin J, Liaw M, Seelbach C, Gunderson C and Denning D 1990 IEEE Trans. on Electron Dev. $\underline{11}$ 253-255

Rodder M and Aur S 1991 IEEE Electron Device Lett. $\underline{12}$ 233-235

Rosler R 1977 Solid State Technology $\underline{20\text{-}4}$ 63-70

Roth S, Ray W, Mazure C, Kirsch H, Gunderson C and Ko J 1992 IEEE Trans. on Electron Dev. $\underline{39}$ 1085-1089

Ryssel H, Iberl H, Bleier M, Prinke G, Haberger K and Kranz H 1981 Appl. Phys. $\underline{24}$ 197-200

Sakao M, Kasai N, Ishijima T, Ikawa E, Watanabe H, Terada K and Kikkawa T 1990 IEDM Tech. Digest 655-658

Greenberg B and Marshall T 1988 J. Electrochem. Soc. 135 2295- 2299

Groeseneken G and Maes H 1986 IEEE Trans. on Electron Dev. 33 1028-1042

Guckel H, Christenson T, Skrobis K,Sniegowski J, Kang J, Choi B and Lovell E 1990 IEDM Tech. Digest 613-616

Hamasaki M, Adachi T, Wakayama S and Kikuchi M 1978 J. Appl. Phys. 49 3987-3992

Hamdy E, McCollum J, Chen S-O, Chiang S, Eltoukhy A.S, Chang J, Speers T and Mohsen A 1988 IEDM Tech. Digest 786-789

Harame D *et al* 1992 IEDM Tech. Digest 19-22

Hatalis M and Greve D 1988 J. Appl. Phys. 63 2260-2266

Hirose M, Taniguchi M and Osaka Y 1979 J. Appl. Phys. 50 377-382

Hite L, Sundaresan R, Malhi S, Lam H, Shah A, Hester R and Chatterjee P 1985 IEEE Electron Device Lett. 6 548-550

Horiuchi S and Blanchard R 1975 Solid St. Electron. 18 529-532

Hoyt J, Crabbe E, Gibbons J and Pease R 1987 Appl. Phys. Lett. 50 751-753

Hsieh T, Chun H and Kwong D 1989 Appl. Phys. Lett. 55 2408-2410

Huang T, Yao W, Martin R, Lewis A, Koyanagi M and Chen J 1986 IEDM Tech. Digest 742-745

Hwang B-Y, Bushey T, Kirchgessner J, Foertsch S, Stipanuk J, Marshbanks L, Hernandez J and Herald E 1989 IEEE J. Solid-State Circuits 24 504-510

Ikeda S, Hashiba S, Kuramoto I, Katoh H, Ariga S, Yamanaka T, Hashimoto T, Hashimoto N and Meguro S 1990 IEDM Tech. Digest 469-472

Irene E, Tierney E and Dong D 1980 J. Electrochem. Soc. 127 705-713

Jackson W, Johnson N and Biegelsen D 1983 Appl. Phys. Lett. 43 195-197

Kamins T, Manoliu J and Tucker T 1972 J. Appl. Phys. 43 83-91

Kamins T 1971 J. Appl. Phys. 42 4357-4365

Kamins T 1988 Polycrystalline Silicon for Integrated Circuit Applications (Boston, Kluwer)

Kamins T 1989 IEEE Electron Device Lett. 9 401-404

Keyes E and Tarr N 1987 IEEE Electron Device Lett. 8 312-314

King T-J, Pfiester J, Shott J, McVittie J and Saraswat K 1990 IEDM Tech. Digest 253-256

King T-J, Saraswat K and Pfiester J 1991 IEEE Electron Device Lett. 12 584-586

Klein R, Owen W, Simko R and Tchon W 1979 Electronics Oct. 11 111-116

Kozicki M and Robertson J 1989 J. Electrochem. Soc. 136 878-881

Kwark Y, Sinton R and Swanson R 1984 IEDM Tech. Digest 742-745

Lau C, See Y, Scott D, Bridges J, Perna S and Davies R 1982 IEDM Tech. Digest 714-717

Lifshitz N 1985 IEEE Trans. on Electron Dev. 32 617-621

Liu D, Chen K, Tigelaar H, Paterson J and Chen S 1991 IEEE Electron Device Lett. 12 151-153

Lu N, Rajeevakumar T, Bronner G, Ginsberg B, Machesney B and Sprogis E 1988 IEDM Tech. Digest 588-591

Malhotra V, Mahan J and Ellsworth D 1985 IEEE Trans. on Electron Dev. 32 2441-2449

Saraswat K and Singh H 1982 J. Electrochem. Soc. 129 2321-2326

Saraswat K, Brors D, Fair J, Monnig K and Beyers R 1983 IEEE Trans. on Electron Dev. 30 1497-1505

Scott D, See Y, Lau C and Davies R 1981 IEDM Tech. Digest 538-541

Seager C and Pike G 1979 Appl. Phys. Lett. 35 709-711

Seager C 1982 J. Appl. Phys. 53 5968-5971

Seto J 1975 J. Appl. Phys. 46 5247-5254

Soerowirdjo B, Ashburn P and Cuthbertson A 1982 IEDM Tech. Digest 668-671

Sokolich M, Adler E, Longo R, Goebel D and Benton R 1990 IEDM Tech. Digest 159-162

Stork J, Arienzo M and Wong C 1985 IEEE Trans. Electron Dev. 32 1766-1770

Sun S, Woo M and Kawasaki H 1991 J. Electrochem. Soc. 138 619-620

Sunami H, Kure T, Hashimoto N, Itoh K, Toyabe T and Asai S 1982 IEDM Tech. Digest 806-809

Suzuki K and Kataoka Y 1991 J. Electrochem. Soc. 138 1794-1798

Swaminathan B, Saraswat K, Dutton R and Kamins T 1982 Appl. Phys. Lett. 40 795-798

Takagi M, Nakayama K, Tevada C and Kamioko H 1972 J. Jap. Soc. Appl. Phys. (Suppl.) 42 101-109

Tanaka H, Aikawa I and Ajioka T 1990 J. Electrochem. Soc. 137 644-647

Tang Y and Wilkinson C 1991 Appl. Phys. Lett. 58 2898-2900

Taylor M, Flowers D and Cosway R 1987 J. Electrochem. Soc. 134 2935-2937

Tsang P, Ogura S, Walker W, Shepard J and Critchlow D 1982 IEEE Trans. on Electron Dev. 29 590-596

Tsaur B and Hung L 1980 Appl. Phys. Lett. 37 648-651

Unagami T and Kogure O 1988 IEEE Trans. on Electron Dev. 35 314-319

Vaidya S, Fuls E and Johnston R 1986 IEEE Trans. on Electron Dev. 33 1321-1328

Wada Y and Nishimatsu S 1978 J. Electrochem. Soc. 125 1499-1504

Warnock et al 1989 IEDM Tech. Digest 903-905

Warnock J, Lu P,Cressler J, Jenkins K, Sun J 1990 IEDM Tech. Digest 301-304

Wolstenholme G, Jorgensen N, Ashburn P and Booker G 1987 J. Appl. Phys. 61 225-233

Wolstenholme G, Browne D, Ashburn P and Landsberg P 1988 IEEE Trans. on Electron Dev. 35 1915-1923

Yamanaka T *et al* 1990 IEDM Tech. Digest 477-480

Yoshimaru M, Miyano J, Inoue N, Sakamoto A, You S, Tamura H and Ino M 1990 IEDM Tech. Digest 659-662

Yu Z, Ricco B and Dutton R 1984 IEEE Trans. on Electron Dev. 31 773-784

Yu C and Mathews V 1992 Appl. Phys. Lett. 60 1501-1503

Biographical Details

Editor' s curriculum vitae

Johan F. Nijs

Johan Nijs received his electronic engineer degree from the Katholieke Universiteit of Leuven (K.U.Leuven), Belgium, in 1977. After a trainee period of 2 months at Philips, he joined the Electronics, Systems, Automation, and Technology (ESAT) Laboratory of K.U.Leuven on a research program for the fabrication of silicon solar cells. He obtained his Ph.D. degree in applied science from K.U.Leuven in 1982. During 1982-1983 he was a postdoctoral visiting scientist at the IBM Thomas J. Watson Research Centre in Yorktown Heights, New York, working on the subject of amorphous silicon technology.

In 1983 he became a research assistant at the ESAT laboratory of K.U.Leuven and in 1984 he joined the Interuniversity Micro-Electronics Centre (IMEC) in Leuven, where he is now head of the silicon materials groups. His interests are and/or were in solar cells, bipolar transistors, low-temperature silicon epitaxy, and polysilicon thin-film transistors on glass.

In 1990 he was appointed part-time assistant professor at K.U.Leuven. He has authored or co-authored more than 100 papers that have been published in conference proceedings and technical journals. Dr. Nijs is a member of the Materials Research Society (MRS) and the IEEE.

Curriculum vitae of authors

Chris A. Baert

C. Baert was born in Leuven, Belgium on October 16, 1960. He took his
Master's Degree of Electrical Engineering and Ph.D. in Micro-electronics and
Materials Science from Leuven University in 1984 resp. 1990. The topic of his
Ph.D. work was low temperature plasma-enhanced CVD of Si materials and their
applications in solar cells and TFT's. From 1990 till 1992 he was with the
Materials and Electronic Devices Laboratory of Mitsubishi Electric, Amagasaki,
Japan where we worked on poly-Si TFT's for active-matrix Liquid Crystal
Displays. At present, he is with the InterUniversity Micro-Electronics Centre
(IMEC) in Leuven, Belgium, where he is responsible for the activities on
microsystems.

Matty R. Caymax

After obtaining his Ph.D. degree in Chemistry from the Katholieke Universiteit
Leuven (K.U.Leuven), M. Caymax worked in the field of solar cells at the ESAT
Laboratory of the Electronics Department of K.U.Leuven from 1984 to 1986.
Then he joined IMEC where he continued his research on solar cells made in
epitaxial layers on cheap polycrystalline Si substrates. In 1988, he started
research on low temperature CVD growth and characterisation of thin epitaxial
Si-layers, followed by work in the SiGe-field. He has authored or co-authored
over 30 papers in these fields.

Cor Claeys

Cor Claeys received the electronic engineering degree in 1974 and the Ph.D.
degree in 1979, both from the Katholieke Universiteit Leuven, Belgium.
From 1974 to 1984 he was respectively, research assistant and staff member of
the ESAT Laboratory of the Katholieke Universiteit Leuven, and is now
Professor there. In 1984, he joined the Interuniversity Micro-Electronics Centre
(IMEC), Leuven, as Head of Silicon Processing. His main interests are in
general IC technology, including MOS, CCD, and CMOS-SOI, device physics
and low temperature operation, defects engineering and material characterisation.
He is author and co-author of three book chapters and more than 200 technical
papers and conference contributions related to he above fields.
Dr. Claeys is a member of the Electrochemical Society, the European MRS and
SEMI.

Charles M. Falco

Charles M. Falco holds joint appointments as professor in the Physics Department, Optical Sciences Center, and Arizona Research Laboratories at the University of Arizona, Tucson. He received his Ph.D. in Physics in 1974 from the University of California, Irvine, and then worked the next eight years at Argonne National Laboratory, where he was group leader of the superconductive and novel materials group. He conducted research at the University of Paris-Sud in 1979 and 1986, and spent the fall of 1989 at the RWTH Aachen as an Alexander vonHumboldt Senior Distinguished Scientist. Falco's research involves studies of physical properties of metallic superlattices, including magnetism, magneto-optical effects, x-ray optical properties, elastic constants, and nucleation and growth of these materials by molecular beam epitaxy and sputtering. He is a Fellow of the APS, Senior Member of the IEEE, and Member of the MRS.

Moustafa Ghannam

Moustafa Ghannam received the B.Sc. in Electrical Engineering from Cairo University, Egypt in 1975 and the Ph.D. degree from the Katholieke Universiteit Leuven, Belgium, in 1985. He has been a Post-Doctoral fellow at Stanford University, USA, during 1985/1986. From 1987 he has been assigned academic and research positions as Assistant, then Associate Professor at Cairo University and Kuwait University, and as Senior Scientist at the University Microelectronics Centre IMEC, Leuven, Belgium. He is author and co-author of more than 60 journal and conference papers and two book chapters, has instructed in several international workshops and holds one patent. He served as committee member of the international IEDM and BCTM conferences of the IEEE. M. Ghannam is a Senior member of the IEEE.

Suresh C. Jain

Professor S.C. Jain obtained his Ph.D. degree in 1955 from Delhi University. He has held several senior positions in India: Head of Solid State Physics Division, National Physical Laboratory at New Delhi, Head of the Physics Department and Dean of the Faculty of Science of I.I.T. Delhi, and Director of Solid State Physics Laboratory of the Ministry of Defence, Govt. of India. (except for about 3 years from 1974 to 1977, when he was a visiting professor at University of Illinois in the U.S.A., at U.K.-A.E.R.E. at Harwell and at Imperial College London). He has spent last 10 years abroad as a visiting professor mostly at K.U.Leuven and IMEC and Clarendon Laboratory at Oxford. Presently he is Clarendon Fellow in Oxford. He has published more than 250 papers on defects in crystals, photovoltaics, short channel MOSFETs, nuclear detectors and heterostructure devices. His present interests are strained layers and heterostructure devices.

Erich Kasper

Prof.Dr. E. Kasper is head of the Institute of Semiconductor Engineering, University of Stuttgart, since autumn 1993. The main activities there will be directed toward novel semiconductor technologies, advanced device concepts and multifunctional circuits. Before he gained more than 20 years experience in industrial research (AEG-Telefunken, Daimler-Benz) dealing with topics like molecular beam epitaxy (Si-MBE), ultra-thin superlattices (Si/Ge), microwave circuits on Si (SIMMWIC) and high frequency transistors. Recently a transit frequency of 100 GHz was obtained by his group with a SiGe/Si-HBT. Born in Eisenerz, Austria (1943), he studied experimental physics in Graz, Austria and is living now with his family (Ann, Johannes, Sebastian, Wolfgang) in Pfaffenhofen, Germany

W.Y. Leong

Dr. Leong completed his Ph.D. studies in 1985 on reduced temperature Si epitaxial growth and doping processes in Si-MBE. He joined DRA Malvern (formerly RSRE) in 1987 where he has since been involved with research in Si and SiGe epitaxy by the UHV-CVD technique for the application in high speed bipolar transistors and long wavelength infra-red detectors. His other interests include novel Si material and device physics.

Santo Martinuzzi

Santo Martinuzzi is full professor at the University of Marseilles III - France. After a Doctorate thesis devoted to the preparation and study of GaAs thin films, he worked in the field of II-VI compounds for photovoltaics, especially Cu_xC-CdS heterojunctions. After 1980, he has investigated the electrical properties of large grain polycrystalline silicon wafers, contributing to demonstrate the paramount influence of the interaction between extended defects and segregated impurities. He has also shown that hydrogenation and external gettering techniques are able to passivate the defects, to remove metallic impurities and to improve drastically the wafers.

Robert P. Mertens

Robert Mertens obtained the electronic degree and the Ph.D. in applied sciences from the K.U.Leuven in 1969 and 1972 respectively. In the period 1974-1974 he was a postdoctoral researcher at the University of Florida. From 1974 to 1984 he was a Senior Research Associate of the National Fund for Scientific Research, Belgium and a staff member at the ESAT Laboratory of the K.U.Leuven. Since 1979 he is also professor at the University of Leuven. In 1984 he joined IMEC in Leuven as a Vice-President of the Materials and Packaging Division. His research interests concern physics of semiconductor devices and technology of solar cells.

Piero Migliorato

Piero Migliorato received his Doctorate in Physics from the University of Rome in 1969. He started his career in the Italian Research Council (CNR) at the Solid State Electronics Institute in Rome, then moving to the Solid State Group in Rome University. He also worked for extended periods at Bell Laboratories, Holmdel, NJ, and at the Royal Signals and Radar Establishment in the U.K. From 1983 until 1989 he was with the GEC-Hirst Research Centre, where he led the Polysilicon Active Matrix Display Programme. He was also Visiting Professor at the Department of Electrical Engineering and Electronics of Liverpool University. Since 1989, he has been with Cambridge University where he is presently Reader in Physical Electronics at the Department of Engineering, and a Fellow of Trinity College. His research activities have included light-emitting diodes, hetero-junction detectors and solar cells, infrared detectors and studies of impurities and defects in semiconductors. He has written many scientific papers, including a number of invited talks at international conferences, and has filed 7 patents.

Johan F. Nijs

Johan Nijs received his electronic engineering degree in 1977 and his Ph.D. degree in 1982 (on the subject of silicon solar cells), both from the Katholieke Universiteit of Leuven (K.U.Leuven), Belgium,. During 1982-1983 he was a postdoctoral visiting scientist at the IBM Thomas J. Watson Research Centre in Yorktown Heights, NY. In 1983 he became a research assistant at the ESAT laboratory of K.U.Leuven and in 1984 he joined the Interuniversity Micro-Electronics Centre (IMEC) in Leuven, where he is now head of the silicon materials groups. His interests are and/or have been solar cells, bipolar transistors, low-temperature silicon epitaxy, and polysilicon thin-film transistors on glass. In 1990 he was also appointed part-time assistant professor at K.U.Leuven. He has authored or co-authored more than 100 papers that have been published in conference proceedings and technical journals. Dr. Nijs is a member of the Materials Research Society (MRS) and the IEEE.

Sergio Pizzini

Sergio Pizzini is professor of Physical Chemistry of Solids at the University of Milano, where he is also teaching Materials Chemistry at the School of Material Science and Technology of the Faculty of Sciences.

He is editor of two books and author or co-author of more than 120 scientific papers published in international Journals, mostly addressed to problems concerning physical and physico-chemical properties of ionic and covalent semiconductors, including silicon and zinc selenide.

In 1991 he won the Philip Morris award in Science and Technology for his contribution to the development of the advanced polycrystalline silicon casting process, now adopted, within others, by Eurosolare and Crystalox.

Jef Poortmans

Jozef J. Poortmans was born in Turnhout, Belgium, on April 24th, 1962. He received the Engineer's degree in electronics from the Katholieke University Leuven, Belgium in 1985 and the Ph.D. degree at the same university in 1993.

Since 1985 he worked at the Interuniversity Microelectronics Centre in Leuven. From 1985 until 1987 his research has been in the field of recrystallization of polycrystalline and amorphous silicon for SOI-applications and Thin Film Transistors. Since 1988 he was engaged in research on the growth of epitaxial Si and SiGe-layers, grown at low temperature and their application in the base of bipolar devices. He has authored or co-authored 30 publications in journals or conference proceedings.

His current interests also include Si-based high efficiency solar cells.

Michael Quinn

Michael Quinn received the B. Eng. degree with First Class Honours from the University of Limerick, Ireland, in 1992 where he specialised in Microelectronics. He spent one year working at Philips Research Laboratories, Eindhoven, on power dissipation in VLSI circuits and Swith Level simulation, before pursuing his Ph.D. at the University of Cambridge. He is the recipient of an Internal Graduate Studentship at Trinity College, Cambridge and is a British Council - Foreign and Commonwealth Office Scholar. He is currently working on the investigation and modelling of conduction mechanisms in polysilicon TFTs, within the framework of the ECAM/INTEGRAL ESPRIT project.

Toshihiro Sugii

Toshihiro Sugii was born in Kyoto, Japan, in November 17, 1956. He received the B.S., M.S. and Ph.D. degrees in electronics engineering from Tokyo Institute of Technology, Tokyo, Japan, in 1979, 1981 and 1991, respectively. In 1981, he joined Fujitsu Laboratories Ltd., Kawasaki, Japan, where he has been engaged in the research and development of VLSI materials, processes and devices. His research interests are in high-speed Si-heterojunction bipolar transistors and related materials, and high-speed sub-quarter micron CMOS devices. Dr. Sugii is a member of the Japan Society of Applied Physics.

Jan Vanhellemont

Jan Vanhellemont graduated in 1978 from the University of Gent, Belgium, in Theoretical Physics. He obtained his Ph.D. in Physics at the University of Antwerp in 1990.
After his military service he joined Siemens Belgium in 1980, where he worked as scientific responsible engineer in the field of laser treatment of solid state materials. From 1981 until 1987, he was a research assistant at the University of Antwerp where he was involved in characterisation of semiconductors with transmission electron microscopy techniques. In 1987 he joined the Interuniversity Micro-Electronics Centre (IMEC) as a scientific collaborator.
Jan Vanhellemont is author and co-author of over 150 scientific papers and three book chapters dealing with characterisation of semiconductor materials using transmission electron microscopy, optical and electrical techniques.

Piet Van Mieghem

Piet Van Mieghem obtained the degree in electrical engineering from the Katholieke Universiteit Leuven (K.U.Leuven, Belgium) in 1987. In 1991 he received the Ph.D. in electrical engineering on the subject Heavy Doping Effects in Semiconductors. From 1992-1993 he joined the Massachusetts Institute of Technology (USA) as a visiting postdoctoral scientist where he studied the influence of ferromagnetic materials on semiconductors. Currently he works in the central research of Alcatel Bell in Antwerp (Belgium).

Roger J. Van Overstraeten

Baron Roger J. Van Overstraeten was born on December 1937 in Vlezenbeek-Belgium. He obtained his engineers degree in electronics and in mechanics from the Katholieke Universiteit Leuven, Belgium, 1960. In 1963 he obtained a Ph.D. degree in physical electronics from Stanford University. He is professor at the Katholieke Universiteit Leuven since 1965 and was the Founder and Director (until 1984) of the laboratory ESAT (Electronics, Systems, Automation, Technoloy) at the Katholieke Universiteit Leuven.

Since 1984, he is President of the Interuniversity Microelectronics laboratory IMEC vzw in Leuven, Belgium with a personnel of 480. This independent R&D laboratory is active in the development of design methodologies for VLSI systems, the development of submicron processes and interconnecting techniques, supporting the next generations of VLSI chips, opto-electronic components, sensors, solar cells, ... and in the training of VLSI designs and of processing engineers for industry.

Professor R. Van Overstraeten is author or co-author of more than 100 papers in scientific journals. His contributions are mainly in the field of physical electronics and of photovoltaics.

He is fellow of IEEE and member of the Belgian Royal Academy of Sciences.

Insead Innovator prize, 1986. He obtained the Honorary doctoral degree from I.N.P.G. Grenoble, France, 1987. He won the Becquerel prize 1989 at the "9th European Photovoltaic Solar Energy Conference" in 1989 and the Flanders Technology International Innovation Award in 1991.

Magnus Willander

Magnus Willander was born in Varberg, Sweden. He received M.S. degrees in Physics and Technical Engineering from Lund University, Lund, Sweden, and Uppsala University, Uppsala, Sweden, in 1974 and 1976 respectively. During 1976 and 1980 he worked with electronic design for the Philips Corp. in Stockholm. He worked as Research Assistant in Physics in Royal Institute of Technology, Stockholm, Sweden, from 1980 till 1984. In 1984 he received Ph.D. degree in Applied Physics from that place. In 1984 and 1985 he was employed at Noble Industries, Stockholm, Sweden, as specialist in electronics. In 1985 he joined Linköping University, Linköping, Sweden, as Associate Professor in Device Physics. His research interests include experimental and theoretical aspects in semiconductor materials and devices. He has authored or co-authored numerous papers in this field, and has authored several book chapters.

Keyword Index